Energy Policy, Regulation and Sustainable Development

Energy Policy, Regulation and Sustainable Development

Editors

Grzegorz Mentel
Sebastian Majewski

MDPI • Basel • Beijing • Wuhan • Barcelona • Belgrade • Manchester • Tokyo • Cluj • Tianjin

Editors
Grzegorz Mentel
Rzeszow University of Technology
Poland

Sebastian Majewski
University of Szczecin
Poland

Editorial Office
MDPI
St. Alban-Anlage 66
4052 Basel, Switzerland

This is a reprint of articles from the Special Issue published online in the open access journal *Energies* (ISSN 1996-1073) (available at: https://www.mdpi.com/journal/energies/special_issues/Energy_Policy_Regulation_and_Sustainable_Development).

For citation purposes, cite each article independently as indicated on the article page online and as indicated below:

LastName, A.A.; LastName, B.B.; LastName, C.C. Article Title. *Journal Name* **Year**, *Volume Number*, Page Range.

ISBN 978-3-0365-6800-3 (Hbk)
ISBN 978-3-0365-6801-0 (PDF)

Contents

About the Editors

Grzegorz Mentel

Grzegorz Mentel, DSc. PhD has been working at the Rzeszow University of Technology at the Faculty of Management since 2000 and is Head of the Department of Quantitative Methods. Between 2010–2012, he was the director of Bank Pocztowy S.A. in Rzeszow. As an author, he has published extensively on a wide variety of topics including the energy market, finance and capital markets, as well as having produced monographs on the risk of financial instruments. Furthermore, he is a specialist in security risk management, fundamental analysis, technical analysis, multidimensional analysis and forecasting.

Sebastian Majewski

Sebastian Majewski (Prof. Dr) currently works at the Institute of Economics and Finance, Department of Sustainable Finance and Capital Markets, University of Szczecin. His broad research interests lie in Sport Finance, Behavioral Finance, Sustainable Economics, Financial Economics and Econometrics. His current project is 'The football club valuation'.

Preface to "Energy Policy, Regulation and Sustainable Development"

Today's economy is faced with a very important challenges—responding to dynamic environmental degradation, dwindling energy resources, public opposition to the changes taking place and also satisfying businesses operating in the market. Taken together, these challenges are driving the development of a sustainable economy.

We are delighted to present the results of the research conducted and published as part of our Special Issue. The papers published here primarily explore renewable energy sources, environmental pollution and the energy transition. The Special Issue's dynamic scope has made it possible to include other topics, such as those relating to the stock market. The papers presented here showcase academic excellence and contribute to a compelling scientific discussion.

Grzegorz Mentel and Sebastian Majewski
Editors

energies

Article

Determinants of Reducing Greenhouse Gas Emissions in European Union Countries

Beata Zofia Filipiak [1,*] and Dorota Wyszkowska [2]

1 Institute of Economics and Finance, University of Szczecin, 71-101 Szczecin, Poland
2 Faculty of Economics and Finance, University of Bialystok, 15-062 Bialystok, Poland
* Correspondence: beata.filipiak@usz.edu.pl; Tel.: +48-914442198

Abstract: In the literature on the subject, it is argued that tax policy is one of the tools stimulating the transition toward sustainable economies. Public authorities can use two functions for this purpose: fiscal and non-fiscal functions. High emission rates and the rising rapid atmospheric changes that come with them are serious threats to the climate and sustainable development. Reducing greenhouse gas emissions is one of the goals towards which the world strives (including the EU), so as to keep a balance between people's expectations, economic aspects, and the needs of the environment. Therefore, it is necessary to explain whether, along with other factors, environmental policy and its component "green taxes" can act as a factor in limiting greenhouse gas emissions. The purpose of this article is to seek an answer to the question of whether "green taxes" as an instrument of tax policy are a significant factor influencing climate change by contributing to reducing greenhouse gas emissions. This article attempts to identify the determinants of greenhouse gas emissions (the dependent variable) using the method of linear regression analysis. Multiple linear regression models are used to predict the value of the dependent variable based on the values of the independent variables (identified from the literature). Trading of CO_2 emissions was not included in the analysis due to lack of data. The regression analysis was carried out using specialized statistical software (SPSS). The authors negatively verified the hypothesis that environmental taxes are a significant determinant of greenhouse gas emission reductions compared to the analyzed determinants. "Population", "current and capital transfers for environmental protection", and "supply, transformation and consumption of solid fossil fuels" are the most important factors influencing greenhouse gas emissions. Changing consumer behavior (as an effect of the non-fiscal function of taxes) appears to be an extremely important factor in reducing greenhouse gas emissions. Hence, the public authorities should promote behaviors conducive to their reduction by means of incentives, and not mainly taxation of negative behavior or fiscal incentives.

Keywords: green taxes; greenhouse gas emissions; indicators of reduction in greenhouse gas emissions; sustainable development

Citation: Filipiak, B.Z.; Wyszkowska, D. Determinants of Reducing Greenhouse Gas Emissions in European Union Countries. *Energies* **2022**, *15*, 9561. https://doi.org/10.3390/en15249561

Academic Editors: David Borge-Diez and Behnam Zakeri

Received: 2 November 2022
Accepted: 14 December 2022
Published: 16 December 2022

1. Introduction

Industrial growth throughout the 19th and 20th centuries caused an increase in greenhouse gas emissions as well as emissions of other gases into the atmosphere, resulting in global warming. There have been numerous changes to the global climate, and international organizations and governments have started to pursue policies to combat climate change [1]. The scientific community has acknowledged climate change as a research subject but, unfortunately, not all countries perceive climate change and the need to reduce greenhouse gas emissions as a political priority [2].

Reducing global carbon emissions is complex and challenging for scientists and politicians. However, the literature on the subject shows the importance of various instruments of influence (e.g., tradable quotas, greenhouse gas emission allowances, a system of accounting for greenhouse gas emissions, recognizing liabilities due to greenhouse gas emissions and their settling, and tax systems with their fiscal and non-fiscal functions) [3,4] and recognizes green taxes—which are also often referred to as environmental taxes—as an important instrument of influencing behavior towards sustainability [5–7].

Environmental taxes have a real impact on the adjustment activities of the enterprises, society, and public entities that pay taxes for the use of the environment [8]. It should also be emphasized that, in addition to tax policy, EU countries have implemented a variety of economic instruments for environmental regulation. These flexible and cost-effective tools are used to correct market failures and internalize externalities in a cost-effective way, unlike administrative or regulatory measures that tackle climate problems by setting maximum allowable emissions limits, banning the use of some materials, or enforcing the application of advanced pollution abatement technologies [9].

The conditions for the use of tax policies containing environmental tax instruments by governments differ depending on the country, membership of a community of states (e.g., the European Union), common policy, environmental value, awareness, or identification with the SDGs' objectives. In countries based on a highly developed awareness of the sustainability and goals of SDGs, there are requirements and tax policies that have a very strong impact on changes with a positive impact on the climate and limiting greenhouse gas emissions. The climate targets for reducing greenhouse gas emissions are becoming increasingly ambitious [10]. In 2019, the "European Green Deal" [11] was adopted, on the way to energy neutrality for the European continent. In 2021, EU countries agreed to reduce emissions by 55% by 2030 compared to the 1990 levels [12]. A key determinant of the promotion of the above actions is their financing and tax incentives. This dimension of actions is regulated by a number of formal and legal instruments, including, among others, Commission Implementing Regulation (EU) 2020/1294 of 15 September 2020 on an EU Renewable Energy Financing Mechanism [13]. "Green" taxes create incentives for business investments to develop and use alternative low-carbon fuels and technologies [10]. This is an important argument because the efficiency of actions in achieving climate targets may be limited by the existing mechanism for the allocation of free emission allowances [14]. The increasing price of "brown" energy (especially polluted) caused by the energy tax leads to a decreased demand for polluted energy sources, caused by a substitution effect [15]—a common argument that is used by society and is strongly associated with non-financial factors (ESG factors), especially with social responsibility for climate change. The literature shows that the impact of normative information varies across climatic mitigation contexts, depending on familiarity and behavioral factors [16]. The new regulations on social and environmental risk management are becoming more and more important in terms of climate responsibility and are related to the non-fiscal function of taxes.

The research shows that green taxes play a very important role for the environment, and that we risk a backlash of increased greenhouse gas emissions if they are abolished [17]. Numerous analyses also show green taxes as an instrument of sustainability [15,16,18]. Dulebenets (2018) binds taxes and advances environmental sustainability. It also indicates the importance of green taxes in the context of estimating the external environmental costs [19]. The literature on the subject also shows that governments must ensure that their green taxation policies are strong enough and are able to mitigate the risk and, thus, affect sustainability [20].

The review of the literature shows that "environmental taxes"—and especially "green" taxes—play a fiscal role, and very rarely a stimulating one. In view of the growing importance of ESG factors (especially in EU member states), there is a need to examine whether "green taxes" have a non-fiscal function. There is also no comprehensive analysis of many factors against the background of the tax factor to see which factors affect the reduction in greenhouse gas emissions and whether "green" taxes are really a significant factor influencing the reduction in greenhouse gas emissions.

Therefore, it is necessary to explain whether, along with other factors, environmental policy and its component "green taxes" can act as a factor in limiting greenhouse gas emissions.

The purpose of this article is to seek an answer to the question of whether "green taxes" as an instrument of tax policy are a significant factor influencing climate change by contributing to reducing greenhouse gas emissions. Since the fiscal function of green taxes is commonly known and confirmed, we sought answers as to whether green taxes can perform a motivational function (non-fiscal) as an instrument of sustainability. The research hypothesis assumes that environmental taxes are a significant determinant of greenhouse gas emission reductions compared to the analyzed determinants. The detailed goals can be defined as follows:

1. The accomplishment of a review and order of terminology relating to environmental taxes and the purposes of their application;
2. Determination of the determinants of greenhouse gas emissions;
3. Designation of the determinants of greenhouse gas emissions in European Union (EU) countries using econometric modeling (multiple linear regression models);
4. Determining whether green taxes are a significant contributor to greenhouse gas emissions.

This paper is organized as follows: The introduction is presented in Section 1. Section 2 contains a literature review and arguments pointing to theoretical research on the impacts of various factors in reducing greenhouse gas emissions. Section 3 presents the methodological approach, data collection procedure, and description of the methods. Section 4 presents the research results. Section 5 presents a discussion of the research results obtained in the context of the existing view that green taxes are used as an instrument for changes towards sustainability. Section 6 presents the conclusions.

2. Literature Review

In addition to orders and bans as well as public spending on pro-ecological adaptation, "environmental taxes"—also known in practice as "green taxes"—are an important instrument. The implementation of the "green taxes" system, as an instrument of government policy for sustainability, is aimed at triggering adjustment reactions both on the market and among the public, in accordance with the needs of environmental protection. The literature review shows that we are dealing with heterogeneous terminology, and "environmental taxes" [21] are very often referred to as "green taxes" [7] or "climate change mitigation taxes" [22,23]. The lack of clarity in defining the concept of "environmental taxes" does not change the fact that they serve the objectives of climate change mitigation policy and contribute to the achievement of environmental objectives—especially the SDGs [21,24]. Eurostat [22] defines environmental taxes as taxes that can be classified by environmental categories, dividing them into four basic categories of taxes: energy, transport, pollution, and resources. According to Eurostat [22], these four groups of taxes can mitigate the negative impacts of various factors—including greenhouse gas emissions—on the environment. The outline of the meaning of environmental taxes, taking into account their impact on water and air as basic resources, is presented in Figure 1. The analysis of the meaning of the data collected by "Eurostat" from the point of view of Eurostat accounts, as well as the definitions of the concepts of "green taxes" and "environmental taxes", indicate that their scope is the same, while from the point of view of the concepts they differ in terms of risk factors. Risk is not taken into account in Eurostat accounts, as it is difficult to quantify objectively. Therefore, the notions of "environmental taxes" and "green taxes" are often equated.

The importance of risk is of particular importance in relation to sustainability, especially through the prism of non-financial factors. For sustainable development, financing sustainable change policies [23], and measurement of ESG risk—and in particular for financing—its inclusion in tax structures will be important in the future for "greening", using tax breaks or even tax expenditures.

The goal of introducing such green taxes is not to achieve fiscal objectives, but to achieve environmental objectives, which are non-fiscal [25]. The theoretical groundwork for the green tax concept was prepared by A.C. Pigou in 1920 [26], who postulated to include so-called externalities in the tax system—in this case, environmental ones. This means that the most important goal of environmental taxes is to encourage entities that create pollution to act in more environmentally responsible ways and, in essence, to "go green" and "go to reduce the climate change" through ESG risk reduction. Therefore, from the point of view of achieving the goals that the government wants to achieve, the terms "environmental taxes" and "green taxes" are often used interchangeably [27].

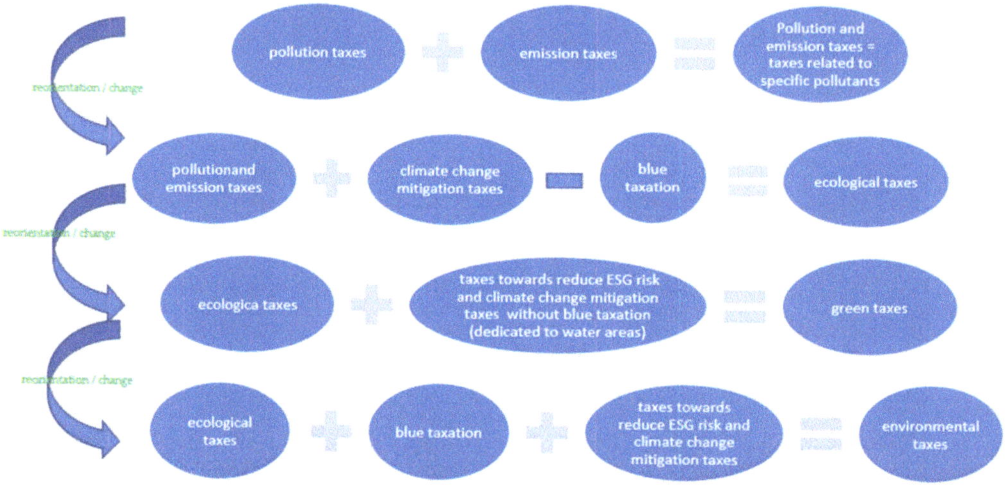

Figure 1. The meaning of environmental taxes. Source: own elaboration.

In the context of this literature review, we wish to present a description of the most relevant directions of study that can allow the identification of factors affecting the reduction in global carbon emissions through politics and tax instruments, the assessment of their impact, and the answer to the question of whether taxes are an important instrument contributing to the reduction in greenhouse gas emissions. Various authors in the literature on the subject represent the following views on "environmental taxes":

1. They are based on a new philosophy of ecological economy, taxing negative externalities—not goods [28–33];
2. The essence of the action of the tax itself affects the behavior of entities that are obliged to pay a public tribute [8–10,15,17,18,34–39];
3. They can be a significant source of budget revenues and stimulate dynamic technological innovations [40–43];
4. They change the face of the economy, allowing "green growth" to be achieved, and changing the face of the economy from traditional to circular, taking into account ESG factors [37,43–53];
5. They should be considered in the context of the theory of tax optimization and the effects of optimization [25,43,54,55];
6. Taxes can be considered in the context of prices and their role in the EU-ETS [4,9,14,15,56,57];
7. They imply a reaction to taxation (social assessment, adaptive responses of enterprises and society, tax evasion, acceptance of tax incentives, strategies towards taxation, desirable adaptive behavior, etc.) [55,58–64];
8. They enable progress to be made in the integration and sustainable development of their countries (bringing measurable effects such as greenhouse gas emission reductions or counteracting climate change); in other words, the research development trends related to the use of environmental taxes as tools to understand the determinants of acceptability for taxes directed at sustainability [47,65–72].

The basic condition for the effectiveness of environmental policy instruments is the possibility of internalizing negative environmental externalities arising from the activities of enterprises or entities. In practice, this means transforming these effects into internal costs (in the form of fees, e.g., ecological taxes) of the functioning of economic entities causing pressure on the environment. It can be indicated that financial instruments should have such an effect in order to force economic entities to reduce the pressure that they place on the environment. The concept of negative effects was defined by Baumol and Oates [28] as the results of the actions of some entities on the utility or production functions of other entities. Dasgupta and Heal [29] indicated that externalities are the result of insufficient incentives to create efficient markets and production factors, causing the market equilibrium to fail to meet the Pareto optimality conditions. External effects are subject to correction, and the most common forms of correcting market misallocation are direct regulations—most often in the form of orders to limit the production of pollutants to optimal levels. However, their role is limited in situations where a large number of entities participate in the creation of the same external effect. There is a problem of how to distribute the limitations resulting from this effect among individual entities [30].

Achieving sustainable development goals; connecting the environment, economy, and society; eliminating externalities; and taking into account the ESG factors are dependent on the use of sufficient capital to finance the long-term transition and long-term changes of the real economy—especially in enterprises and financial institutions [31]. The basic condition is to provide complementary forms of finance for low-carbon investment (the so-called transition to sustainable and responsible finance), which combines what is most effective with the private and public financial system [32,33].

In the public financial system, environmental taxes constitute a combination of prices and standards, designed not to achieve a Pareto-efficient allocation but to achieve a preset arbitrary environmental target contributing to the achievement of the SDG objectives and solving specific sustainability problems [17,18]. Furthermore, this detour from the theory of optimal taxation leads to a more pragmatic approach precisely because "the level of acceptable pollution is not a question of economics, but of environmental as well as of social (particularly intergenerational) justice considerations and can be set by the government" [34–37].

The literature review shows many trends and studies on the role of environmental taxes in reducing greenhouse gas emissions, as well as the role of green taxation and its impacts on the economy and society [8–10,15]. In addition to serving a non-fiscal function related to mitigating environmental problems—such as reducing greenhouse gas emissions, pollution, and degradation of nature—green taxes are regarded as market-based, incentive-driven mechanisms to stimulate desired sustainability and mitigate climate change. They are identified in the literature as being more efficient than so-called regulation and control mechanisms, and their acquisition costs are usually low [38]. The literature analyzes the impact of factors such as GDP per capita, population, renewable energy, energy intensity, and the economic crisis on GHG emissions [39]. The findings show that green taxation can also help promote sustainable growth, support intergenerational fairness, and maintain tax revenue levels for EU member states. Such research confirms the direction of the action of the tax itself on the behavior of entities that are obliged to pay a public tribute.

In many countries, green taxes differ in many ways, and their use as a source of budget revenues also differs [40]. The following applications are indicated in the literature on the subject: the green tax revenue contributes to the general public budget without being tied to environmental or sustainability goals (fiscal functions); the revenue from green taxes is recycled (partly or wholly) as reductions in social security taxes [41]; some of the green taxes are used to compensate polluters or to subsidize sustainable investments in technology favoring the fight against climate change, and their use is indicated for innovations conducive to climate change [42,43].

The macroeconomic theory indicates that different types of taxes fulfil the function of a repressing function in the economy and society [44]. Tax instruments can prompt people to adapt to specific government policies with varying degrees of force. The idea of using taxation to correct negative externalities [43]—especially those relating to climatic factors—is generally credited to Pigou's theory (1920) [26]. However, it should be remembered that the introduction of new instruments and techniques of taxation (e.g., new tax rates, exemptions, subsidies), while striving for growth of the tax income of the government, puts a downturn risk on the economy [45]. Hence, numerous studies refer to the impacts of environmental taxes on GDP and the economy itself [46–48].

Green taxes are discussed in the literature regarding their potential effects on resource savings [49,50]. However, they are also examined from the point of view of affecting types of funding [51]. Green taxes internalize the environmental and social externalities of resource extraction [52] and belong to the instruments of policies for addressing resources [53], which contribute to reducing negative environmental effects from the use of resources [54]. Thus, on the one hand, environmental taxes influence GDP, and on the other, by influencing resources, they contribute to the shift towards a circular economy. Vence and Pérez indicated their low efficiency and low level of achievement of environmental goals (including greenhouse gas emissions) [37]. The authors discussed the impact of taxes on the circular economy. To strengthen the circular economy and sustainable growth, activity should focus on reform and make use of available tax expenditure measures, including the many tax benefits, exemptions, deductions, and allowances applicable to existing large taxes. The tax expenditure in the general tax policy, according to Vence and Pérez (mainly with non-environmental purposes), could be reshaped and used to promote the transition towards a circular economy (replacing all environmentally harmful subsidies and tax benefits with a tax treatment favoring all circular and sustainable activities) [37].

In the context of the influence of ESG factors, the presence of negative externalities should be emphasized, which is related to market failure. This applies especially to greenhouse gas emissions, as indicated in the literature on the subject [36]. Imposing taxes on externality-generating goods can correct the externality, but in the case of greenhouse gas emissions there are complication in the operation of this theory. These complications are related to the fact that the marginal external damage caused by a good varies based on who produces the good or how it is produced. Greenhouse gas emissions from a power plant that burns natural gas are much lower than the greenhouse gas emissions from electricity produced by burning coal [43].

In the literature on the subject, environmental and green taxation is discussed in the context of the theory of tax optimization [25]. Optimal taxes correspond to the shadow prices that are generated in the social optimum, as indicated by Ploeg and Withagen [54]. They also argue that, on a global scale, optimal taxes are difficult to use in public policy. In order to influence greenhouse gas emissions, new technologies are indispensable, and it is necessary to influence the development of clean, "green" technologies with fiscal instruments. Moreover, only a few countries choose green taxes, whereas most nations want to implement a flexible approach based on subsidies for renewable energy, which are seen as an instrument for influencing technological change toward reducing greenhouse gas emissions [56]. Carattini et al. (2017) showed that the implementation of green taxation has proven difficult in many nations [43,55], due to the resistance of large companies, but also to rising commodity prices due to rising energy prices caused by the imposition of CO_2-related taxes.

The literature also indicates, in the context of the EU ETS, the effects of their introduction may transcend climate mitigation and, thus, extend beyond the impact on greenhouse gas emissions [14]. In addition to cutting emissions, there are other goals—such as environmentally friendly investments, security of energy supply, local pollution reduction, or industrial development. The challenge lies in calibrating policies, which means broader inclusion of green taxation policy in the EU ETS [56,57]. In addition, an economically efficient instrument is to provide a carbon price signal that increases over time [4]. This price and tax signal makes it possible to calibrate the cost of GHG emissions to society as a whole and to encourage governments to reduce the consumption of those energy sources that cause the highest pollution. Thus, the "polluter pays" principle was consolidated [9,15].

Another trend represented in theory relates to voters' reactions to green taxation [55,58]. The literature review shows the importance of considering voters' preferences, as well as the consideration of the economic objectives of voters over public interests and the public good, including environmental preferences. In addition, the use of green tax opinion by the media becomes a problem, possibly leading to erroneous opinion formation and incorrect understanding of political decisions among voters [59]. The public's responses to green taxes aimed at reducing greenhouse gas emissions can include membership and activity of citizens in climate protection advocacy groups and active lobbying of these groups in the political arena. Such groups, through lobbying, can apply pressure for the use of other, non-tax instruments that are more effective [60]. Baranzini et al., in 2017, indicated that green taxes represent a basis to study the consequences of informational asymmetries between citizens, policymakers, and experts [61]. In conclusion, it should be noted that the information asymmetry theory plays a key role in explaining the reactions of society and various entities to "green taxation" related to reducing greenhouse gas emissions.

In theoretical considerations, there is also a visible trend indicating the application of diversified tools and instruments to understand the determinants of the acceptability of "green taxes". Conducting qualitative assessments has helped in understanding of the obstacles to tax reform and to the introduction and modification of environmental taxes [62,63]. Using quantitative and qualitative methods, the high level of distrust in environmental tax reforms among the general public has been demonstrated [62,64]. Research has led to the conclusion that society may only be willing to support the introduction of "green taxes" if their revenues are clearly earmarked for environmental purposes.

Vence and Pérez point to the proliferation of new, specific, and relatively marginal taxes related to environmental goals. The indicated studies of Vence and Pérez should also be considered in the context of the current research, which shows that depending on the design of the environmental taxation, a double dividend can be obtained (i.e., a benefit to both the environment and the economy) [37,65–67]. The study of Freire-González et al. (2022) reinforces the theory that authorities need to impose green taxes to stimulate and reinforce the circular economy, reduce greenhouse gas emissions and environmental burdens, and act for the climate, but also demonstrates that they can improve their design without additional costs [68].

An important aspect of the activities undertaken within the framework of tax policy is the pursuit of the restriction of climate change. Both the studies presented and the activities of many countries show the need for tax reforms—in particular, environmental tax reforms (ETRs) in the national legislation of EU countries. Criticism of the application of environmental tax mechanisms in both the literature and the assessment of tax policy has led to a transfer of the tax burden from factors of production to polluters themselves, summarized as a step from economic "goods" to environmental "bads" [47,65,69,70]. This idea, in particular, provides the basis for the introduction of "energy taxes" and a "green taxes system", with the aim of stimulating the reduction in greenhouse gas emissions.

In this context, the link between environmental taxes and air pollution (and other emissions) has also been recognized by various researchers. The literature on the subject [61,70–72] shows the impact of fiscal spending patterns on the environment by taking into account CO_2, greenhouse gas emissions, and other emissions.

"Green taxes" are those for which the tax base is a physical unit (or a substitute for a physical unit) of a good that has a proven, specific, and highly negative impact on the climate and environment [27], i.e., emissions that meet the negative impact on environment criterion, in particular on natural resources, air, water, animals, or humans and society.

Research by Rybak et al. (2022) [24] showed that the direction and strength of the impacts of green taxes differ depending on the greenhouse gas. Environmentally responsible tax policies can guide entities to circular production and make them more environmentally efficient, in an effort to limit greenhouse gas emissions, reduce negative environmental and climate impacts, and further the development of new markets. This is the battle between the so-called "crowding-out effect" and the "Porter effect". In the literature on the subject, it has been shown that green tax incentives play a role as stimulators of technological changes towards green transformation and sustainability [44,73–76].

It should be emphasized that the eight groups of tendencies highlighted in the views on "environmental taxes" do not exhaust the discussion, which continues and is still being developed. The intention of the authors was to show the fiscal and non-fiscal functions of green taxes. In fact, the non-fiscal functions are developed only in the views of groups 4 and 8.

Against the background of the discussion on the specificity, role, and meaning of "environmental taxes", the idea of research on the factors determining greenhouse gas emissions was developed. The purpose of this article is to search for the answer to the question of whether "green taxes", as an instrument of tax policy, are an important factor affecting climate change by limiting greenhouse gas emissions. This goal prompts us to analyze the factors influencing greenhouse gas emissions. Considering the factors contributing to reducing greenhouse gas emissions based on the literature on the subject, we have seen that these factors are considered on a case-by-case basis. This study considers many factors, and the methodological approach results in an empirical model in the form of a regression equation.

Table 1 presents the factors influencing greenhouse gas emissions discussed in the literature on the subject, as well as the relationships between them that have been analyzed in the literature.

Table 1. Determinants of greenhouse gas emissions—literature review.

Authors	Investigated Determinants
K.M. Azis, T. Widodo (2019); B. Jóźwik, A. Gavryshkiv (2020)	GDP, environmental tax revenues [48,77]
L.N. Hao et al. (2021)	Environmental tax, human capital, development of environmental technology, consumption of renewable energy, GDP [78]
G. Liobikiene, M. Butkus (2017)	Economic growth, energy consumption, and renewable energy sources [79]
R. Hashmi, K. Alam (2019)	Environmental tax, environmental technologies, patents, prices for CO_2 emission allowances [80]
G. Lapinskiene et al. (2017)	Economic growth, energy consumption, energy taxes, and research and development [81]
L. Meng, B. Huang (2018)	GDP per capita, industrial sector, population, and energy consumption, among a set of qualitative variables (i.e., policy, government effectiveness, and location) [82]
I.M. de Alegría et al. (2016)	GDP per capita, population, renewable energy, energy intensity, financial and economic crisis [39]
L. Andrés, E. Padilla, (2018)	Population, economic activity, transport volume, transport energy intensity, and transport activity composition in terms of modal share and of energy source mix [83]
M. González-Sánchez, J. L. Martín-Ortega (2020)	Population, GDP per capita, temperature—heating degree days, temperature—cooling degree days, primary energy consumption, final energy consumption, primary and final energy intensity, fuel prices, carbon intensity, RWE, structure of the economy, fuel mix [40]
European Environment Agency (2019)	Energy efficiency, renewables, structural changes in the economy towards a circular economy, temperature, GDP, emission reduction measures, carbon intensity, switch from gasoline to diesel and use of biofuels, fuel mix [84]
Federal Environment Agency (2019)	Higher technical efficiency due to the closure of plants (e.g., replacement of old lignite plants), increasing use of renewable and nuclear energy, electricity demand, production evolution (e.g., petroleum refining), production level (GDP), emission reduction measures, refueling in other countries (fuel prices), substitution of diesel fuel for gasoline, use of admixtures with biodiesel, fuel changeovers (fuel mix), higher energy and technical efficiencies, and temperature [85]
Environmental Protection Agency (2019)	GDP, fuel mix, changes in the fuel mix (e.g., displacement of oil by natural gas), energy efficiency, GDP, share of renewables in gross electricity consumption, wind and hydro electricity generation (precipitation and wind), production levels (GDP), fuel mix (large increases in the use of petroleum coke), closure of high-energy-intensity production plants, road transport volumes (vh kn, passenger fleet and goods vehicles), fuel tourism (fuel price), impact of registration tax and road tax introduced in 2008, biofuels obligation scheme, and population [86]
Agencia Portuguesa do Ambiente (2019)	GDP, energy demand, mobility, investment in renewable sources, energy efficiency, precipitation, GDP, energy demand, production levels (GDP), vehicle fleet, income and investment in road infrastructure [87]
Witkowski et al. (2021)	The cumulative rate of return—the criterion of missing emission allowances to allocate to companies between portfolios, thereby taking into account companies' exposure to changes in the price of CO_2 emission allowances—was analyzed [14]
Rybak et al. (2022)	Energy, m. EUR Transport, m. EUR Resources, m. EUR Pollution, m. EUR CO_2 Emission, Mg CH_4 Emission, Mg [24]

Source: own study.

The direct and indirect impacts of environmental taxes on climate change may not happen instantaneously. There are many factors to consider and key factors to be established.

Greenhouse gas emissions are undoubtedly a very important factor causing climate change but, as shown in Table 1, the indicated factors were analyzed and their impacts on climate change were demonstrated. On the one hand, Table 1 shows the specific variables analyzed in the literature on the subject, the influence of which has previously been confirmed. On the other hand, we have industrial research with the use of specific variables describing the industries most associated with greenhouse gas emissions. Table 1 indicates the significant variables that are often presented in the literature on the subject: production level (GDP), emission reduction measures, use of various energy sources (e.g., RWE, gasoline, diesel), energy productivity, factors influencing energy efficiency, transport, and population.

Therefore, we can see the desirability of the factors listed in Table 1 to show their impact on reducing greenhouse gas emissions. As shown by the literature review, single variables have been studied, while the impact of many variables in the context of their relationship with environmental taxes has not been analyzed, which is the authors' own contribution and an extension of the empirical research conducted so far.

3. Materials and Methods

When starting to elaborate the methodology of research on the factors influencing greenhouse gas emissions, on the basis of the analysis of the literature concerning this topic, we determined the measures illustrating these factors, as well as the method of cause–effect analysis, where the explained variable was the amount of greenhouse gas emissions and the explanatory variables were amounts showing the diagnosed factors. The stages of this work are presented in Figure 2.

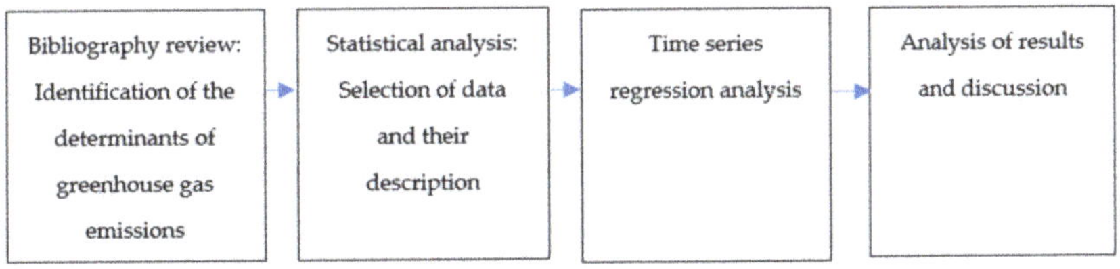

Figure 2. Methodological framework. Source: own study.

In order to select the explanatory variables, the evolution of greenhouse gases in the European Union countries was additionally analyzed by emission sector (Table 2).

Table 2. Greenhouse gas emissions in the European Union countries, by sector (%).

Sector	2010	2015	2019
Energy, including:	79.42	77.97	77.08
Energy industries	38.08	36.88	31.55
Manufacturing industries and construction	14.30	14.44	15.53
Transport	24.75	27.13	30.52
Other sectors	22.88	21.56	22.39
Industrial processes and product use	8.27	8.77	9.06
Agriculture	8.83	9.99	10.54
Waste	3.48	3.27	3.32
Total	100	100	100

Source: own calculations based on the OECD database (https://data.oecd.org/) https://data.oecd.org/searchresults/?q=green+gas (accessed on 10 July 2022).

In order to determine the coexistence of both analyzed categories (i.e., environmental taxes and greenhouse gas emissions), we calculated the Pearson's correlation coefficients, and their statistical significance was verified using the test of significance for Pearson's product–moment correlation coefficients.

When making the final selection of variables for the description of the greenhouse gas emission quantity, on the one hand, we took into account their best fit to the description of the analyzed factors (categories), and on the other hand, the availability of data was considered (e.g., comparability of data; availability of long time series for many EU countries). Selected variables reflecting the identified factors are presented for selected years. In the event of the absence of data for a country for a specific year, the value for the next year was used.

Determining the relationships between the indicated variables made it possible to determine the impact of environmental taxes on greenhouse gas emissions in comparison with the other defined variables. Modeling allowed us to determine whether environmental taxes are an important determinant of greenhouse gas emissions in the European Union countries. Assuming such an aim of the research, we decided to use the analysis of interdependence of variables in multiple linear regression [88–92]. This enabled us to study the existence of correlation between the categories under consideration, which is a condition for a cause-and-effect relationship between them. This method enables the construction of models of linear dependence between many variables. The result is an empirical model in the form of a regression equation that takes the following form:

$$y_t = b_0 + b_1 x_{1t} + b_2 x_{2t} + \ldots + b_n x_{nt} + \varepsilon_t, \quad (1)$$

where the meaning is as follows:

y_t—the explained variable at time t;
x_t—the explanatory variable at time t (predictor);
b_i—unknown regression coefficients $i = 0, 1, 2, \ldots, n$, where b_0 denotes a free term;
n—the number of explanatory variables;
ε_t—a random component expressing the influence of all those factors that were not included in the model on the dependent variable.

Linear regression requires the assumption that the relationship between the variables is linear. In practice, the validity of this assumption is almost impossible to prove; however, multiple regression procedures are quite resistant to slight derogations from this assumption [93]. Regression allows us to estimate how the dependent variable changes as the independent variables change. The main advantages of using this modeling are its simplicity, interpretability, and speed. The indicated advantage—simplicity—is at the same time its greatest disadvantage, because the surrounding reality does not consist of simple linear relationships. The disadvantages of using a multiple regression model are also related to the use of data, which can lead to false conclusions. In addition, outliers greatly distort the results. Hence, it is necessary to carry out activities minimizing the occurrence of these problems (e.g., collecting and preparing data, examining their quality, initial model construction, verification/validation). Before modeling, it was established that the variables were continuous, the observations were independent of one another, and there were no significant outliers.

When building subsequent models, the determination coefficient was used as a measure of the models' fit to empirical data, and significance tests (F-test) and the empirical significance level (the so-called p-value) were used to determine the significance of the changes. The analysis of residuals was also used to evaluate the models. After the models were estimated, they were verified in order to confirm the accuracy of the assumptions of the least squares method.

Regression coefficients are estimated using the least squares method (LSM). The constructed multiple linear regression models must conform to the assumptions regarding their quality. The intention was to search for such models that would allow the explanation of the explained variable—greenhouse gas emissions (GGEs)—to the highest degree, with simultaneous consideration of the low value of the variance inflation factor (VIF) for particular explained variables.

The basic measure of regression matching is the coefficient of determination R^2, which describes the strength of the linear relationships between variables, i.e., the match of the regression line to empirical data. The coefficient of determination takes values in the range [0,1] and indicates what part of the variability of the GGE variable is explained by the estimated model. The higher the level of the determination coefficient, the more of the variability in the dependent variable is explained by the model. Another important parameter used to assess the quality of the regression model is the significance coefficient, the value of which should not exceed 0.05. The model should be matched with independent (explanatory) variables that are strongly correlated with the dependent (explained) variable but weakly correlated with one another, so that the phenomenon of collinearity does not occur, which would weaken the quality of adjusting the model to reality. The measure of collinearity is the tolerance coefficient or its reciprocal—called the match error inflation or variance inflation factor (VIF). Explanatory variables with a large VIF value should be eliminated from the model [94].

The analysis of interdependence of variables in multiple linear regression is used to a rather limited extent in social research [95]. So far, it has not been used to study the relationships between environmental taxes and greenhouse gas emissions. The constructed model is an original proposal to study the abovementioned relationships. The use of regression analysis allows determination of the strength of the influence of individual explanatory variables on the explained variable.

When starting the modeling, all correlation coefficients between the variables were calculated. When looking for regression models that explain the explained variable as best as possible, the following factors were taken into account: the amount of the regression coefficient, the value of the coefficient of variance (VIF), the possibility of interpreting the results, and the level of significance. Variables that were strongly correlated with one another were eliminated from the models.

Before starting the multiple regression analysis, the variables (predictors) influencing greenhouse gas emissions were selected regarding substantive considerations, the universality of the measures that were used, and their comparability, variability, and importance, as well as their availability [96]. It should be emphasized that the selection of variables (predictors) was influenced by the analysis of the literature on the subject, where we checked which variables (predictors) were analyzed and why (Table 1). From the set of statistical features used to build the greenhouse gas emissions regression model in the EU countries, 14 variables were selected. Among the described predictors are indicators identified in the theoretical part as determinants of greenhouse gas emissions. The complete set of variables is presented in Table 3.

The variables indicated in Table 3 are mostly destimulants (i.e., their impact on greenhouse gas emissions should be negative), except for HC, CO_2, and FWA, which are stimulants (i.e., their impact should be positive). We also decided to introduce two nominative variables: GDP and ETR. When searching for regression models that best explain the dependent (explained) variable—greenhouse gas emissions (GGEs; thousands tonnes of CO_2 equivalents)—the following factors were taken into account: the value of the regression coefficient, the value of the variance inflation factor (VIF), the possibility of interpreting the results, and the level of significance. Variables that were strongly correlated with one another were eliminated from the models. The calculations were made with the use of statistical data analysis software (SPSS–PS-Imago version 7.0). A stepwise regression analysis was used, allowing us to enter into the model only those variables (predictors) that significantly affect the dependent variable [96,97]. This allowed us to eliminate unnecessary variables that did not contribute anything to the model and, thus, to obtain only those variables affecting the prediction of the dependent variable. At the same time, the stepwise method allows for the elimination of the problem of collinearity. Successively introduced predictors also take into account the mutual correlation between them. In the regression analysis, we used statistical data for 2020 (as the last year for which all data were available), along with data for 2018, 2016, 2014, 2012, and 2010 for comparison. The data for the analysis were derived from the resources of Eurostat, the OECD, the European Environment Agency, and the Food and Agriculture Organization of the United Nations. Models were built on the basis of data for each EU country in the selected years.

Table 3. Set of variables shaping the greenhouse gas emissions in EU countries.

Symbol	The Predictor	Unit	N/D/S *	Data Source **
GDP	GDP and main components (output, expenditure, and income)	Million EUR per capita	N	Eurostat
EP	Energy productivity	EUR per kilogram of oil equivalent (KGOE)	D	Eurostat
HC	Supply, transformation, and consumption of solid fossil fuels—final consumption, hard coal	Thousand tonnes	S	Eurostat/EEA
REC	Final energy consumption in households by fuel—renewables and biofuels	Percentage of total consumption	D	OECD
FA	Forest land in % of the country's land area	Percentage of total area	S	FAO
OF	Area under organic farming	Percentage of total utilized agricultural area	D	OECD
CO_2	Average CO_2 emissions per km from new passenger cars	Average carbon dioxide (CO_2) emissions per km from new passenger cars in a given year	S	Eurostat
EHC	The share of electric and hybrid passenger cars in total passenger cars	Percentage of total passenger cars	D	Eurostat
ETR	Environmental tax revenues	Million EUR	N	Eurostat
EPT	Environmental protection transfers by environmental protection activity and institutional sector—current and capital transfers for environmental protection	Million EUR	D	Eurostat
EPI	Environmental protection investments of total economy	Million EUR	D	Eurostat
RRW	Recycling rate of municipal waste	Percentage of total municipal waste	D	Eurostat
FWA	Annual freshwater abstraction by source and sector	Million cubic meters	S	Eurostat
P	Population	Persons	S	Eurostat

* N/D/S: nominee/destimulant/stimulant. ** Eurostat: https://ec.europa.eu/eurostat/web/main/data/database (accessed on 10 May 2022); OECD: https://stats.oecd.org/ (accessed on 10 April 2022); FAO: https://www.fao.org/faostat/en/#data/RL (accessed on 10 April 2022); EEA: https://www.eea.europa.eu/data-and-maps (accessed on 10 February 2022). Source: own study.

At the stage of preparing the data for analysis, the descriptive statistics of all variables were calculated. An example of the results of the analysis for 2020 are included in Appendix A. The aim of the study was to identify factors influencing greenhouse gas emissions, including green taxes. At the beginning, we aimed to build models for individual EU countries using data for the last available year in databases. For comparison, we did the same for additional selected years to confirm the results that we obtained.

4. Results

Table 4 presents the values of Pearson's linear correlation coefficients, which can be used for the initial assessment of the level of linear dependence (co-occurrence) between greenhouse gas emissions and environmental taxes.

Table 4. Correlation between greenhouse gas emissions and environmental taxes.

2010	2012	2014	2016	2018	2020
0.928	0.906	0.895	0.890	0.901	0.905

Source: own study with the use of SPSS software.

All computed Pearson's correlation indicators were statistically significant (p-value < 0.01). When analyzing the values of the correlation index, it can be concluded that there is a strong positive relationship between the analyzed variables. In the authors' opinion, this relationship should take negative values, because the introduction of environmental taxes should lead to a reduction in greenhouse gas emissions. The primary aim of introducing environmental taxes is to reduce the emissions of pollutants, not to obtain budget revenues. Nevertheless, it is justified, because an increase in gas emissions leads to an increase in fiscal sanctions (e.g., environmental taxes/green taxes) on the issuers.

When starting the modeling, we searched for multiple regression models that would correspond to the assumptions made regarding their quality. We searched for models that would allow the explanation of the dependent variable to the greatest extent, while taking into account the low value of the variance inflation factor (VIF) of individual variables. Attention was also paid to the level of the coefficient of determination and its significance level (F-test). All of the variables in the models were significant, as evidenced by their *p*-values (below 0.001). The characteristics of the obtained models are presented in Table 5.

Table 5. Analysis of variance (ANOVA).

Model	The Sum of the Squares	df	The Mean Square	F	Relevance
			2020		
Regression	7.439×10^{11}	3	2.480×10^{11}	280.062	0.000
Remainder	19,479,662,816	22	885,439,218.9		
Overall	7.634×10^{11}	25			
			2018		
Regression	1.170×10^{12}	5	2.339×10^{11}	328.038	0.000
Remainder	1.569×10^{10}	22	713,153,995.8		
Overall	1.185×10^{12}	27			
			2016		
Regression	1.242×10^{12}	4	3.105×10^{11}	289.183	0.000
Remainder	2.985×10^{10}	23	1,297,984,106		
Overall	1.272×10^{12}	27			
			2014		
Regression	1.240×10^{12}	3	4.134×10^{11}	277.982	0.000
Remainder	3.569×10^{10}	24	1,487,162,308		
Overall	1.276×10^{12}	27			
			2012		
Regression	1.375×10^{12}	3	4.583×10^{11}	268.442	0.000
Remainder	4.089×10^{10}	24	1,707,296,500		
Overall	1.416×10^{12}	27			
			2010		
Regression	1.457×10^{12}	3	4.856×10^{11}	243.794	0 < 001
Remainder	3.768×10^{10}	24	19,917,117,777		
Overall	1.505×10^{12}	27			

Source: own study with the use of SPSS software.

Tables 6–11 present the results of the estimation of the regression coefficients along with the errors in their estimation, as well as the statistics on their significance and the levels of collinearity.

Table 6. Coefficients in the regression model of greenhouse gas emissions in 2020.

Variable	Standardized Coefficients—Beta	t	Relevance	Collinearity Statistics	
				Tolerance	VIF
Constant	7340.632	−0.238			
P	0.763	15.068	<0.001	0.452	2.211
ETP	0.226	4.460	<0.001	0.488	2.050
HC	0.147	4.100	<0.001	0.898	1.113

Source: own study with the use of SPSS software.

Table 7. Coefficients in the regression model of greenhouse gas emissions in 2018.

Variable	Standardized Coefficients—Beta	t	Relevance	Collinearity Statistics	
				Tolerance	VIF
Constant	−5130.058	−0.554	0.585		
P	0.772	19.211	<0.001	0.373	2.683
HC	0.131	4.849	<0.001	0.825	1.212
EHC	−0.105	−3.405	<0.001	0.634	1.578
EPT	0.232	5.981	<0.001	0.400	2.502

Source: own study with the use of SPSS software.

Table 8. Coefficients in the regression model of greenhouse gas emissions in 2016.

Variable	Standardized Coefficients—Beta	t	Relevance	Collinearity Statistics	
				Tolerance	VIF
Constant	−54,645.663	−2.482	0.021		
P	0.726	13.215	<0.001	0.339	2.950
HC	0.107	3.381	<0.001	0.807	1.239
EPT	0.271	5.306	<0.001	0.390	2.562

Source: own study with the use of SPSS software.

Table 9. Coefficients in the regression model of greenhouse gas emissions in 2014.

Variable	Standardized Coefficients—Beta	t	Relevance	Collinearity Statistics	
				Tolerance	VIF
Constant	−15,013.018	−1.585	0.126		
P	0.680	11.509	<0.001	0.334	2.966
HC	0.118	3.156	<0.001	0.833	1.200
EPT	0.315	5.627	<0.001	0.371	2.694

Source: own study with the use of SPSS software.

Table 10. Coefficients in the regression model of greenhouse gas emissions in 2012.

Variable	Standardized Coefficients—Beta	t	Relevance	Collinearity Statistics	
				Tolerance	VIF
Constant	−15,436.685	−1.509	0.144		
P	0.797	15.129	<0.001	0.435	2.300
HC	0.088	2.345	<0.001	0.850	1.176
EPT	0.210	4.206	<0.001	0.484	2.066

Source: own study with the use of SPSS software.

Table 11. Coefficients in the regression model of greenhouse gas emissions in 2010.

Variable	Standardized Coefficients—Beta	t	Relevance	Collinearity Statistics	
				Tolerance	VIF
Constant	−14,779.854	−1.314	0.201		
P	0.786	12.235	<0.001	0.321	2.119
HC	0.089	2.186	<0.001	0.805	1.242
EPT	0.207	3.391	0.02	0.357	2.804

Source: own study with the use of SPSS software.

Concluding, as a result of the modeling, with the use of the variables defined in the previous point, six multiple regression models were obtained for each year. Of the proposed variables, only four were introduced into the models, namely, P, HC, EHC, and EPT (depending on the year of analysis).

In the case of all of the constructed models, the coefficient of determination R^2 was obtained at a level exceeding 0.95, which means that the explanatory variables accounted for over 95% of the total variability in the development of greenhouse gas emissions. This proves the high quality of the constructed models. On the basis of the F-tests, it can also be concluded that the coefficient of determination was statistically significant. All of the variables adopted for the models were characterized by a variance inflation factor (VIF) value below 3.

The obtained results of the regression analysis indicate that greenhouse gas emissions in individual years are essentially shaped by the same variables. Table 12 presents a comparison of the variables and the strength of their impacts on greenhouse gas emissions in the individual years of the analysis.

Table 12. Explanatory variables and their strength in the regression model of greenhouse gas emissions.

Explanatory Variables	2010	2012	2014	2016	2018	2020
P	0.786	0.797	0.680	0.726	0.772	0.763
HC	0.089	0.088	0.118	0.107	0.131	0.147
EHC	-	-	-	-	−0.105	-
EPT	0.207	0.210	0.315	0.271	0.232	0.226
R^2	0.964	0.967	0.969	0.972	0.980	0.974

Source: own study.

We analyzed the impact of all variables (P, HC, EHC, and EPT), the values of which are presented in Table 12 for each year, using an empirical model in the form of a regression equation (Equation (1)) and supplementing the scheme with data from Table 12. The basic conclusion is that green taxes are not included in the models for individual years and that the values for individual variables for individual years do not differ significantly. This means that, in the years analyzed, environmental taxes did not significantly affect greenhouse gas emissions, and the values of the variables (factors) did not change significantly.

Among the variables introduced into the models, the population variable had the strongest impact on the explained variable—greenhouse gas emissions—in all years of the analysis (as marked in the darkest gray in Table 12). Apart from this, the explained variable was also significantly influenced by the variable environmental protection transfers according to environmental protection activity and institutional sector. In the case of any analyzed year, the model did not include the environmental taxes variable, despite the previously indicated coexistence of this variable with greenhouse gas emissions (significant correlation indicators). This proves that, despite the fact that there is a statistical relationship (correlation) between them, in combination with other explanatory variables of greenhouse gas emissions, environmental taxes do not contribute to the shaping of the analyzed dependent (explained) variable. The constructed regression models of one variable (explained variable = GGE, explanatory variable = ETR) indicate the existence of a cause-and-effect relationship between them. The models are characterized by a high coefficient of determination (R^2).

In each of the analyzed years, the main determinant of greenhouse gas emissions was the variable P. Populations of particular countries report the demand for all kinds of goods and services that generate gas emissions, especially when it comes to the need for all kinds of energy carriers. Currently, people are characterized by a high level of consumerism, especially in highly developed countries—which include EU countries. It seems necessary to undertake research on the consumption behavior of societies in the context of their awareness of and responsibility for reducing greenhouse gas emissions by limiting purchases and making choices about pro-environmental goods.

It is puzzling that environmental protection transfers based on environmental protection activities undertaken in countries do not result in reducing gas emissions but, rather, increase them. When starting the modeling, we assumed that these transfers would positively affect the reduction in gas emissions. However, the achieved results did not confirm this. This could be for a number of reasons. Firstly, their values are too low, meaning that they do not reduce emissions. Secondly, these transfers are used to reduce other environmental pollution, such as water pollution. The explanation of the observed relationship merits additional research.

In summary, the selected variables, which were initially adopted after studies of the literature to explain the development of greenhouse gas emissions in EU countries, allowed us to construct regression models for the selected years. The ETR variable was not introduced in any of the models, indicating that other factors have a greater impact on GGE.

5. Discussion

This study explored existing environmental taxes as a tool for counteracting climate change. We analyzed whether green taxes are an incentive tool for limiting greenhouse gas emissions and asked whether there is really an impact of green taxes on sustainable development in reducing greenhouse gas emissions. The literature on this subject analyzes various greenhouse gas emissions policies and instruments that could help curb the effects of climate change [37,40,58,65]. The assessment of these policies and tools in the literature on the subject varies and depends on the continent and country [76,98,99]. The factors contributing to reducing greenhouse gas emissions were considered individually. Therefore, in order to test the effectiveness of tools such as environmental taxes (among other), it is necessary to consider all relevant factors that may affect the reduction in greenhouse gas emissions.

The conducted analysis and research results in the first stage of our study are mostly consistent with the results that can be found in the literature [76,99] in relation to the most important variables influencing climate change. Table 13 presents the differences and similarities between the findings of this study and those of other scholars. The main difference was in the use of a research method that allowed for multivariate analysis, which is a new approach, as previous research did not include the analysis of 14 variables.

The main difference was in the use of a research method that allowed for multivariate analysis, which is a new approach, as previous research did not include the analysis of 14 variables. Despite the demonstration of the coexistence of the variable "Environmental taxes" with greenhouse gas emissions (significant correlation indicators), the variable illustrating "green taxes" was not included in the developed model. Therefore, our study showed that despite the existence of a statistical dependence between "green taxes" and greenhouse gas emissions, in combination with other explanatory greenhouse gas emission variables, green taxes are not a significant instrument for reducing greenhouse gas emissions.

The lesson learned from existing situations in the scope of greenhouse gas emissions and climate change is to develop a more effective policy framework, as also postulated in previous studies [25,49–51,68]. Our research shows that the model lacks "green taxes" as an important factor influencing greenhouse gas emissions and, thus, the climate. The lack of the "green taxes" variable in the model supports the argument that a more effective policy framework is necessary.

The EU countries are not only looking for solutions and tools, but also have to look for the determinants of greenhouse gas emissions so that their policy becomes effective. Until now, determinants have been considered and tested individually (Table 13). This is the first study to compare the determinants of greenhouse gas emissions. Although many factors were considered, some of them did not have a significant impact on greenhouse gas emissions. In the first stage, potential variables were selected for the linear regression model of greenhouse gas emissions, taking into account the availability of data and their substantive importance by specifying stimulants, destimulants, and nominees (Table 3). In the second stage, the selected variables were analyzed (based on descriptive statistics), and the modeling method was determined. In the last stage, the significance of the selected variables (predictors) in shaping greenhouse gas emissions was modeled (Table 12). The indicated studies of other scholars and their achievements show the importance of single determinants or just a few factors. Our model shows the dependence over time and verifies the meaning of as many as 14 variables.

Table 13. The differences and similarities between the findings of the present study and those of other scholars.

Authors	The Differences among the Findings	The Similarities among the Findings
Aidt (2010) [41]; Zhang et al. (2020) [42]; Zhang et al. (2021) [43]	The use of green taxes as a tool of change in the direction of technological changes and the study of the subsidization effect (public policy)	The impact of taxes was examined, and they were found to be a significant variable, but green taxes (as an instrument of public policy) were not a significant element influencing the behavior of entities
Vence and Pérez (2021) [37]; Williams (2016) [46]; Zhang et al. (2021) [43]	Different types of taxes perform a repressive function in the economy and society, using green taxation to correct negative externalities; the study did not cover the impact of greenhouse gas emissions	A similar research position that environmental taxes should serve to correct negative externalities, i.e., the effects of greenhouse gas emissions; the research focused on the impact of greenhouse gas emissions
Carattini et al. (2017) [55]; Kirchgassner and Schneider (2003) [58]; Millner, A.; Ollivier (2016) [59]; Marchiori et al. (2017) [60]; Barranzini et al. (2017) [61]	Analysis of green taxes in the context of the theory of tax optimization	Lack of analysis of green taxes in the context of the theory of tax optimization; research on the relationship between taxes and greenhouse gas emissions
Rudolph et al. (2017) [36]; Dresner et al. (2006) [62]; Kallbekken et al. (2010) [63]; Baranzini and Carattini (2017) [64]	The application of diversified tools and instruments to understand the determinants of the acceptability of "green taxes"; research on the people factor	The problem of the human factors was not analyzed; the variable "Population–Persons" was analyzed as one of the factors of the model
Vence and Pérez (2021) [37]; Morley and Abdullah (2010) [47]; Freire-González (2018) [65]; Pereira et al. (2016) [66]; Sajeewani et al. (2015) [67]; Freire-González et al. (2022) [68]; Morley (2012) [57]; Kotniks et al. (2014) [70]	Specific demands were made to change the taxation system related to environmental goals	Verification from the point of view of significance—green taxes are an important factor
Kotniks et al. (2014) [70]; López et al. (2008) [71]; López et al. (2011) [72]	The impact of fiscal spending patterns on the environment by taking into account CO_2, greenhouse gas emissions, and other emissions	The influence of 14 variables (Table 3) on greenhouse gas emissions was analyzed
Rybak et al. (2022) [24]	The direction and strength of the impact of green taxes differ depending on the greenhouse gas emissions	Green taxes, as one of 14 variables, were analyzed in the context of their influence on greenhouse gas emissions
Kalendienė and Pukelienė (2011) [44]; Li et al. (2019) [73]; Li and Zhu (2019) [74]; Hu et al. (2020) [75]; Yu et al. (2021) [76]	Green tax incentives fulfilled their role as a stimulator of technological changes toward green transformation and sustainability	The study did not consider green tax incentives

Source: own study.

This is the first study to compare the determinants of greenhouse gas emissions. First, the aims of this study are to motivate governments of EU countries to formulate a national carbon abatement policy and reduce greenhouse gas emissions, to change technology, and to invest in clean technology to grow the circular economy.

Second, policies should be implemented not only in EU countries, but also in regions with high levels of greenhouse gas emissions. Then, such policies would be more effective. The problem, as our research has shown, is the policy of environmental taxes. The governments of the EU countries, as a model organization, should remodel their tax policy so as to link environmental taxes with other elements of the policy of counteracting climate change, e.g., subsidies for clean technology to grow the circular economy. Such solutions would serve as a model for the less developed regions as far as possible and help them transform their resource use based on efficient instruments and policies based on environmental taxes.

As the literature on the subject shows [27], environmental taxation is merely a tool—or rather, an incentive—as our research confirms, leading to lower pollution emissions. As part of changing tax systems in EU countries, we should consider the role of environmental taxes within the larger tax system, considering both the potential for using environmental tax revenue to lower other taxes (or prevent raising them) and how environmental taxes interact with the rest of the tax system.

The implications of these results are that the current use of environmental taxes to reduce the EU's present levels of greenhouse gas emissions appears to be having some effect, although their relationships with other taxes and instruments need to be considered. The lack of a significant effect on greenhouse gas emissions (which was confirmed by modeling) suggests that environmental taxes are not reducing greenhouse gas emissions, implying that pollution is being reduced through the use of cleaner technologies and other activities or policies.

We support the postulates in the subject literature to change the tax policy [47,65,69,70], so as to give more meaning to "green taxes" as a non-fiscal instrument stimulating changes towards sustainability. It should also be emphasized that since we have shown a lack of the "green taxes" in the econometric model, changes in the tax policy—which is connected with public policy—should have such an effect in order to force economic entities to reduce the pressure that they place on the environment in terms of greenhouse gas emissions. We propose the use of both financial and non-financial instruments, but they should not contribute to the budget; rather, they should imply technological and adaptive changes aimed at reducing greenhouse gas emissions. In this way, "green taxes" will play the role of an important sustainability factor and influence climate change. Thus, our survey complements the conclusions concerning the mitigation of externalities, using the "green taxes system", formulated in the literature on the subject [30,32,33].

The literature on the subject indicates the effectiveness of using alternative activities, such as investments in new technologies that are environmentally friendly [42,49,50]. Therefore, it is worth considering combining ecological taxes with other tools supporting the reduction in greenhouse gas emissions. This requires remodeling the existing tax system, where environmental taxes should be more strongly linked with investment subsidies for entities introducing new environmentally friendly technologies. As indicated in the literature on the subject [5], combining taxation with other instruments may bring benefits; hence, our postulate regarding the achieved results of the model.

Optimal taxation levels and tax structures have been an issue for discussion and empirical research [44] for a long time but, as practice shows, it is very difficult to apply relatively new phenomena, such as the use of taxation as an instrument supporting climate policy (as shown by our studies on greenhouse gas emissions).

Our models showed the lack of a lasting impact of green taxes on greenhouse gas emissions. However, the first part of our research confirmed that the variable "green taxes" reduced greenhouse gas emissions and, thus, contributed to mitigating the effects of climate change. Our hypothesis was negatively verified. Therefore, we can conclude that the existing tax policy needs to be verified. Our research confirmed that the following:

1. Environmental taxes perform a fiscal function, because both greenhouse gas emissions and revenues from green taxes are growing (very strong dependency);
2. Environmental taxes do not have a motivational function, because there is no dependency showing that environmental taxes accelerate the decrease in greenhouse gas emissions.

The occurrence of the fiscal function has been confirmed in the literature for various countries [24,99]. As our research shows, a multivariate analysis is necessary to show the occurrence of the motivational function and to establish the variability of factors. This variability of factors makes it possible to determine which factors will significantly affect the implementation of the motivational function of environmental taxes.

The implementation of the second and third objectives of the study—i.e., determination of the determinants of greenhouse gas emissions, and designation of the determinants of greenhouse gas emissions in EU countries using modeling (linear multiple regression models)—made it possible to achieve these objectives, and a review of potential significant determinants is included in Tables 1 and 3. The determinants presented and considered in this study are measurable, being sourced from the Eurostat databases. However, there are determinants for which there are no comparable data in the analyzed period (e.g., agriculture, access to clean water, trading in greenhouse gas emissions). In our study, despite the fact that we managed to identify the determinants of greenhouse gas emissions, we were not able to quantify the behaviors of consumers and pollutant emitters. There are no data in this regard in Eurostat's databases. Therefore, our study may be extended by these two indicated factors in the future.

Increasingly, the risk of an increase in greenhouse gas emissions in connection with the ongoing war between Russia and Ukraine is indicated. At present, it is very difficult to quantify this factor for 2022, but its impact and significance should be examined [100]. In our model, we also did not take into account factors such as politics, or even the trading of CO_2 emissions itself. This important factor may be a component of tax policy, and the actions of governments themselves may also lead to a change in attitudes towards trading in greenhouse gas emissions. In the literature on the subject, there is still a discussion on the future significance of this factor, and it is also considered from the point of view of its impact on the growth of greenhouse gas emissions [101].

When undertaking research on the factors influencing greenhouse gas emissions, one should mainly consider the energy sector, which is responsible for almost 80% of greenhouse gas emissions. However, the available statistics for individual countries (economies) prevent a more detailed analysis in this regard. There are single variables available in public statistical resources that can be used in modeling the development of greenhouse gas emissions. In the obtained models (with high R^2), it is the size of the country's population that determines the emissions of the analyzed gases to the greatest extent. As already noted, it is the population and the need to meet their needs in the EU countries that generate greenhouse gas emissions.

6. Conclusions

The use of environmental taxes to reduce the EU's present levels of greenhouse gas emissions, in the context of energy transformation and in the era of climate change, is an important research issue. This article attempts to determine the factors influencing greenhouse gas emissions. Environmental taxes were examined as a special variable in the context of their contribution to reducing greenhouse gas emissions. It was assumed that the tax incentive is an important instrument for influencing the reduction in greenhouse gas emissions. The methods of analysis used in this paper made it possible to trace changes in the area under consideration, taking into account 14 diagnostic features describing factors related to greenhouse gas emissions in the EU countries.

The method of linear regression analysis (multiple variables) was used to determine the relationships between the explained variable—greenhouse gas emissions—and other variables shaping it, defined on the basis of the literature. Linear regression models, in spite of their disadvantages (as mentioned in the text of this article), provide a fairly simple way to determine the impacts of independent variables on the dependent variable (a broad description of the method is included in Section 3).

Our research shows that only 4 out of the 14 diagnostic features are relevant for greenhouse gas emissions, and they do not include "environmental taxes". There is a strong need to change the tax system and introduce both non-financial and financial tax incentives for influencing the reduction in greenhouse gas emissions.

Our study complements the existing research, as it shows the effects of taking into account many factors that have been previously studied, their verification from the point of view of significance (in this case, environmental taxes are an important factor), and an indication of the strength of their impact on greenhouse gas emissions in individual years when analyzing the most important factors.

Our research also prompted us to consider the key determinants of greenhouse gas emissions that shape the trends of changes in policy in EU countries. We believe that the indicated factors determined in relation to EU countries are similar in countries outside the EU. Therefore, in light of the first stage of our study (determining the significant impacts on greenhouse gas emissions), we postulate that EU countries should change their tax policy so as to include "environmental taxes" as an important determinant of reducing greenhouse gas emissions. We propose a modification of the environmental tax system, taking into account tax expenditures, as they may affect specific, needed changes in the direction of reducing greenhouse gas emissions. The governments of the EU member states should use the fiscal function of "green taxes" as well as redirecting their actions towards the use of non-fiscal functions of "green taxes" and good practices.

Many publications indicate the need to change the tax policy towards sustainability, taking into account the significant impacts on the behavior of enterprises and society. Not only can our observations be of use to EU governments to justify a change in this policy, but our research also indicates specific factors influencing the reduction in greenhouse gas emissions.

The obtained results could be used in subsequent years to check the directions of observed changes in the individual EU member states. They can also be used to monitor the changes in factors and, thus, whether the changes in tax policy have been successful in the pursuit of sustainability. This could be the basis for determining the feasibility of forecasts and making economic decisions aimed at environmentally friendly technologies with the use of tax incentives. We can also see that the results obtained could be helpful to decision-makers as an informative element for countries' positioning, as well understanding the current state to which previous decisions have led.

In summary, taking into account the results of the study, it is necessary to pay attention primarily to the following needs:

- To refocus the fiscal policy of countries on the use of environmental taxes, so that they achieve the goals assigned to them—including the reduction in greenhouse gas emissions;
- To monitor the effects of environmental protection transfers in order to analyze the principles and legitimacy of spending public funds.

We should also acknowledge some limitations in our research. The first limitation of this research is the inability to verify the various income groups obtained through green taxes. The possibility of introducing these data into the model would be highly desirable and would make it possible to verify that one (or more) of the income groups obtained through green taxes could be the explanatory variable of the model. Additionally, the authors noticed that this research could be extended to include the "trading in greenhouse gas emission" factor, which may distort the results of public intervention and impact through the tax incentive.

Additionally, as another limitation of this research, we can see that the research initiated in the field of searching for factors influencing greenhouse gas emissions should be continued and extended with qualitative (soft) variables that would illustrate consumer behavior. Our research shows that the size of the population is the main determinant of greenhouse gas emissions. This is quite obvious, as it affects many other variables that determine—for example—the consumption of electricity, heat, and other goods and services, which contribute to increased emissions. Hence, it seems extremely important to undertake research on the behavior of the population, which would be conducive to limiting their consumerism and could be used as an instrument of environmental tax policy. Should governments focus on "punishing" emission-generating consumption, or should they use incentives?

It seems that the ongoing conflict in Ukraine and the restrictions on trade in energy carriers (especially natural gas) between the EU and Russia will be significant not only for greenhouse gas emissions, but also for their determinants. This is likely to increase the use of hard coal, which will be reflected in the amounts of greenhouse gas emissions (see model estimates). How large the increase will be will depend on the behavior of the population (e.g., acceptance of lower temperatures in buildings and, in the long term, the use of renewable energy sources).

We can also see the possibility of researching the importance of taxes as a factor in limiting greenhouse gas emissions in the context of the EU ETS. It is important that the climate policy is effective, and the analysis of the EU ETS together with the assessment of the effectiveness of environmental taxes would undoubtedly be justified to change the existing climate policy.

Author Contributions: Conceptualization, B.Z.F. and D.W.; methodology, D.W.; software, D.W.; validation, B.Z.F. and D.W.; formal analysis, B.Z.F. and D.W.; investigation, B.Z.F. and D.W.; resources, B.Z.F. and D.W.; data curation, D.W.; writing—original draft preparation, B.Z.F. and D.W.; writing—review and editing, B.Z.F. and D.W.; visualization, B.Z.F. and D.W.; supervision, B.Z.F.; project administration, B.Z.F.; funding acquisition, B.Z.F. All authors have read and agreed to the published version of the manuscript.

Funding: This research was financed within the framework of the program of the Minister of Science and Higher Education under the name "Regional Excellence Initiative" in the years 2019–2022 (project number 001/RID/2018/19); the amount of financing was PLN 10,684,000.00.

Data Availability Statement: Not applicable.

Conflicts of Interest: The authors declare no conflict of interest. The funder had no role in the design of the study; in the collection, analyses, or interpretation of data; in the writing of the manuscript; or in the decision to publish the results.

Appendix A

Table A1. Data for 2020.

	GDP	EP	HC	REC	FA	OF	CO_2	EHC	ETR	EPT	EPI	RRW	FWAp	P	GE
Arithmetical mean	26,284.62	7.91	850.93	25.45	35.34	8.82	111.28	0.02	11,087.80	565.93	1986.15	38.98	351.66	16,767,800	124,770.95
Standard error	3495.11	0.86	571.30	2.39	3.40	1.15	2.30	0.00	3232.53	223.23	675.99	2.83	49.03	4,454,054	34,323.28
Median	20,025.00	7.14	179.64	24.33	34.55	8.38	113.30	0.01	4358.23	195.85	596.95	39.15	276.96	7,926,273	54,254.80
Standard deviation	17,821.66	4.39	2913.05	12.16	17.32	5.85	11.74	0.02	16,482.72	1138.26	3446.88	14.43	249.99	22,711,309	175,015.09
Curtosis	2.89	4.39	24.98	−0.76	0.27	−0.14	0.08	1.80	3.14	17.41	7.10	−0.34	−0.17	3	5.40
Skewness	1.63	1.86	4.96	−0.21	0.40	0.58	−0.56	1.46	2.05	3.96	2.70	−0.13	0.87	2	2.26
Min	6380.00	2.47	7508.00	2.58	1.44	0.47	82.30	0.00	296.75	4.20	25.90	10.50	79.50	514,564	2321.40
Max	82,250.00	22.61	15,007.13	46.00	73.73	22.41	133.00	0.07	57,528.00	5674.00	13,977.80	67.00	943.44	83,166,711	742,490.79

Source: own study with the use of SPSS software.

References

1. Kamieniecki, S.; Kraft, M. *The Oxford Handbook of U.S. Environmental Policy*; Kamieniecki, S., Kraft, M., Eds.; University Press: Oxford, UK, 2012.
2. Milanes-Montero, P.; Esteban Perez-Calderon, E.; Dias, A.I. Greenhouse gas emission allowances: Financial reporting transparency. *J. Bus. Manag.* **2022**, *62*, 1–19. [CrossRef]
3. Gramkow, C. Green fiscal policies: An armoury of instruments to recover growth sustainably. In *Studies and Perspectives Series-ECLAC Office in Brasilia, 5 (LC/TS.2020/24) (LC/BRS/TS.2019/7)*; Economic Commission for Latin America and the Caribbean (ECLAC): Santiago, Chile, 2020. Available online: https://repositorio.cepal.org/bitstream/handle/11362/45418/1/S2000005_en.pdf (accessed on 5 October 2022).
4. Bellec, C.; Bouthevillai, C. European Instruments Aimed at Reducing Greenhouse Gas Emissions. Macroeconomics, Microeconomics and Structures. European and Multilateral Policies Directorate, Bulletin de la Banque de France 2022, 242/1, pp. 1–14. Available online: https://www.banque-france.fr/sites/default/files/medias/documents/822303_bdf242-1_en_ges_vfinale.pdf (accessed on 10 April 2022).
5. Bird, R.; Davis-Nozemack, K. Tax Avoidance as a Sustainability Problem. *J. Bus. Ethics* **2018**, *151*, 1009–1025. [CrossRef]
6. Organisation for Economic Cooperation and Development. Available online: https://www.oecd.org/env/tools-evaluation/48164926.pdf (accessed on 1 July 2022).
7. Mpofu, F.Y. Green Taxes in Africa: Opportunities and Challenges for Environmental Protection, Sustainability, and the Attainment of Sustainable Development Goals. *Sustainability* **2022**, *14*, 10239. [CrossRef]
8. Sterner, T. Fuel taxes: An important instrument for climate policy. *Energy Policy* **2007**, *35*, 3194–3202. [CrossRef]
9. Streimikiene, D.; Siksnelyte, I.; Zavadskas, E.K.; Cavallaro, F. The Impact of Greening Tax Systems on Sustainable Energy Development in the Baltic States. *Energies* **2018**, *11*, 1193. [CrossRef]

10. Chomać-Pierzecka, E.; Sobczak, A.; Urbańczyk, E. RES Market Development and Public Awareness of the Economic and Environmental Dimension of the Energy Transformation in Poland and Lithuania. *Energies* **2022**, *15*, 5461. [CrossRef]
11. European Union. Available online: https://eur-lex.europa.eu/legal-content/PL/TXT/?uri=CELEX:52019DC0640 (accessed on 10 June 2022).
12. European Union. Available online: https://ec.europa.eu/info/strategy/priorities-2019-2024/european-green-deal/delivering-european-green-deal_en (accessed on 10 June 2022).
13. European Union. Available online: https://eur-lex.europa.eu/legal-content/PL/TXT/PDF/?uri=CELEX:32020R1294&from=PL (accessed on 10 June 2022).
14. Witkowski, P.; Adamczyk, A.; Franek, S. Does Carbon Risk Matter? Evidence of Carbon Premium in EU Energy-Intensive Companies. *Energies* **2021**, *14*, 1855. [CrossRef]
15. Liobikienė, G.; Butkus, M.; Matuzevičiūtė, K. The Contribution of Energy Taxes to Climate Change Policy in the European Union (EU). *Resources* **2019**, *8*, 63. [CrossRef]
16. Rinscheid, A.; Pianta, S.; Weber, E. What shapes public support for climate change mitigation policies? The role of descriptive social norms and elite cues. *Behav. Public Policy* **2021**, *5*, 503–527. [CrossRef]
17. Dulebenets, M.A. Green vessel scheduling in liner shipping: Modeling carbon dioxide emission costs in sea and at ports of call. *Int. J. Transp. Sci. Technol.* **2018**, *7*, 26–44. [CrossRef]
18. Green Taxation–In Support of a More Sustainable Future. European Green Deal. European Commision. Available online: https://taxation-customs.ec.europa.eu/green-taxation-0_en (accessed on 1 July 2022).
19. Al-Alawi, H.; Nobanee, H.; Nobanee, H. A Study on Green Taxation and its Impact on Economic Development: A Mini-Review. *SSRN Electron. J.* 2020. [CrossRef]
20. OECD. Available online: https://www.oecd.org/env/tools-evaluation/environmentaltaxation.htm (accessed on 1 July 2022).
21. Opodatkowanie Ekologiczne Narzędziem Ochrony środowiska Państw Nordyckich. Available online: repozytorium.ka.edu.pl/bitstream/handle/11315/24644/TOMALA_Opodatkowanie_ekologiczne_narzedziem_ochrony_srodowiska_2018.pdf?sequence=1&isAllowed=y (accessed on 10 July 2022).
22. Eurostat. Available online: https://ec.europa.eu/eurostat/web/main/data/database (accessed on 15 July 2022).
23. Ziolo, M.; Filipiak, B.Z.; Tudys, B. *Sustainability in Bank and Corporate Business Models. Ling between ESG Risk Assessment and Corporate Sustainability*; Palgrawe: Cham, Switzerland, 2021.
24. Rybak, A.; Joostberens, J.; Manowska, A.; Pielot, J. The Impact of Environmental Taxes on the Level of Greenhouse Gas Emissions in Poland and Sweden. *Energies* **2022**, *15*, 4465. [CrossRef]
25. Mankiw, N.G.; Weinzierl, M.; Yagan, D. Optimal Taxation in Theory and Practice. *J. Econ. Perspect.* **2009**, *23*, 147–174. [CrossRef]
26. Pigou, A.C. *The Economics of Welfare*; Macmillan: London, UK, 1920.
27. Labeaga, J.M.; Labandeira, X. Economics of Environmental Taxes and Green Tax Reforms. *Sustainability* **2020**, *12*, 350. [CrossRef]
28. Baurnol, J.; Oates, W.E. *The Theory of Environmental Policy (Extemalities, Public Outlays, and the Quality of Life)*; Prentice-Hall: Englewood Clifs, NJ, USA, 1975; p. 17.
29. Dasgupta, P.S.; Heal, G.M. *Economic Theory and Exhaustible Resources*; Cambridge University Press: London, UK, 1979; 516p.
30. Filipiak, B.Z.; Wyszkowska, D. Impact of green taxes on the public financial system–An example of European Union countries. In *Social, Economic, and Environmental Impacts Between Sustainable Financial Systems and Financial Markets*; Ziolo, M., Ed.; IGI Global: Hershey, PA, USA, 2019; Chapter 6.
31. Falcone, P.M. Green investment strategies and bank-firm relationship: A firm-level analysis. *Econ. Bull.* **2018**, *38*, 2225–2239.
32. Owen, R.; Brennan, G.; Lyon, F. Enabling investment for the transition to a low carbon economy: Government policy to finance early stage green innovation. *Curr. Opin. Environ. Sustain.* **2018**, *31*, 137–145. [CrossRef]
33. Hänsel, M.C.; van den Bergh, J.C.J.M. Taxing interacting externalities of ocean acidification, global warming, and eutrophication. *Nat. Resour. Model.* **2021**, *34*, e12317. [CrossRef]
34. Domenech, T.; Bahn-Walkowiak, B. Transition towards a resource efficient circular economy in Europe: Policy lessons from the EU and the Member States. *Ecol. Econ.* **2019**, *155*, 7–19. [CrossRef]
35. Milios, L. Towards a Circular Economy Taxation Framework: Expectations and Challenges of Implementation. *Circ. Econ. Sustain.* **2021**, *1*, 477–498. [CrossRef]
36. Rudolph, S.; Kawakatsu, T.; Lerch, A. Regional Market Based Climate Policy in North America: Efficient, Effective, Fair? In *Environmental Taxation and Green Fiscal Reform: Theory and Impact*; Kreiser, L., Lee, S., Ueta, K., Milne, J., Eds.; Edvard Elgar: Cheltenham, UK, 2014; pp. 273–288.
37. Vence, X.; López Pérez, S.d.J. Taxation for a Circular Economy: New Instruments, Reforms, and Architectural Changes in the Fiscal System. *Sustainability* **2021**, *13*, 4581. [CrossRef]
38. Miller, S.J.; Vela, M.A. *Are Environmentally Related Taxes Effective?* Inter-American Development Bank: Washington, DC, USA, 2013.
39. De Alegría, I.M.; Basañez, A.; de Basurto, P.D.; Fernández-Sainz, A. Spain's fulfillment of its Kyoto commitments and its fundamental greenhouse gas (GHG) emission reduction drivers. *Renew. Sustain. Energy Rev.* **2016**, *59*, 858–867. [CrossRef]
40. González-Sánchez, M.; Martín-Ortega, J.L. Greenhouse Gas Emissions Growth in Europe: A Comparative Analysis of Determinants. *Sustainability* **2020**, *12*, 1012. [CrossRef]
41. Aidt, T.S. Green taxes: Refunding rules and lobbying. *J. Environ. Econ. Manag.* **2010**, *60*, 31–43. [CrossRef]

42. Zhang, S.; Yu, Y.; Zhu, Q.; Qiu, C.M.; Tian, A. Green Innovation Mode under Carbon Tax and Innovation Subsidy: An Evolutionary Game Analysis for Portfolio Policies. *Sustainability* **2020**, *12*, 1385. [CrossRef]
43. Zhang, Z.; Chen, Y.-H.; Wang, C.-M. Can CO_2 Emission Reduction and Economic Growth Be Compatible? Evidence From China. *Front. Energy Res.* **2021**, *9*, 693767. [CrossRef]
44. Kalendienė, J.; Pukelienė, V. Taxation and economic sustainability. *Ekonomika* **2011**, *90*, 63–75. [CrossRef]
45. Ko, H. How Does a Welfare State achieves Fiscal Sustainability? A Study of the Impact of Tax Equit. In *Taxes and Taxation Trends*; Iwin-Garzynska, J., Ed.; IntechOpen: London, UK, 2017. [CrossRef]
46. Williams, R.C. *Environmental Taxation*; Discussion Paper 16–24; RFF DP: Washington, DC, USA, 2016.
47. Morley, B.; Abdullah, S. Environmental Taxes and Economic Growth: Evidence from Panel Causality Tests. Bath Economics Research Papers 04. 2010. Available online: https://www.researchgate.net/profile/Bruce-Morley/publication/46468103_Environmental_Taxes_and_Economic_Growth_Evidence_from_Panel_Causality_Tests/links/543fab830cf21227a11a643b/Environmental-Taxes-and-Economic-Growth-Evidence-from-Panel-Causality-Tests.pdf?origin=publication_detail (accessed on 10 July 2022).
48. Azis, K.M.; Widodo, T. The Impact of Carbon Tax on GDP and Environment. MPRA Paper 91314. 2019. Available online: https://mpra.ub.uni-muenchen.de/91314/ (accessed on 10 July 2022).
49. Ekins, P.; Pollitt, H.; Summerton, P.; Chewpreecha, U. Increasing carbon and material productivity through environmental tax reform. *Energy Policy* **2012**, *42*, 65–376. [CrossRef]
50. Milne, J.E.; Andersen, M.S. *Handbook of Research on Environmental Taxation*; Edward Elgar Publishing: Cheltenham, UK, 2012.
51. Cecere, G.; Corrocher, N.; Mancusi, M.L. Financial constraints and public funding of eco-innovation: Empirical evidence from European SMEs. *Small Bus. Econ.* **2020**, *54*, 285–302. [CrossRef]
52. Palmer, K.; Walls, M. *Extended Product Responsibility: An Economic Assessment of Alternative Policies*; Discussion Papers 10830; Resources for the Future: Washington, DC, USA, 1999. [CrossRef]
53. Söderholm, P. Taxing virgin natural resources: Lessons from aggregates taxation in Europe. *Resour. Conserv. Recycl.* **2011**, *55*, 911–922. [CrossRef]
54. Van der Ploeg, F.; Withagen, C. Is there really a green paradox? *J. Environ. Econ. Manag.* **2012**, *64*, 342–363. [CrossRef]
55. Carattini, S.; Baranzini, A.; Thalmann, P.; Varone, F.; Vöhringer, F. Green taxes in a post-Paris world: Are millions of nays inevitable. *Environ. Resour. Econ.* **2017**, *68*, 97–128. [CrossRef]
56. Meckling, J.; Kelsey, N.; Biber, E.; Zysman, J. Winning coalitions for climate policy. Green industrial policy builds support for carbon regulation. *Clim. Chang.* **2015**, *349*, 1170–1171.
57. Verde, S.; Galdi, G.; Alloisio, I.; Borghesi, S. The EU ETS and its companion policies: Any insight for China's ETS? *Environ. Dev. Econ.* **2021**, *26*, 302–320. [CrossRef]
58. Kirchgassner, G.; Schneider, F. On the political economy of environmental policy. *Public Choice* **2003**, *115*, 369–396. [CrossRef]
59. Millner, A.; Ollivier, H. Beliefs, politics, and environmental policy. *Rev. Environ. Econ. Policy* **2016**, *10*, 226–244. [CrossRef]
60. Marchiori, C.; Dietz, S.; Tavoni, A. Domestic politics and the formation of international environmental agreements. *J. Environ. Econ. Manag.* **2017**, *81*, 115–131. [CrossRef]
61. Barranzini, A.; van den Bergh, J.C.J.M.; Carattini, S.; Howarth, R.B.; Padilla, E.; Roca, J. Carbon Pricing in Climate Policy: Seven Reasons, Complementary Instruments, and Political Economy Considerations. *CSLF Artic.* **2017**, *7*, e462. [CrossRef]
62. Dresner, S.; Dunne, L.; Clinch, P.; Beuermann, C. Social and political responses to ecological tax reform in Europe: An introduction to the special issue. *Energy Policy* **2006**, *34*, 895–904. [CrossRef]
63. Kallbekken, S.; Kroll, S.; Cherry, T.L. Pigouvian tax aversion and inequity aversion in the lab. *Econ. Bull.* **2010**, *30*, 1914–1921.
64. Baranzini, A.; Carattini, S. Effectiveness, earmarking and labeling: Testing the acceptability of carbon taxes with survey data. *Environ. Econ. Policy Stud.* **2017**, *19*, 197–227. [CrossRef]
65. Freire-González, J. Environmental taxation and the double dividend hypothesis in CGE modelling literature: A critical review. *J. Pol. Model.* **2018**, *40*, 194–223. [CrossRef]
66. Pereira, A.M.; Pereira, R.M.; Rodrigues, P.G. A new carbon tax in Portugal: A missed opportunity to achieve the triple dividend? *Energy Policy* **2016**, *93*, 110–118. [CrossRef]
67. Sajeewani, D.; Siriwardana, M.; Mcneill, J. Household distributional and revenue recycling effects of the carbon price in Australia. *Clim. Chang. Econ.* **2015**, *6*, 1550012. [CrossRef]
68. Freire-González, J.; Martinez-Sanchez, V.; Puig-Ventosa, I. Tools for a circular economy: Assessing waste taxation in a CGE multi-pollutant framework. *Waste Manag.* **2022**, *139*, 50–59. [CrossRef]
69. Morley, B. Empirical Evidence on the Effectiveness of Environmental Taxes. *Appl. Econ. Lett.* **2012**, *19*, 1817–1820. [CrossRef]
70. Kotniks, Z.; Klun, M.; Skulj, D. The Effect of Taxation on Greenhouse Gas Emissions. *Transylv. Rev. Adm. Sci.* **2014**, *43*, 168–185.
71. López, R.; Vinod, T.; Wang, Y. *The Quality of Growth: Fiscal Policies for Better Results*; IEG Working Paper 6; The World Bank: Washington, DC, USA, 2008.
72. López, R.; Galinato, G.I.; Islam, A. Fiscal Spending and the Environment: Theory and Empirics. *J. Environ. Econ. Manag.* **2011**, *62*, 180–198. [CrossRef]
73. Li, H.L.; Zhu, X.H.; Chen, J.Y.; Jiang, F.T. Environmental regulations, environmental governance efficiency and the green transformation of China's iron and steel enterprises. *Ecol. Econ.* **2019**, *165*, 106397. [CrossRef]

74. Li, D.; Zhu, J. The Role of Environmental Regulation and Technological Innovation in the Employment of Manufacturing Enterprises: Evidence from China. *Sustainability* **2019**, *11*, 2982. [CrossRef]
75. Hu, B.; Dong, H.; Jiang, P.; Zhu, J. Analysis of the Applicable Rate of Environmental Tax through Different Tax Rate Scenarios in China. *Sustainability* **2020**, *12*, 4233. [CrossRef]
76. Yu, L.; Zhu, J.; Wang, Z. Green Taxation Promotes the Intelligent Transformation of Chinese Manufacturing Enterprises: Tax Leverage Theory. *Sustainability* **2021**, *13*, 13321. [CrossRef]
77. Jóźwik, B.; Gavryshkiv, A. Wpływ podatku środowiskowego na emisję gazów cieplarnianych w państwach Grupy Wyszehradzkiej. *Gospod. Nar.* **2022**, *1*, 66–78. [CrossRef]
78. Hao, L.N.; Umar, M.; Khan, Z.; Ali, W. Green growth and low carbon emission in G7 countries: How critical the network of environmental taxes, renewable energy and human capital is? *Sci. Total Environ.* **2021**, *752*, 141853. [CrossRef] [PubMed]
79. Liobikiene, G.; Butkus, M. The European Union possibilities to achieve targets of Europe 2020 and Paris agreement climate policy. *Renew. Energy* **2017**, *106*, 298–309. [CrossRef]
80. Hashmi, R.; Alam, K. Dynamic relationship among environmental regulation, innovation, CO_2 emissions, population, and economic growth in OECD countries: A panel investigation. *J. Clean. Prod.* **2019**, *231*, 1100–1109. [CrossRef]
81. Lapinskiene, G.; Peleckis, K.; Slavinskaite, N. Energy consumption, economic growth and greenhouse gas emissions in the European union countries. *J. Econ. Bus. Manag.* **2017**, *18*, 1082–1097. [CrossRef]
82. Meng, L.; Huang, B. Shaping the relationship between economic development and carbon dioxide emissions at the local level: Evidence from spatial econometric models. *Environ. Resour. Econ.* **2018**, *71*, 127–156. [CrossRef]
83. Andrés, L.; Padilla, E. Driving factors of GHG emissions in the EU transport activity. *Transp. Policy* **2018**, *61*, 60–74. [CrossRef]
84. European Environment Agency. Annual European Union Greenhouse Gas Inventory 1990–2017 and Inventory Report 2019. Submission to the UNFCCC Secretariat. 2019. Available online: https://unfccc.int/documents/194921 (accessed on 7 July 2022).
85. Federal Environment Agency. National Inventory Report for the German Greenhouse Gas Inventory 1990–2017. Submission to the UNFCCC Secretariat. 2019. Available online: https://unfccc.int/documents/194930 (accessed on 6 July 2022).
86. Environmental Protection Agency. Ireland's National Inventory Report 2019. Greenhouse Gas Emissions 1990–2017. Submission to the UNFCCC Secretariat. 2019. Available online: https://unfccc.int/documents/194638 (accessed on 8 July 2022).
87. Agencia Portuguesa do Ambiente. Portuguese National Inventory Report on Greenhouse Gases, 1990–2016. Submitted Under the United Nations Framework Convention on Climate Change and the Kyoto Protocol. 2019. Available online: https://unfccc.int/documents/194464 (accessed on 8 July 2022).
88. Cox, D.R. Regression models and life-tables (with discussion). *J. R. Stat. Soc. Ser. B Methodol.* **1972**, *34*, 187–202. [CrossRef]
89. Sysk, J. *Współczesne Metody Analizy Regresji Wspomagane Komputerowo*; Uniwersytet Śląski: Katowice, Poland, 2014. Available online: http://el.us.edu.pl/ekonofizyka/images/b/b6/MetodyDoboruModeli.pdf (accessed on 1 July 2022).
90. Draper, N.R.; Smith, H. *Analiza Regresji Stosowana*; Państwowe Wydawnictwo Naukowe: Warszawa, Poland, 1973.
91. Węziak, D. Wielopoziomowe modelowanie regresyjne w analizie danych. *Wiadomości Stat.* **2007**, *9*, 1–12.
92. Szaleniec, M. *Sieci Neuronowe i Regresja Wieloraka–Czyli Jak Okiełznać Złożoność w Badaniach Naukowych*; Statsoft Polska: Kraków, Poland, 2008. Available online: https://media.statsoft.pl/_old_dnn/downloads/sieci%20neuronowe.pdf (accessed on 1 July 2022).
93. Rydzewski, P. Metody analizy danych w socjologicznych badaniach rozwodów. *Rocz. Nauk. Społecznych* **2010**, *2*, 276–280.
94. Konarski, R. Regresja Wielokrotna: Diagnostyka i Selekcja Modelu, 9. Available online: http://www.wns-stat.strony.ug.edu.pl/artykuly/art3.pdf (accessed on 1 July 2022).
95. Raczkiewicz, D. Zastosowanie analizy regresji w reprezentacyjnych badaniach społeczno-go¬spodarczych. *Ekonometria* **2016**, *1*, 31–47.
96. Rószkiewicz, M. *Analiza Klienta*; Wydawnictwo SPSS Polska: Kraków, Poland, 2011; pp. 240–247.
97. Nazlioglu, S.; Soytas, U.; Gupta, R. Oil prices and financial stress: A volatility spillover analysis. *Energy Policy* **2015**, *82*, 278–288. [CrossRef]
98. Chiroleu-Assouline, M.; Fodha, M. From regressive pollution taxes to progressive environmental tax reforms. *Eur. Econ. Rev.* **2014**, *69*, 126–142. [CrossRef]
99. Aydin, C.; Esen, Ö. Reducing CO_2 emissions in the EU member states: Do environmental taxes work? *J. Environ. Plan. Manag.* **2018**, *61*, 2396–2420. [CrossRef]
100. What Does the Russia-Ukraine War Mean for the Climate? Available online: https://www.dw.com/en/ukraine-russia-climate-emissions-fossil-fuels/a-61130583 (accessed on 30 October 2022).
101. Guo, Q.; Su, Z.; Chiao, C. Carbon emissions trading policy, carbon finance, and carbon emissions reduction: Evidence from a quasi-natural experiment in China. *Econ. Chang. Restruct.* **2022**, *55*, 1445–1480. [CrossRef]

energies MDPI

Article

Impact of Subsidy Programmes on the Development of the Number and Output of RES Micro-Installations in Poland

Beata Bieszk-Stolorz

Institute of Economics and Finance, University of Szczecin, 71-101 Szczecin, Poland; beata.bieszk-stolorz@usz.edu.pl

Abstract: Renewable energy sources are intended to support the decarbonisation process of the Polish economy. Since 2005, the share of renewable energy in total electricity in Poland has been increasing. The number of photovoltaic panels installed by prosumers as part of micro-installations increased particularly strongly. The aim of this research is an assessment of the impact of government programmes on the development of RES micro-installations in Poland. A regression discontinuity design was used in the analysis. It is a model from the group of average impact effect models used in evaluation studies. The added value of the presented study is its application in the assessment of the impact of implemented programmes on the number and output of micro-installations in Poland. In the study, it is shown that there had been no increase in the number and output of micro-installations at the adopted threshold (2019Q4). On the other hand, there was a sharp increase in them over the whole period starting from 2019Q4.

Keywords: renewable energy sources; government regulations; subsidy programmes; prosumers; micro-installations; regression discontinuity design

Citation: Bieszk-Stolorz, B. Impact of Subsidy Programmes on the Development of the Number and Output of RES Micro-Installations in Poland. *Energies* **2022**, *15*, 9357. https://doi.org/10.3390/en15249357

Academic Editors: Grzegorz Mentel and Sebastian Majewski

Received: 11 November 2022
Accepted: 8 December 2022
Published: 10 December 2022

1. Introduction

In 1996, the European Parliament and the European Council adopted the 'White Paper for a Community Strategy and Action Plan' [1]. The Union's objective at the time was to double the share of renewable energy in total gross internal energy consumption. Unfortunately, the enormous potential of renewable energy resources (RES) has still not been realised [2]. The political aim of the European Union is its transformation into a competitive, modern, prosperous, and climate-neutral economy by 2050. To accelerate this transformation, the European Commission presented 'The European Green Deal' in 2019 [3]. It aim was the reduction of over half of greenhouse gas emissions by 2030. The European Union is now the largest political bloc with policy targets to create a climate-neutral economy by 2050 [4]. The development of renewable energy sources is linked to the issue of energy security. Jonek-Kowalska [5] showed that the level of energy security for most of the 32 assessed European countries was low due to the predominant use of non-renewable energy resources in their energy mixes. Countries that have a high level of energy security own non-renewable resources or use alternative energy carriers in the form of nuclear power or renewables. EU legislation is forcing the reduction of carbon emissions and the abandonment of non-renewable energy resources and is promoting renewable energy sources. This is a real economic and technological challenge. Such a transformation requires an effective and strategic approach, especially in those countries that have so far mainly used hard coal in the energy sector [6].

In Poland, the Energy Law came into force in 1998 and initiated substantial restructuring and regulation of heat markets and electricity. It introduced a free market for energy and regionally diversified rates. The law favours innovative methods of energy saving and supports clean technologies. It also recognises renewable energy sources as one option to achieve environmental targets in Poland. Nevertheless, the Polish energy system is dominated by coal-fuelled power stations [7]. Renewable energy sources are intended to support the process of decarbonising economies. A situation in which the renewables can be equally competitive must be cultivated by institutional considerations and inducements [8]. The legal situation is supportive of introducing renewables energy. Prosumers in Poland are largely interested in photovoltaics. The concept of rooftop PVs, also known as "building integrated/applied PV" (BIPV/BAPV) is drawing significant attention. Jurasz et al. [9] emphasise that PVs are almost the only source of electricity that can be used on a large scale in highly populated areas. PV technology requires low maintenance and is noise- and emission-free during its operation phase. Recently, there has been a dynamic increase in the number and output of micro-installations connected to the distribution grid. The majority of them are photovoltaic (PV) installations, of which the dominant majority are prosumer micro-installations connected to the grid on the basis of notification [10]. At the end of 2021, the installed capacity in EU countries reached 158 GW, representing an annual increase of 21.4 GW. Poland was just behind Germany in terms of PV cell growth [11].

Hypothesis and the Research Objective

In the study, the following hypothesis is put forward:

Hypothesis (H): *Regulations and projects implemented in Poland in 2019 have had a positive impact on the increase in the number and output of micro-installations.*

The aim of this article is to estimate the impact of government programmes on the development of RES micro-installations in Poland. Funds for the modernisation and upgrading of energy installations in Poland from the budget are subject to evaluation. The various reports and studies are based on basic statistical research. They do not usually go beyond the scope of descriptive statistics. In this research, an evaluation is carried out using an econometric model. A regression discontinuity design approach was used in the analysis. It is a model from the group of average impact effect models used in evaluation studies. A model using the Regression Discontinuity Design (RDD) in the social sciences was first introduced by Thistlethwaite and Campbell in 1960 [12]. They studied the impact of merit awards on students' future achievements (career aspirations, postgraduate registration, academic achievement, etc.). The RDD has been known for many years. However, its use in scientific research has been quite rare [13]. In socio-economic analysis, it has been used more and more since the late 1990s, with many articles appearing in the fields of labour market, education, environment, forensics, or health [14]. One strand of literature has emerged on measuring the effects of potential benefit duration on unemployment duration [15–18]. Another strand in which RDD has been applied is the impact of regulatory changes on socio-economic phenomenon [19–29].

The added value of the presented study is its application to assess the impact of implemented programmes on the number and output of micro-installations in Poland. An applied econometric model (RDD) allowed a comparison of changes in the years before the subsidy programme was introduced with the period after the introduction of the programme. It was possible to assess the impact at the time of the introduction of the programme, as well as to assess changes over the whole period after its introduction.

2. The Renewable Energy in Poland

There is an increasing emphasis worldwide on the development of renewable energy sources. Their use is not connected with a long-term deficit, as their resource is renewed in a relatively short time. Such sources include the sun, wind, water (tidal and wave power), closed-cycle nuclear power, biomass, biogas, bioliquids, and biofuels, as well as geothermal, aerothermal, and hydrothermal energy. RES are used, among other things, for the production of electricity.

The production of electricity in Poland depends on the economic situation. Energy demand increases with the economic development of the country. Figure 1 shows two characteristic declines caused by the global crises [30]. Since 2005, the share of renewable energy in Poland's total electricity production has been increasing. Since 2019, the structure of gross electricity production from renewable energy sources in Poland has changed (Figure 2). The share of photovoltaic power plants has begun to increase in this structure. Particularly rapid growth has occurred in the number of electricity prosumers in Poland (Table 1). Their number was 25,623 in 2017 and 845,505 in 2021, almost 100% of which were prosumers using and producing energy from photovoltaic panels.

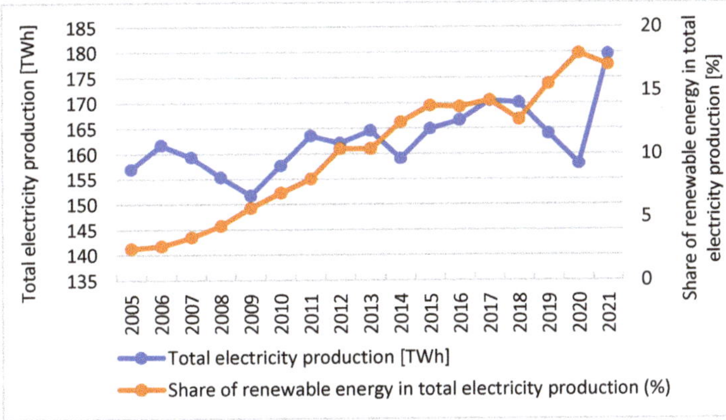

Figure 1. Total electricity production and share of renewable energy in Poland 2005–2021. Source: own elaboration on the basis of data from https://stat.gov.pl, accessed on 11 November 2022.

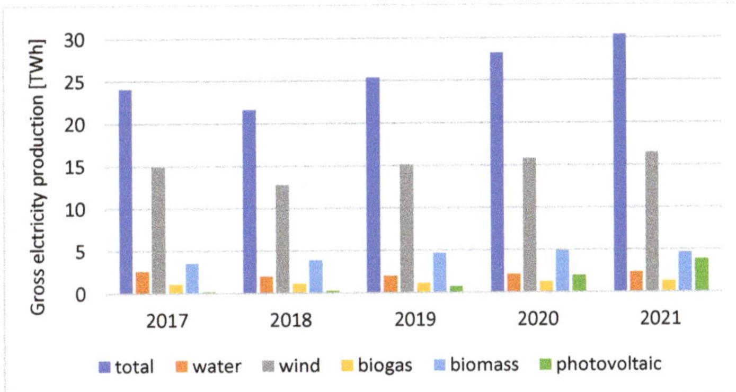

Figure 2. Gross electricity production from renewable energy sources in Poland 2017–2021 (GWh). Source: own elaboration on the basis of data from https://www.are.waw.pl/, accessed on 11 November 2022.

Table 1. Number of electricity prosumers in Poland in 2017–2021. Source: own elaboration on the basis of data from https://www.are.waw.pl/, accessed on 11 November 2022.

	Energy Sources (Electricity)					
Years	**Total**	**Water**	**Wind**	**Photovoltaic**	**Hybrid RES Installations**	**Biogas**
2017	25,623	3	19	25,571	28	2
2018	51,016	8	54	50,933	18	3
2019	144,940	8	56	144,856	17	3
2020	435,455	18	67	435,314	29	10
2021	845,505	75	70	845,259	45	33

One obstacle to the development of renewable energy sources in Poland is the state of the electricity transmission and distribution infrastructure [31]. In order to support investments in RES, it is first necessary to strive to simplify all state administrative procedures concerning investments in this sector. Decisive legal actions creating favourable conditions for RES development ought to be implemented by state authorities. Modification of the RES support system and harmonisation of the interpretation of the regulatory framework would also promote a faster introduction of new renewable energy sources [31]. The potential of renewable energy resources in Poland is high, which indicates real opportunities for further development of the renewable energy sector. According to experts, wind energy, solar energy, and solid biomass processing have the best chances for development in Poland. They point out that the main cause for the slow development of RES in Poland is the coal lobby [32]. Despite many efforts, the public administration in Poland is not conducive to the development of RES, which is often emphasised by owners of RES installations [33]. Administrative simplification is needed to reduce the investment preparation period and the associated costs. Existing procedures must be as simple as possible. The deadlines that individual decision-makers have should be reviewed (shortened if possible) or methods should be found on how to bypass certain systemic preparation stages [34]. Poland is one of the countries whose economy largely depend on energy imports and relies on non-organic energy sources, mainly coal or gas [35].

The rapid development of photovoltaic installations in Poland may exacerbate the problem of the mismatch between evening and daytime electricity demand in households. The Polish electricity system should prepare to stabilise the grid. This is a problem that should be solved in the near future [36].

The literature highlights the fact that the impact of government expenditure on economic development is not unambiguous. It depends on the country studied, the methodology used, the period studied, and the analysed socio-economic phenomenon [37].

A number of programmes are being implemented in Poland with the aim of meeting international obligations in the field of renewable energy, improving air quality, and increasing energy security. Some of them are directly dedicated to renewable energy sources, while others include support for the installation of these sources alongside other measures. Some of these programmes apply to individuals: homeowners or flat owners. Therefore, they concern micro-installations and prosumers.

An RES micro-installation is an installation of RES with a total installed electrical output of no more than 50 kW, connected to an electricity grid with a rated voltage of less than 110 kV or with a cogeneration heat output of no more than 150 kW, where the total installed electrical capacity is no more than 50 kW.

A prosumer is a final consumer purchasing electricity on the basis of a comprehensive contract, generating electricity exclusively from renewable energy sources in a micro-installation, for consumption for their own needs.

On 30 August 2019, the call for applications for the "My Electricity" programme began in Poland. This is the first time such a large nationwide support has been applied for individuals. The programme is a strong impetus for the further development of prosumer energy. The aim of the programme is to increase the production of electricity from photovoltaic micro-installations in the Republic of Poland. Importantly, the obtained subsidy can be combined with the so-called thermo-modernisation relief, which translates into financial efficiency of this support. The fourth edition of this programme was launched in 2022.

Recently, in Poland, the increase in output was mainly due to micro-installations (inter alia, the programme "My Electricity"), and in the near future there should be an increase in the output of large installations supported by the auction system [38].

A similar programme is the "My Heat" programme. Its aim is to support the development of individual heating and prosumer energy development in the area of air, water, and ground source heat pumps in new single-family residential buildings. The programme will run in the period 2022–2026.

The "Clean Air" programme is being implemented in Poland between 2018 and 2029. It is aimed at owners and co-owners of single-family houses, or at separate dwellings in single-family buildings with a separate land register. It is a response to the worsening air quality in Poland. Thanks to the programme, it is possible to apply for subsidies for the replacement of cookers, insulation of buildings, and the installation of photovoltaic panels and heat pumps. The programme aims to increase the energy efficiency of households and reduce or avoid emissions of harmful pollutants introduced into the atmosphere by single-family houses. This is mainly achieved by replacing old solid fuel heat sources with modern heat sources, meeting the highest standards and thermo-modernisation of buildings (e.g., external wall insulation). The "Clean Air" programme can contribute to reducing energy poverty in Poland.

Another programme to be implemented between 2019 and 2027 is "Stop SMOG". It is dedicated to municipalities located in areas where the so-called anti-smog resolution is in force to support the elimination or replacement of heat sources with low-emission ones and thermo-modernisation in single-family residential buildings of the least affluent people.

The "Warm Dwelling" programme will be implemented between 2022 and 2026 and is intended for municipalities, which will then issue an intake in their area for individuals with a legal title arising from ownership or a limited right to a dwelling unit located in a multi-family residential building. It is aimed at replacing all inefficient solid fuel heat sources for heating the dwelling with efficient heat sources or connecting to an efficient heat source in the building.

The transition to an electro-energy system dominated by renewable energy sources requires the consideration of economic, technical, and socio-political, as well as regulatory and institutional, aspects. Any changes in laws and regulations should be made with a full understanding of these aspects, otherwise change processes may be blocked [39]. Currently, a number of regulations are being introduced in Poland as a part of decarbonisation. Their aim is to develop a sustainable energy strategy. Some of these regulations include renewable energy sources. In order to achieve the sustainable development, governments ought to implement cost-effective, environmentally friendly and socially acceptable policies. Czarnecka et al. [40] identified socio-economic characteristics that influence the evaluation of the decarbonisation process in Poland. Respondents were those using central or local funds to conduct this process. The authors showed that, in Poland, financial incentives have the greatest potential for successful decarbonisation. Particularly important are subsidies for changing individual heating infrastructure or for developing prosumer photovoltaics. According to them, these activities should be financed from local resources.

As Rynska [11] points out, the first report on the PV market in Poland was published in 2012. At that time, the output of installations was estimated at 7.9 MW. Initial estimates for the following years turned out to be incorrect. In 2020, production was already higher than proposed for 2030. Even with various policy measures, the strongest incentive can only come from national regulations and financial incentives supporting stakeholders. PV development often does not depend on climatic conditions, but rather on the level of incentives taken by individual countries and the overall policy measures taken at the EU level.

The problem with subsidising renewables in Spain was described by Wuebben and Peters [41]. In 2008, Spain was the world leader in the number and output of photovoltaic installations. During this period, the tariff deficit (the mismatch between government expenditures on subsidies and revenues from taxes and tariffs for electricity) reached record levels in Spain. In the following years, the Spanish government removed incentives for prosumers and even tightened existing regulations. Spain has become one of the most restrictive regulatory regimes of all EU Member States. The setbacks and lack of stable regulation have created an unattractive market for investors. The number of new renewable energy installations has stopped growing [42].

Christoforidis et al. [43] proposed a novel generalised methodology for the techno-economic assessment of different Net-Metering (NEM) policies with respect to profitability for the prosumer. They tested the methodology on the example of Greece. They showed that the macroeconomic impacts of NEM are country specific. They affect each economy differently. In order that a policy is successful, it must create a win–win situation for all stakeholders.

3. Methods

The estimation of the mean impact effect is related to the analysis of dependence between the occurring phenomena. The basic element of these methods is a random variable describing two states: impact (Y_1) and the lack of impact (Y_0). The relationship between the empirical and hypothetical results can be presented as follows: $Y = dY_1 + (1 - d)Y_0$ for $d \in \{0, 1\}$. The dependent variable Y is modelled as a conditional expected value. It has a known realisation of the vector of observed characteristics, X. The mean treatment effect is defined by the formula $ATE = E(Y_1 \mid X) - E(Y_0 \mid X)$. The regression discontinuity design (RDD) is one of the popular methods of evaluation research.

The study uses the RDD model. Two forms of such models are mentioned in the literature: sharp and fuzzy [44,45]. The sharp form can be used if the cut-off point fully identifies the experimental group. If not, all units on a given side of the cut-off point can be assigned to the experimental group (they still have to satisfy an additional condition), then the fuzzy form of the model ought to be used. In the case of this study, due to a single set of conditions, the sharp form of the model has been used

On each side of the cut-off point, c, we estimate the parameters for two separate regression functions. On the left-hand side of the cut-off point, c, the function has the form [14]:

$$Y = \alpha_l + f_l(X - c) + \varepsilon \tag{1}$$

and on the right:

$$Y = \alpha_r + f_r(X - c) + \varepsilon \tag{2}$$

Both models (1) and (2) can be written as one:

$$Y = \alpha_l + \tau\, D + f(X - c) + \varepsilon \tag{3}$$

where:

$$\tau = \alpha_r - \alpha_l \tag{4}$$

$$f(X - c) = f_l(X - c) + D[f_r(X - c) - f_l(X - c)] \tag{5}$$

D is a dichotomic variable whose form depends on the position of the experimental group in relation to the defined eligibility threshold, c. If it is defined:

$$D = \begin{cases} 0 \text{ for } X < c \\ 1 \text{ for } X \geq c \end{cases} \tag{6}$$

then the experimental group is defined by the inequality $X \geq c$, and the control group by $X < c$.

If f_l and f_r have a linear form, the following holds:

$$f_l(X - c) = \beta_l(X - c) \tag{7}$$

$$f_r(X - c) = \beta_r(X - c) \tag{8}$$

and model (3):

$$Y = \alpha_l + \tau\, D + \beta_l(X - c) + \beta D(X - c) + \varepsilon \tag{9}$$

where $\beta = \beta_r - \beta_l$.

In model (9), the parameter τ reports the change in the value of the variable Y when passing the threshold. If this parameter is positive, there has been an increase in the threshold, if negative, there has been a decrease in the value of the dependent variable.

The parameter β, on the other hand, reports whether there has been a change in the rate of change in the growth of the dependent variable after passing the threshold.

The study uses data from Statistics Poland [7]. These are only annual data and do not detail the types of renewable energy sources. For this reason, data collected by The Energy Market Agency (ARE) [46] and Polish Power Transmission and Distribution Association [47] were also used. The quarterly data contained therein enabled the construction of the RDD model. Data on the number of micro-installations and micro-installation output in Poland from 2017Q2 to 2022Q2 were used to construct the model.

4. Data Analysis

At the outset, the research period was established. Due to data availability, quarterly data were included in the modelling. We mark the subsequent years as X_i for $i = 1, 2, \ldots,$ 21; 2017Q2 (the first observed quarter) is defined as 1 ($X_1 = 1$), and quarter 2022Q2 (the last observed) as 21 ($X_{21} = 1$). Preliminary analysis of the data is shown in Figure 2 and Table 1, and a review of the programmes targeting prosumers allowed us to assume the threshold as ($X_{11} = X_0 = 11$) (2019Q4). This estimate of the number and the output of micro-installations, Y_i, was used in two models. Each has the general form of:

$$\hat{Y}_i = \alpha_l + \tau D + \beta_l (X_i - X_0) + \beta D (X_i - X_0) \tag{10}$$

where:

Y_i—number of micro-installations (output of micro-installations);

X_i—quarter number;

τ—average impact of the change of regulations on the number of micro-installations (output of micro-installations) at the threshold $X_0 = 11$.

We define the dummy variable, D, as follows:

$$D = \begin{cases} 0 \text{ for } 1 \le X_i < 11 \\ 1 \text{ for } 11 \le X_i \le 21 \end{cases} \tag{11}$$

Parameter estimates are shown in Table 2.

Table 2. Number of electricity prosumers in Poland in the period 2017–2021.

Parameter	Parameter's Estimator	Standard Error	p-Value	Parameter's Estimator	Standard Error	p-Value
	Number of Micro-Installations $R^2 = 0.9910$			Output of Micro-Installations $R^2 = 0.9829$		
α_l	98,064.73	24,782.23	0.0010	630.03	247.61	0.0209
τ	−19,481.87	32,138.82	0.5524	−333.64	321.12	0.3134
β_l	8744.70	3994.02	0.0428	56.26	39.91	0.1767
β	88,935.20	5283.60	0.0000	675.63	52.79	0.0000

For the number of micro-installations, the following models were obtained:

$$\hat{Y}_i = 1873.06 + 8744.70 X_i \text{ for } D = 0 \tag{12}$$

$$\hat{Y}_i = -995,896.04 + 97,679.90 X_i \text{ for } D = 1 \tag{13}$$

For the output of micro-installations [MW] the following models were obtained:

$$\hat{Y}_i = 11.21 + 56.26 X_i \text{ for } D = 0 \tag{14}$$

$$\hat{Y}_i = -7754.35 + 731.89 X_i \text{ for } D = 1 \tag{15}$$

Equations (12) and (14) describe the linear models of the number and the output of micro-installations before 2019Q4, respectively. Equations (13) and (15) describe the linear models of the number and the output of micro-installations starting from 2019Q4, respectively.

The RDD models (12)–(15) are shown in Figures 3 and 4.

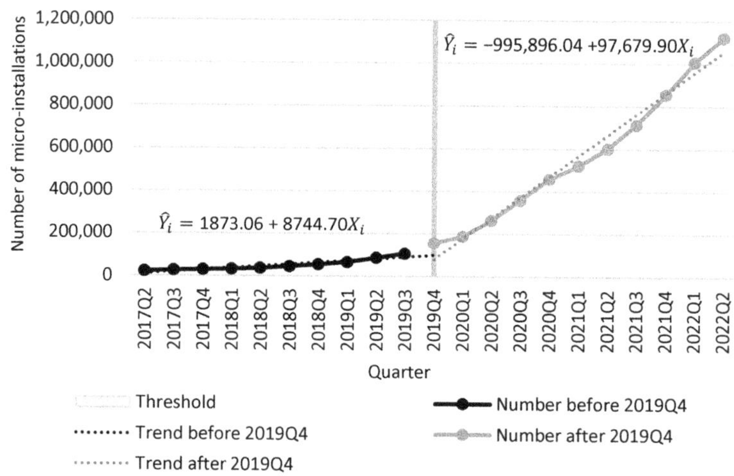

Figure 3. The regression discontinuity design models for the number of micro-installations in Poland.

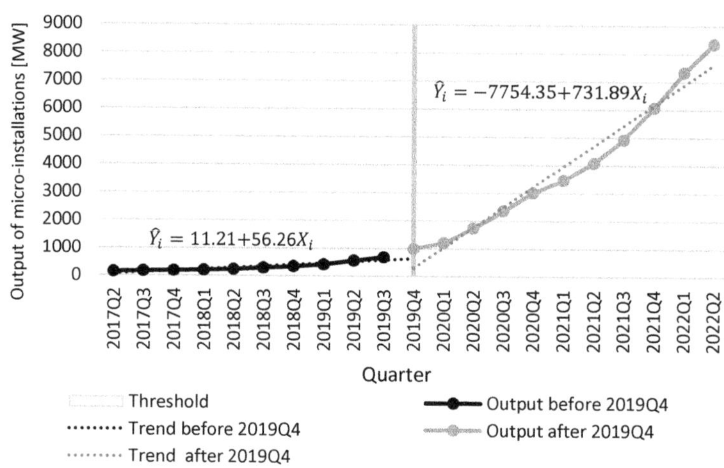

Figure 4. The regression discontinuity design models for the output of micro-installations in Poland.

The following conclusions can be drawn from the estimations in Table 2:

- The launch of the "My Electricity" programme did not cause a significant increase in the number and output of micro-installations at the threshold, i.e., in 2019Q4 (lack of statistical significance of the parameter τ at variable D). This fact is justified by the fact that the installation of photovoltaic panels is an investment that takes more than one quarter to complete. The increase in the number and output of installed micro-installations will be visible more than a quarter after the introduction of the new legislation.
- Both models, (12) and (14), have significant slope parameters. Positive signs of estimators of β_l indicate that before 2019Q4 the number and output of micro-installations

increased. However, the lack of significance of these parameters informs the fact that this growth was slow from quarter to quarter.

- The estimators of the β parameter were statistically significant and positive for both models. Comparing the slope parameters in models (12) and (13), as well as (14) and (15), one can see a clear sharp increase in their values. This means that since 2019Q4 the number and output of micro-installations started to increase rapidly.

5. Discussion

The analysis confirms the results of other studies. The RES subsidy programmes introduced in Poland for prosumers have significantly increased the number of RES and the output of energy production. Kata et al. [48] analysed the role of municipalities in the adaptation of RES installations in residents' households. Municipalities promote renewable energy in the "civic" segment primarily through the introduction of the so-called umbrella projects. They also positively influence RES adaptation through the mimicry effect, also in other households in the nearest neighbourhood. The imitation effect results in more inhabitants taking an interest in RES investments and expecting support from the municipality. These expectations relate to the co-financing of the installation, information, and consultancy support. It is the imitation effect that causes a rapidly growing interest in a particular programme. This was particularly evident after the introduction of the "My Electricity" programme. In Poland, the number of companies offering the installation of photovoltaic panels increased in the period 2020–2021 despite the ongoing pandemic, and so did the number of realised orders. The results obtained with the RDD model also indicate a rapid interest in the programme.

Zdonek et al. [49] attempted to evaluate the "My Electricity" programme for increasing renewable energy production in prosumer photovoltaic (PV micro-installations) sources in Poland. The evaluation was performed from the perspective of business, beneficiaries, and the local community. The authors concluded that the beneficiaries evaluated the surveyed programme rather well. Owners of small PV installations felt that the subsidy covered by the programme was a good motivation. Owners of larger PV installations prefer the subsidy being a percentage, not an amount. The cooperation with companies installing domestic PV systems and the process of handling the application for grants were also positively assessed by the beneficiaries. This positive assessment is reflected in the number of new PV sources installed and the output of new connections. The "My Electricity" programme has contributed to approximately 2 GWp of installed PV capacity between 2019 and 2021. The total cost of the programme on the state side is approximately EUR 390 million in direct subsidies and additional tax credits for prosumers [50]. The research in this paper is in line with these findings.

The process of growth in the overall share of energy from renewable sources in Poland follows the upward European trend (Figure 5). The exception in Poland is the year 2017, when there was a decrease and then an increase in the share due to changes in legislation and subsidies favourable to prosumers. However, despite these favourable changes, Poland's situation differs significantly from that of other European countries. Although the average overall share of energy from renewable sources in Poland is not much lower than the average for EU countries, Poland ranks among the last in Europe in 2020 (Figure 6). The situation of Poland is slightly better in relation to other European states in terms of the share of RES in total electricity production (Table 3). In Poland, most electricity from RES is produced in wind power plants. The changes analysed in the research part of the article concern electricity produced in prosumer micro-installations, which mainly concern solar installations. If the favourable changes in legislation and investment subsidies do not worsen, the number of these installations and their capacity will certainly continue to grow. This is because the overall aim is to increase the overall share of energy from renewable sources and to become independent of foreign supplies of energy resources.

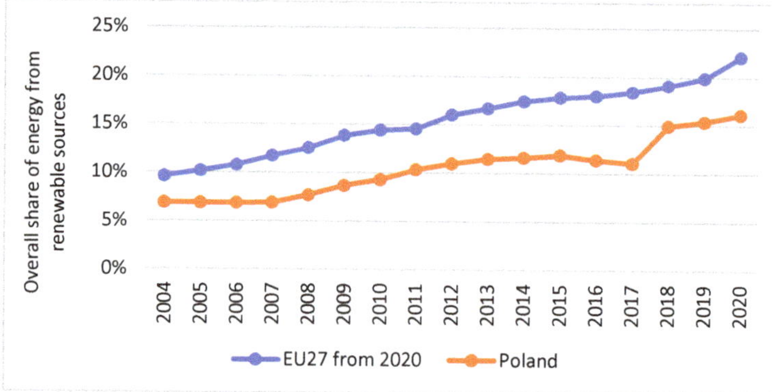

Figure 5. Overall share of energy from renewable sources, 2004–2020, EU27 from 2020 (%). Source: own elaboration on the basis of data from https://ec.europa.eu/eurostat/web/energy/data/shares, accessed on 11 November 2022.

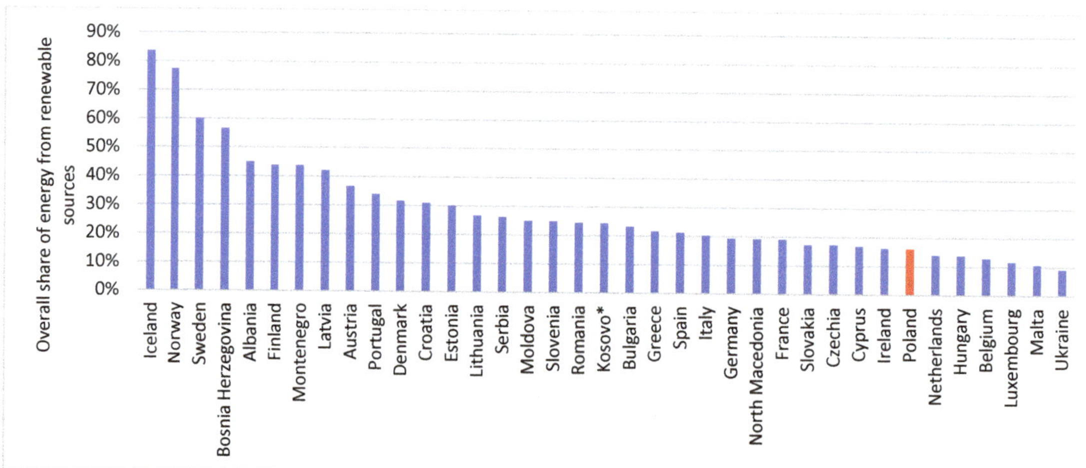

Figure 6. Overall share of energy from renewable sources in 2020 (%) in selected European countries. Source: own elaboration on the basis of data from https://ec.europa.eu/eurostat/web/energy/data/shares, accessed on 11 November 2022.

The development of RES in Poland depends on the level of public awareness in all consumer groups. The results of a study by Maciaszczyk et al. [51] showed that there is a relationship between the respondents' age and education and their readiness to consider renewable energy issues and implement them in their households. Awareness of the importance of RES increases slightly with the age of the respondents and their level of education. However, these variables did not significantly influence the decision to install RES in the household. The implication is that potential prosumers in Poland are still in the phase of following the RES market.

As Pietrzak and Kuc-Czarnecka [52] emphasise for the RES sector, there should be a focus, not on independent RES, but on the energy mix. Any regulations and implemented projects should allow the creation of cogeneration from different renewable sources. This direction of development of energy markets should increase the level of energy security, both in analysed countries and worldwide.

Table 3. Share of RES in total production of electricity and share of main RES sources in total RES in European countries in 2020. Source: own elaboration on the basis of data from https://ec.europa.eu/eurostat/web/energy/data/shares, accessed on 11 November 2022.

Countries	Electricity					
	RES %	**Hydro %**	**Wind %**	**Solar %**	**Solid Biofuels %**	**All Other Renewables %**
Norway	113.8	93.8	6.0	0.0	0.0	0.1
Iceland	102.7	69.6	0.0	0.0	0.0	30.3
Albania	100.0	99.6	0.0	0.4	0.0	0.0
Austria	78.2	75.6	12.5	3.7	6.5	1.7
Sweden	74.5	64.4	23.7	1.0	9.2	1.6
Denmark	65.3	0.1	68.6	5.1	18.5	7.7
Montenegro	61.5	85.2	14.7	0.1	0.0	0.0
Portugal	58.0	40.0	41.5	5.5	10.4	2.6
Croatia	53.8	70.4	17.5	1.0	5.8	5.3
Latvia	53.4	74.0	3.9	0.1	13.3	8.8
Bosnia & Herzegovina	49.3	94.2	4.8	0.7	0.1	0.1
Germany	44.7	8.2	50.6	20.0	4.6	16.5
Romania	43.4	63.9	27.0	6.9	2.0	0.2
Spain	42.9	26.8	49.8	18.0	4.0	1.4
Finland	39.6	43.6	20.9	0.7	32.4	2.5
Ireland	39.1	6.1	85.8	0.5	3.5	4.1
Italy	38.1	40.5	16.8	21.1	3.8	17.8
Greece	35.9	27.5	47.4	23.3	0.1	1.7
Slovenia	35.1	87.7	0.1	7.0	3.0	2.2
Serbia	30.7	89.5	8.6	0.1	0.2	1.5
Estonia	28.3	1.3	26.1	4.5	64.3	3.9
The Netherlands	26.4	0.3	43.7	27.5	18.1	10.4
Belgium	25.1	1.4	51.7	23.1	15.0	8.9
France	24.8	50.7	30.1	11.2	3.3	4.7
Bulgaria	23.6	47.2	16.3	17.0	16.9	2.6
North Macedonia	23.5	89.6	5.9	1.3	0.0	3.2
Slovakia	23.1	64.8	0.1	10.0	16.8	8.3
Lithuania	20.2	17.1	54.6	5.1	14.6	8.7
Poland	**16.2**	**8.4**	**54.4**	**7.1**	**25.0**	**5.1**
Czechia	14.8	21.2	6.5	22.0	24.1	26.2
Ukraine	13.9	51.6	15.0	29.7	1.4	2.3
Luxembourg	13.9	11.2	31.3	17.3	28.6	11.6
Cyprus	12.0	0.0	39.0	50.6	0.0	10.4
Hungary	11.9	4.3	12.2	44.3	30.0	9.2
Malta	9.5	0.0	0.0	97.5	0.0	2.4
Kosovo	5.3	79.2	17.7	3.1	0.0	0.0
Moldova	3.1	41.0	34.9	2.9	0.0	21.1

6. Conclusions

The conducted analysis confirmed the hypothesis. Regulations and projects implemented in Poland 2019 have had a positive impact on the increase in the number and output of micro-installations. The main contributor to this was the "My Electricity" programme. Due to the character of the investment, this change did not occur immediately, but several months after the introduction of the favourable programme. There was a sharp increase in the number of micro-installations and their output. A new Renewable Energy Sources Act came into force in Poland on 1 April 2022. The profitability of photovoltaics under the new system will be lower. The new system is not as favourable for prosumers as the previous one. The return on investment will extend by several years. This does not mean the end of the prosumer. Photovoltaics will still be more profitable than paying the usual electricity bills. The change in legislation will provide the impetus for further research into the impact of regulation on the development of RES in Poland. However, it will be necessary to wait for new data on the energy market in Poland.

One limitation of the research is the access to data. They are usually only available on an annual basis. This is because such a high degree of data aggregation makes it impossible to directly track changes over the year. Another problem is the lack of data by different energy sources, as well as the lack of data distinguishing micro-installations. In addition, data published by different institutions differ. Hence, the analyses appearing in the literature differ in the results obtained. However, the assessments of changes in the access and use of renewable energy sources are the same. Studies confirm the increase in the use of RES in Poland, especially by prosumers.

This article can be an important source of information for policy makers and creators of energy policy in Poland. Any new legal regulations may have a positive or negative impact on prosumer behaviour. This is especially important in the case of economic crises. Measures to mitigate the current energy crisis in Europe should be based on scientific research in the particular area. Thus, future legal regulations are largely likely to contribute to the rapid growth of micro-installation, or its decline.

Funding: The project is financed within the framework of the program of the Minister of Science and Higher Education under the name "Regional Excellence Initiative" in the years 2019–2022; project number 001/RID/2018/19; the amount of financing was PLN 10,684,000.

Data Availability Statement: Contained within the text and cited references.

Conflicts of Interest: The author declares no conflict of interest.

References

1. Energy for the Future: Renewable Sources of Energy—White Paper for a Community Strategy and Action Plan, COM(97)599 Final (26/11/1997). Available online: http://europa.eu.int/comm/energy/library/599fi_en.pdf (accessed on 25 October 2022).
2. Jäger-Waldau, A.; Scholz, H. EU renewables: Energy end-use efficiency and electricity from biomass, wind and PV in the EU. *Refocus* **2005**, *6*, 58–60. [CrossRef]
3. European Commission Communication. *The European Green Deal; COM (2019) 640 Final*; European Commission: Brussels, Belgium, 2019. Available online: https://ec.europa.eu/info/publications/communication-european-green-deal_en (accessed on 23 October 2022).
4. Jäger-Waldau, A. The Untapped Area Potential for Photovoltaic Power in the European Union. *Clean Technol.* **2020**, *2*, 440–446. [CrossRef]
5. Jonek-Kowalska, I. Assessing the energy security of European countries in the resource and economic context. *Oecon. Copernic.* **2022**, *13*, 301–334. [CrossRef]
6. Jonek-Kowalska, I. Transformation of energy balances with dominant coal consumption in European economies and Turkey in the years 1990–2017. *Oecon. Copernic.* **2019**, *10*, 627–647. [CrossRef]
7. GUS. Energy 2022. Warszawa 2022. Available online: https://stat.gov.pl/obszary-tematyczne/srodowisko-energia/energia/energia-2022,1,10.html (accessed on 2 November 2022).
8. Pietruszko, S.M. Photovoltaics in Poland. *Appl. Energy* **2003**, *74*, 169–175. [CrossRef]
9. Jurasz, J.K.; Dąbek, P.B.; Campana, P.E. Can a city reach energy self-sufficiency by means of rooftop photovoltaics? Case study from Poland. *J. Clean. Prod.* **2020**, *245*, 118813. [CrossRef]
10. Mikroinstalacje w Polsce. Available online: http://www.ptpiree.pl/energetyka-w-polsce/energetyka-w-liczbach/mikroinstalacje-w-polsce (accessed on 10 October 2022).
11. Rynska, E. Review of PV Solar Energy Development 2011–2021 in Central European Countries. *Energies* **2022**, *15*, 8307. [CrossRef]
12. Thistlethwaite, D.L.; Campbell, D.T. Regression Discontinuity Analysis: An Alternative to ex post facto Experiment. *J. Educ. Psychol.* **1960**, *51*, 309–317. [CrossRef]
13. Cook, T.D. "Waiting for Life to Arrive": A history of the regression-discontinuity design in Psychology, Statistics and Economics. *J. Econom.* **2008**, *142*, 636–654. [CrossRef]
14. Lee, D.S.; Lemieux, T. Regression Discontinuity Designs in Economics. *J. Econ. Lit.* **2010**, *48*, 281–355. [CrossRef]
15. Lalive, R. Unemployment Benefits, Unemployment Duration, and Post-Unemployment Jobs: A Regression Discontinuity Approach. *Am. Econ. Rev.* **2007**, *97*, 108–112. [CrossRef]
16. Lee, H.; Munk, T. Using regression discontinuity design for program evaluation. In Proceedings of the 2008 Joint Statistical Meeting, Denver, CO, USA, 3–7 August 2008; American Statistical Association: Alexandria, VA, USA, 2008; pp. 3–7. Available online: http://www.amstat.org/sections/srms/proceedings/y2008/Files/301149.pdf (accessed on 1 October 2022).
17. Brügger, B.; Lalive, R.; Zweimüller, J. Does Culture Affect Unemployment? Evidence from the Röstigraben, IZA Discussion Papers 428, 2009. Available online: https://www.iza.org/publications/dp/4283/does-culture-affect-unemployment-evidence-from-the-rostigraben (accessed on 25 October 2022).

18. Schmieder, J.F.; von Wachter, T.; Bender, S. The Effects of Extended Unemployment Insurance Over the Business Cycle: Evidence from Regression Discontinuity Estimates Over 20 Years. *Q. J. Econ.* **2012**, *127*, 701–752. [CrossRef]
19. Decramer, S.; Vanormelingen, S. The effectiveness of investment subsidies: Evidence from a regression discontinuity design. *Small Bus. Econ.* **2016**, *47*, 1007–1032. [CrossRef]
20. Markowicz, I.; Bieszk-Stolorz, B. The Application of Survival Analysis to Evaluating the Efficiency of a Professional Effectiveness Enhancement Programme. *Acta Univ. Lodziensis. Folia Oeconomica* **2016**, *1*, 85–91. [CrossRef]
21. Cadil, J.; Mirosnik, K.; Petkovova, L.; Mirvald, M. Public Support of Private R&D—Effects on Economic Sustainability. *Sustainability* **2018**, *10*, 4612. [CrossRef]
22. Cansino, J.M.; Sánchez-Braza, A.; Espinoza, N. Evaluating the Impact of an Active Labour Market Policy on Employment: Short- and Long-Term Perspectives. *Soc. Sci.* **2018**, *7*, 58. [CrossRef]
23. Wang, S.; Li, Y.; Haque, M. Evidence on the Impact of Winter Heating Policy on Air Pollution and Its Dynamic Changes in North China. *Sustainability* **2019**, *11*, 2728. [CrossRef]
24. Wang, X.; Chen, C.; Du, Y.; Zhang, Y.; Wu, C. Analysis of Policies Based on the Multi-Fuzzy Regression Discontinuity, in Terms of the Number of Deaths in the Coronavirus Epidemic. *Healthcare* **2021**, *9*, 116. [CrossRef]
25. Lotsu, S.; Yoshida, Y.; Fukuda, K.; He, B. Effectiveness of a Power Factor Correction Policy in Improving the Energy Efficiency of Large-Scale Electricity Users in Ghana. *Energies* **2019**, *12*, 2582. [CrossRef]
26. Dong, D.; Xu, X.; Wong, Y.F. Estimating the Impact of Air Pollution on Inbound Tourism in China: An Analysis Based on Regression Discontinuity Design. *Sustainability* **2019**, *11*, 1682. [CrossRef]
27. Ogunmodede, A.M.; Ogunsanwo, M.O.; Manyong, V. Unlocking the Potential of Agribusiness in Africa through Youth Participation: An Impact Evaluation of N-Power Agro Empowerment Program in Nigeria. *Sustainability* **2020**, *12*, 5737. [CrossRef]
28. Bieszk-Stolorz, B.; Dmytrów, K. Impact of Changes to Procedures on the Evaluation of the Effectiveness of Forms of Professional Activation in Poland. *Economies* **2021**, *9*, 88. [CrossRef]
29. Zhao, X.; Cheng, Z.; Jiang, C. Could Air Quality Get Better during Epidemic Prevention and Control in China? An Analysis Based on Regression Discontinuity Design. *Land* **2021**, *10*, 373. [CrossRef]
30. Dmytrów, K.; Landmesser, J.; Bieszk-Stolorz, B. The Connections between COVID-19 and the Energy Commodities Prices: Evidence through the Dynamic Time Warping Method. *Energies* **2021**, *14*, 4024. [CrossRef]
31. Serowaniec, M. Sustainable Development Policy and Renewable Energy in Poland. *Energies* **2021**, *14*, 2244. [CrossRef]
32. Pietrzak, M.B.; Igliński, B.; Kujawski, W.; Iwański, P. Energy Transition in Poland—Assessment of the Renewable Energy Sector. *Energies* **2021**, *14*, 2046. [CrossRef]
33. Iglinski, B.; Flisikowski, K.; Pietrzak, M.B.; Kiełkowska, U.; Skrzatek, M.; Zyadin, A.; Natarajan, K. Renewable Energy in the Pomerania Voivodeship—Institutional, Economic, Environmental and Physical Aspects in Light of EU Energy Transformation. *Energies* **2021**, *14*, 8221. [CrossRef]
34. Jankowska, B.; Staliński, A.; Trąpczyński, P. Public policy support and the competitiveness of the renewable energy sector–The case of Poland. *Renew. Sust. Energ. Rev.* **2021**, *149*, 111235. [CrossRef]
35. Pietrzak, M.B.; Olczyk, M.; Kuc-Czarnecka, M.E. Assessment of the Feasibility of Energy Transformation Processes in European Union Member States. *Energies* **2022**, *15*, 661. [CrossRef]
36. Olczak, P.; Jaśko, P.; Kryzia, D.; Matuszewska, D.; Fyk, M.I.; Dyczko, A. Analyses of duck curve phenomena potential in polish PV prosumer households' installations. *Energy Rep.* **2021**, *7*, 4609–4622. [CrossRef]
37. Nyasha, S.; Odhiambo, N. The Impact of Public Expenditure on Economic Growth: A Review of International Literature. *Folia Oeconomica Stetin.* **2019**, *19*, 81–101. [CrossRef]
38. Gnatowska, R.; Moryń-Kucharczyk, E. The Place of Photovoltaics in Poland's Energy Mix. *Energies* **2021**, *14*, 1471. [CrossRef]
39. Wilkinson, S.; John, M.; Morrison, G.M. Rooftop PV and the Renewable Energy Transition; a Review of Driving Forces and Analytical Frameworks. *Sustainability* **2021**, *13*, 5613. [CrossRef]
40. Czarnecka, M.; Chudy–Laskowska, K.; Kinelski, G.; Lew, G.; Sadowska, B.; Wójcik-Jurkiewicz, M.; Budka, B. Grants and Funding for the Processes of Decarbonization in the Scope of Sustainability Development—The Case from Poland. *Energies* **2022**, *15*, 7481. [CrossRef]
41. Wuebben, D.; Peters, J.F. Communicating the Values and Benefits of Home Solar Prosumerism. *Energies* **2022**, *15*, 596. [CrossRef]
42. Mir-Artigues, P.; Cerdá, E.; del Río, P. Analyzing the Impact of Cost-Containment Mechanisms on the Profitability of Solar PV Plants in Spain. Renew. *Sustain. Energy Rev.* **2015**, *46*, 166–177. [CrossRef]
43. Christoforidis, G.C.; Panapakidis, I.P.; Papadopoulos, T.A.; Papagiannis, G.K.; Koumparou, I.; Hadjipanayi, M.; Georghiou, G.E. A Model for the Assessment of Different Net-Metering Policies. *Energies* **2016**, *9*, 262. [CrossRef]
44. Trochim, W. *Research Design for Program Evaluation: The Regression-Discontinuity Approach*; Sage Publications: Beverly Hills, LA, USA, 1984.
45. Hahn, J.; Todd, P.; Van der Klaauw, W. Identification and Estimation of Treatment Effects with a Regression-Discontinuity Design. *Econometrica* **2001**, *69*, 201–209. [CrossRef]
46. ARE. Available online: https://www.are.waw.pl/ (accessed on 14 October 2022).
47. PTPiREE. Available online: http://ptpiree.pl/ (accessed on 14 October 2022).
48. Kata, R.; Cyran, K.; Dybka, S.; Lechwar, M.; Pitera, R. The Role of Local Government in Implementing Renewable Energy Sources in Households (Podkarpacie Case Study). *Energies* **2022**, *15*, 3163. [CrossRef]

49. Zdonek, I.; Tokarski, S.; Mularczyk, A.; Turek, M. Evaluation of the Program Subsidizing Prosumer Photovoltaic Sources in Poland. *Energies* **2022**, *15*, 846. [CrossRef]
50. Kulpa, J.; Olczak, P.; Surma, T.; Matuszewska, D. Comparison of Support Programs for the Development of Photovoltaics in Poland: My Electricity Program and the RES Auction System. *Energies* **2022**, *15*, 121. [CrossRef]
51. Maciaszczyk, M.; Czechowska-Kosacka, A.; Rzepka, A.; Lipecki, T.; Łazuka, E.; Wlaź, P. Consumer Awareness of Renewable Energy Sources: The Case of Poland. *Energies* **2022**, *15*, 8395. [CrossRef]
52. Pietrzak, M.B.; Kuc-Czarnecka, M. Transformation of Energy Markets: Description, Modeling of Functioning Mechanisms and Determining Development Trends. *Energies* **2022**, *15*, 5493. [CrossRef]

energies

MDPI

Article

How Do FDI and Technological Innovation Affect Carbon Emission Efficiency in China?

Qizhen Wang [1] and Suxia Liu [2,*]

1 Business School, Nanjing Xiaozhuang University, Nanjing 210017, China
2 Business School, Zhengzhou University, Zhengzhou 450001, China
* Correspondence: liusuxia@zzu.edu.cn

Abstract: China's economic development is characterized by openness, and trade and investment are important engines for promoting economic development. China's economy is now in a transitional period, during which excessive carbon emission reduction would inevitably hinder economic development. In this context, improving carbon emission efficiency is an effective way to achieve sustainable development. This paper deals with the relationships among foreign direct investment, technological innovation and carbon emission efficiency. Our research findings include the following. First, carbon efficiency shows regional differences. East China has the highest mean value of carbon emission efficiency, followed by central China and west China over the sample period. Second, FDI exerts both direct and indirect impacts on carbon emission efficiency through technological innovation, which confirms the intermediate effect of technological innovation. Finally, sub-sample analysis indicates that the impact of FDI and technological innovation on carbon emission efficiency show regional heterogeneity. According to these findings, we offer policy recommendations as follows. The government should stimulate independent innovation, promote technological progress in renewable energy and green energy, and attract environmentally friendly foreign investment to improve carbon emission efficiency and boost green development.

Keywords: FDI; technological innovation; carbon emission efficiency; carbon emission

Citation: Wang, Q.; Liu, S. How Do FDI and Technological Innovation Affect Carbon Emission Efficiency in China?*Energies* **2022**, *15*, 9209. https://doi.org/10.3390/en15239209

Academic Editors: Grzegorz Mentel and Sebastian Majewski

Received: 22 October 2022
Accepted: 2 December 2022
Published: 5 December 2022

1. Introduction

Environmental problems are among the major challenges facing the world in the 21st century. As the largest energy consumer, China faces great pressure in reducing carbon emissions (CE). At the 75th UN General Assembly in 2020, China put forward the commitment to reach peak carbon emissions by 2030 and realize carbon neutrality by 2060. Two methods can be adopted to reduce carbon emissions. One is to reduce the use of fossil fuels, and the other is to improve carbon emission efficiency (CEE) so that maximum economic benefits are generated with less resource consumption and less environmental cost. There is an inverse relationship between CE and CEE. The less the CE, the higher the CEE. China is currently undergoing economic transformation towards green development, a low-carbon economy, and sustainable development. Excessive emission reduction would inevitably hinder economic development. Against this background, improving CEE is an effective means to accomplish sustainable development.

China has made opening up a basic state policy since 1978. A great deal of foreign companies have relocated to the Chinese market and accelerated China's modernization. On the whole, China's use of foreign direct investment (FDI) has been expanding in scale and optimizing in structure. FDI has two-sided influence on China's economy. On the one hand, it brings advanced technology and management experience, optimizes industries, and promotes technological innovation (TI). On the other hand, it increases energy consumption, expands the utilization of resources, and increases carbon emissions and environmental pollution.

FDI can affect regional CEE. On the one hand, it directly affects regional CEE. Firstly, the effect depends on the TSE, structure effect and scale effect of FDI inflow. The TSE and structure effect positively affect CEE and negatively affect CE, while the scale effect negatively affects CEE and positively affects CE. Secondly, the effect is also decided by the stage of local economic development. In the initial stage, the host country would weaken environmental protection in order to develop the economy. In addition, the lack of environmental awareness and technology leads to low intensity of environmental regulations (ERG), which attracts a large amount of foreign investment with high CE. This leads to the agglomeration of FDI enterprises featuring high energy use and high CE, which results in the decline of CEE. At the stage of rapid economic growth, environmental deterioration and improved environmental awareness impel local governments to increase the intensity of ERG and give priority to environmentally friendly foreign enterprises. Local TI is improved through the demonstration effect, TSE, and market competition effect of FDI inflow, which boost technological progress in environmental protection and raise CEE.

On the other hand, FDI indirectly affects regional CEE in different ways, including TI. FDI brings advanced environmental protection technology and pollution control experience. Through the demonstration effect and competition effect, it encourages enterprises to optimize production, improves environmental protection technology, and increases CEE. This covers upstream and downstream enterprises as well as related industries. Advanced experience in environmental protection technology and pollution control is spread among related industries and enterprises through the personnel flow effect and industrial correlation effect. As the technologies become more and more mature, they will be spread to other regions through the spatial spillover effect and eventually lead to overall improvement of environmental protection technology and CEE in the host country. It is worth noting that foreign investment may squeeze out domestic investment, so the TSE of FDI enterprises may be weakened.

We conducted research on the relationships among FDI, TI, and CEE to test whether FDI has a pollution paradise effect or a pollution halo effect on China's environment. Compared with existing literature, major contributions of this paper are as follows. (1) Previous literature has carried out a lot of research on CE, energy efficiency, etc., but the research on CEE is less. In the research on China's CEE, more attention is paid to the spatio-temporal heterogeneity of CEE. Few studies pay attention to the factors affecting CEE. (2) Previous literature pays attention to CEE and its influencing factors, but as one of the important factors, FDI is rarely considered; the impact of FDI on CEE through TI is even less. (3) We conduct a sub-sample analysis to check the regional heterogeneity of the effect of FDI on CEE through TI, having expanded the present research.

2. Literature Review

2.1. FDI and Carbon Emissions

Scholars have conducted many studies on FDI and CE. They put forward two main hypotheses. The "pollution paradise" hypothesis (PPH) assumes that developed countries have strict environmental standards. In order to obtain more profits, some pollution-intensive enterprises will relocate to developing countries which have a low intensity of ERG. The inflow of foreign investment not only influences economic growth [1], but also increases environmental pollution. Therefore, FDI is an important variable in PPH [2–4]. Bae et al. [5] explored the factors influencing CE in 15 post-Soviet countries. They found that FDI exerted a positive effect on CEE, which indicated that PPH existed in these countries. Therefore, it was very important for the countries to attract FDI in order to obtain renewable energy, since most of them were not developed and faced high costs in deploying new technologies. Meanwhile, they should focus on the detrimental influence of FDI before implementing regulations on attracting FDI inflows. Arain et al. [6] found that FDI and CE were closely related in terms of wavelet scales, which showed close relations between FDI and CE in the short term. The coherence analysis revealed that the relationship was significant. FDI was favorable for China's economic growth because renewable technologies and low-carbon technologies were adopted by multinational corporations. On the other hand, these multinational corporations led to the increase of CO_2 emissions. Therefore, the government should introduce appropriate policies to develop the economy while also reducing CE. Zakarya et al. [7] found that FDI positively affected carbon emissions in emerging countries, verifying the existence of PPH. Behera and Dash [8] came to the same conclusion.

The "pollution halo" hypothesis (PHH) assumes that the inflow of FDI is beneficial for optimizing energy structure and generating green technology spillovers, thereby leading to the decline of CE in the host country [9,10]. Zhu et al. [11] found that FDI had a heterogeneous effect on CE. The coefficient of FDI was positive but insignificant at the 5th quantile, whereas other coefficients were significantly negative at high quantiles. Also, the influence of FDI on CE was significantly negative in highly polluted countries, which supported the halo effect hypothesis. One of the plausible reasons was that multinationals had more advanced technologies compared with enterprises in highly polluted countries. When investing in high-emission countries, multinationals brought clean technology and innovative skills which helped improve environmental quality, while in low-emission countries, they tended to invest in non-polluting sectors. Hence, the inflow of FDI had an insignificant impact on CE.

Many scholars assumed FDI's effect on CE was unstable [12]. Yildrim [13] found that the PPH was valid in Mozambique, Oman, and the United Arab Emirates, while PHH was invalid in Zambia, Iceland, Panama, and India. Khurram et al. [14] found that a positive shock in FDI increased CE, especially in the long term. A negative shock in FDI had an insignificant impact on CE. However, the negative shock had a positive impact on CE in the short run. This indicated that FDI and CE had a non-linear relationship, which was consistent with the study of Zia et al. [15]. Zia et al. [15] employed an autoregressive distribution lag model to study PPH in Pakistan. They also found that the increase in FDI positively affected CE, and the decrease of FDI insignificantly affected CE.

Regarding carbon emission efficiency, Wu et al. [16] employed data envelopment analysis (DEA) to investigate the influence of energy subsidies on CEE based on the role of FDI in competition. They found that fiscal decentralization positively affected CEE, which meant local governments had strong financial resources and low willingness to enrich the tax base through economic growth when the fiscal decentralization index was high, which led to high inputs in low-carbon industries and improvement in CEE. The interaction between fiscal decentralization and government competition negatively affected CEE, which meant that local governments were not willing to save energy and reduce emissions in order to attract FDI inflow. This results in the decline of CEE. It is evident that FDI inflows make local government lower their environmental standards, which is not conducive to improving CEE [17].

2.2. FDI and Technological Innovation

The earliest research on how FDI affects TI began with the theory of TSE of foreign direct investment on host countries proposed by MacDougall [18]. This theory holds that FDI can produce technology spillovers on relevant industries in the host country. Keller and Yeaple [19] made a comparative analysis on international TSE on U.S. enterprises through imports and FDI. According to them, FDI allowed domestic firms to obtain great productivity gains, accounting for approximately 14 percent of total productivity growth in the U.S. Imports had technology spillovers to domestic firms, but the effect was weaker than that of FDI.

Some studies showed that TSE of FDI positively affected TI in the host country [20,21]. Munteanu [22] divided knowledge spillovers into two categories: supplemental spillovers and complementary spillovers. Both play an important role in increasing the value of technology transfer through FDI and are influenced by learning ability and technology gap. TSE and propagation effects, especially for the knowledge of technology, significantly influence economic growth at both horizontal and vertical levels. Fernandes and Paunov [23] found that FDI in services positively affected the TFP of manufacturing firms in Chile. It was evident that the horizontal spillovers of FDI were weaker than its vertical spillovers. The reason was that foreign-owned firms did not intend to produce technology spillovers to domestic firms within the industry. They were more willing to produce technology spillovers to downstream providers and upstream clients. To explore the influence of different technological sources on energy conservation technology, Yang et al. [24] analyzed six basic technological sources. They found that forward TSE of FDI competition forced domestic firms to improve energy-saving technology, while backward TSE, horizontal TSE, learning by export, and innovation had no significant effect on energy-saving technology. Negash et al. [25] found that Chinese-invested firms had higher productivity than local firms in Ethiopia. Hence, Chinese-invested firms brought advanced technology, capital, and knowledge to local firms, which stimulated the latter's TI. Whether local firms could gain significant positive TSE from Chinese-invested firms was decided by their absorptive capacity.

Some studies demonstrated that FDI negatively affected TI in the host country [26,27]. Feng et al. [28] found that FDI negatively affected China's urban innovation. Therefore, the government is advised to optimize the structure of FDI: central and west regions should attract high-quality FDI instead of expanding the size of FDI and should not lower the intensity of ERG in order to attract foreign investments. What's more, local government should implement strict environmental regulations to attract multinationals with advanced technology and energy-saving ability and restrict enterprises with weak technological strength and high energy consumption. Hu et al. [29] classified FDI into labor-based FDI and capital-based FDI. The research results demonstrated that both types of FDI negatively affected green total factor productivity (GTFP), but only the effect of labor-based FDI was significant. The plausible reason was that labor-intensive industries produced less carbon emissions and featured cleaner production [30], and labor-based FDI may improve the quality of human capital and generate TSE in the host country. Meanwhile, there were weak ERG in labor-intensive industries, and these industries had no strong technological innovation, which led to the decline of GTFP. The intensity of ERG on capital-intensive industries was higher, so capital-based FDI brings less pollution. Hence, the influence of capital-based FDI on GTFP in the host country was uncertain.

Some scholars also found that FDI had mixed effects on TI in the host country. Anwar and Nguyen [31] investigated the effect of FDI on TFP and found that it was different in different regions. For instance, FDI produced positive horizontal TSE in the northeast, but negative horizontal TSE in the Red River Delta. FDI produced positive backward TSE in the Red River Delta, but negative backward TSE in the Mekong River Delta. Hu et al. [32] found that labor-based FDI had a significantly negative TSE, while capital-based FDI had a significantly positive TSE in industries with a low intensity of ERG. In industries with a high intensity of ERG, labor-based FDI had an insignificantly negative TSE, and capital-based FDI still had a significant TSE. This indicated that the negative TSE of labor FDI was decided by the intensity of ERG. The "pollution haven" hypothesis was valid in industries with a low intensity of environmental regulations.

2.3. Technological Innovation and Carbon Emissions

Many studies focused on environmental variation and its determinants [33–36]. Among them, the link between technological innovation and CE has been abundantly investigated. Most studies assume that technological innovation reduces CE [37–40]. Using firm level data, Lee and Min [41] examined the impact of investment in green R&D on the environment and financial performance. They found that green R&D negatively affected CE, but positively affected financial performance. Kong et al. [42] found that all efficient energy technologies could greatly improve pulp and paper technology. Therefore, advanced technologies should be used to decrease energy consumption and CE in the industry. Sgobbi et al. [43] found that the improvement of technological efficiency was very important to tidal energy. Improving efficiency was more important than reducing costs, because technological upgrades can increase energy supply and reduce CE. Zeeshan et al. [44] found that R&D negatively affected CE. In the short run, there was an insignificant negative relationship between R&D and CE. As far as China is concerned, spending on R&D was also related to CE.

Few studies concluded that technological innovation increased CE. R&D plays a major role in new technologies and new products. Through R&D, more and more competitive products are produced. Shaari et al. [45] employed the panel DOLS and FMOLS to analyze the relations between R&D and CO_2 emissions. The results of FMOLS suggested that R&D positively affected CO_2 emissions. The results of DOLS suggested that R&D positively affected CE. This implies that expenditures on R&D should be reduced to improve environmental quality and boost economic growth.

There were also some mixed or different results regarding the relationship between TI and CE. Demir et al. [46] analyzed this relationship using the ARDL approach. They found that the relationship between home patents and CE followed an inverted U-shape in Turkey. This meant that home patents were positively related to CO_2 emissions in the early stages of economic development. When economic development reached a certain level, home patents were negatively related to CO_2 emissions. Like Demir et al. [46], Gu et al. [47] found the relationship between energy-technological progress and CE in China was in an inverted U-shape. Yii and Geetha [48] found that technological innovation negatively affected CE in the short term and positively affected the latter in the long term in Malaysia. Dauda et al. [49] revealed that technological innovation negatively affected CE in the G6 countries but positively affected CE in the MENA (Middle East and North Africa) and BRICS (Brazil, Russia, India, China, and South Africa) countries. Erdoğan et al. [50] studied the influence of innovation on CE in G20 countries. They found that the increase in innovation resulted in the reduction of CE in the industrial sector and resulted in the increase of CE in the construction sector.

As for the connection between energy and environment, CEE which reflects the level of green development has attracted a lot of attention. Zhang and Chen [51] found that technological progress negatively affected CEE. This is the same as the conclusion of Wang et al. [52] and Huang et al. [53]. The reason behind this conclusion was the rebound effect. Usually, technological progress positively affected energy efficiency and negatively affected CE. But technological progress would lead to production expansion and more energy consumption, so carbon efficiency declined because of the rebound effect. In addition, technological progress changed lifestyles and led to the use of a large number of electronic products, which greatly increased electrical energy consumption and thus improved CE. Hence, the positive effect of technological progress may be offset by a negative rebound effect. Different from Zhang and Chen [51], Yan et al. [54] found that technological progress was a main driving force for improving CEE. Santra [55] studied the effect of environmental innovation on energy efficiency and CEE in BRICS countries. They concluded that green technological innovation reduced energy absorption and CE and improved energy efficiency and CEE in each member country.

2.4. Research Gap

CO_2 is a kind of heat-trapping gas and the largest contributor global warming. Hence, curbing or reducing CO_2 emissions is crucial to sustainable development [56]. According to PPH and PHH, FDI is a double-edged sword that can both increase and decrease CE, which is decided by the sum of TSE, the structure effect, and the scale effect of FDI in different stages of economic development. There are abundant studies on the relationship between FDI and CE, but few deal with the relationship between FDI and CEE. Is there a linear or non-linear relationship between the two variables? If FDI is related to CEE, what is the degree of FDI's impact on CEE? Few studies have focused on the channels through which CEE is affected, and possible channels of technological innovation have been ignored. In addition, regional analysis is also missing. Since China has many regions, heterogonous characteristics should be investigated. In view of this situation, this paper examines the effect of FDI on CEE, including direct and indirect channels, and takes into account regional differences. The results are quotable for policymakers to increase CEE through FDI.

3. Methods and Materials

3.1. Super Efficiency DEA Method

Data envelopment analysis (DEA) is a special tool based on linear programming proposed by Charnes et al. [57]. It is mainly used to evaluate the efficiency of a decision-making unit (*DMU*) using an input–output method. There are two traditional DEA methods: CCR-DEA model [57] and BCC-DEA model [58]. Both can be used to calculate the efficiency score and test the effectiveness of the efficiency score for each DMU. One of their disadvantages is that they cannot compare and analyze different DMUs when DMUs are effective at the same time. In addition, they do not take into consideration the impact of random errors and are easily affected by sample data. Hence, the efficiency score may be biased. When the error item is large, the results estimated by the DEA model will be serious biased. The super efficiency DEA model which was proposed by Anderson and Petersen [59] is different from traditional DEA models and can compare different effective DMUs.

After calculating efficiency with the traditional DEA method, we obtain two kinds of efficiency values: those less than 1 and those is equal to 1. The former means that the efficiency value is invalid, while the latter shows the efficiency value is valid. The traditional DEA method is unable to compare and analyze effective *DMUs* because all efficiency values are 1. In fact, there are differences among effective *DMUs*. According to Anderson and Petersen [59], the formula for this study is:

$$\begin{cases} \min(\theta) - \varepsilon\left(\sum\limits_{i=1}^{m} s_i^- - \sum\limits_{r=1}^{s} s_r^+\right) \\ s.t. \sum\limits_{j=1}^{n} x_{ij}\lambda_j + s_r^- = \theta x_{ij_0} \\ \sum\limits_{j=1}^{n} y_{ij}\lambda_j - s_r^+ = y_{ij_0} \\ \lambda_j, s_r^-, s_r^+ \geq 0, i = 1, 2, \ldots, m \\ j = 1, 2, \ldots, j_0 - 1, j_0 + 1, \ldots, n \end{cases} \tag{1}$$

This is a super efficiency model used to evaluate j_0 decision-making unit DMU_0. It removes the evaluated unit DMU_0 from the reference set and obtains its own value by referring to the frontiers of other *DMUs*. This fills in the blank of the traditional DEA model, which cannot be used for further analysis when the efficiency value is 1. The multi-input–multi-output evaluation system has n *DMUs*, including m input indicators and s output indicators. x_{ij} is the i-th input of the j-th *DMU*, and y_{ij} is the i-th output of the j-th *DMU*. θ is the super efficiency value of *DMU* j_0. ε indicates the non-Archimedean infinitesimal, and s^- and s^+ are slack variables.

3.2. Data Resources

The data are from the China Statistical Yearbook, China Economic Network Statistical Database, China Science and Technology Statistical Yearbook, World Bank, United Nations Conference on Trade and Development (UNCTAD) Database, and China's Carbon Emissions Database from 2004 to 2019. Tibet, Macao, Hong Kong, and Taiwan are not included owing to incomplete data.

3.2.1. Carbon Emission Efficiency

The super efficiency DEA model is employed to estimate CEE. Based on previous research literature and taking into account the characteristics of this study, the input variables are capital (X_1), labor (X_2), and energy consumption (X_3), and the output variables are GDP (Y_1) and CO_2 (Y_2). The input and output variables are presented in Table 1.

Table 1. Index system of regional carbon efficiency.

	Index Category	Index Form
Input index	Capital (X_1) Labor (X_2) Energy Consumption (X_3)	Capital stock Employees Total energy consumption
Output index	GDP (Y_1) CO_2 (Y_2)	Regional GDP CO_2 emissions

In Table 1, capital refers to the stock of fixed capital, which is the weighted sum of the previous investment flows measured at constant prices. Total current capital is equal to total capital of the previous period—depreciation + current capital, which is expressed as $K_{jt} = K_{jt-1}(1 - \delta_{jt}) + I_{jt}$. K_{jt} is the capital stock of province j in period t, K_{jt-1} means the capital stock of province j in period $t - 1$, δ_{jt} is the depreciation rate of province j in year t, and I_{jt} represents the investment in province j in year t by the prices of the current year. The base year is 2004. Labor input is the number of employees in each province, and energy consumption is the total energy consumption of each province. The GDP is calculated based on the GDP deflator of 2004, and CO_2 refers to the CO_2 emissions of each province.

The values of CEE are shown in Figure 1. It is clear that the value of CEE in east China was the highest before 2012, and central China had the highest value after 2012. Both national and regional values showed an increasing trend, and the value of CEE showed regional imbalances. Since 2004, the national average value of CEE has been growing. It reached a peak in 2015 and then decreased. The CEE value of east China had a downward trend, while central China had an upward trend. The lowest value of CEE was in west China, but the value showed an upward trend.

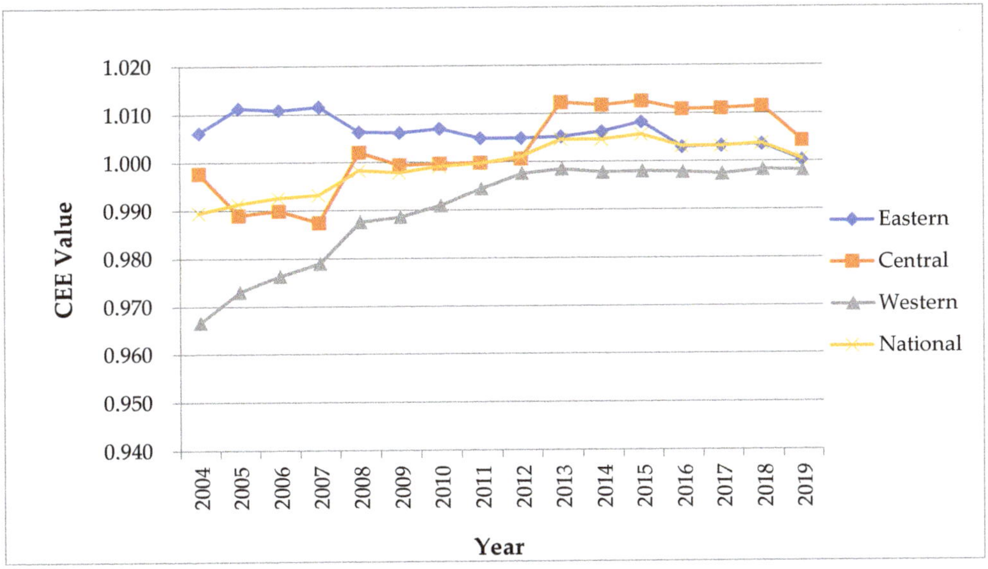

Figure 1. Average CEE values, 2004–2019.

3.2.2. Other Variables

In this study, FDI is an investment behavior in which investors in one country use their capital for production or operations in other countries and exert control. FDI (*fdi*) is a main explanatory variable of this study, which is measured by the total amount of foreign direct investment by logarithm. Technological innovation (*rd*) is the application of new knowledge, new technology, and new processes by enterprises to improve product quality, develop new products, and ultimately occupy a certain market share and realize market value, which refers to the R&D expenditure of each province during the sample period. Based on previous studies [60,61] and the characteristics of this study, we add the degree of openness (*open*), industrial structure (*industry*), and fiscal expenditure (*fine*) as control variables to our models. *open* is measured by the proportion of the import and export in GDP. Industrial structure (*industry*) is equal to the proportion of the secondary industry in the total output. Fiscal expenditure (*fine*) is the ratio of fiscal expenditure to the GDP.

3.3. Econometric Models

The inflow of FDI helps expand local economy. However, it also increases energy consumption and CO_2 emissions in the host country. Since FDI enterprises usually have better technological strengths than local enterprises, its inflow will cause technology spillover to local enterprises, which improves local industrial technology and increases CEE. Therefore, FDI will directly and indirectly affect CEE, and the indirect effect is through TI. In order to test the relationships among FDI, technological innovation, and CEE, the following three models are built:

$$co_{it} = \gamma_0 + \gamma_1 rd_{it} + \gamma_2 fdi_{it} + \gamma_3 open_{it} + \gamma_4 industry_{it} + \gamma_5 fine_{it} + \varepsilon_{it} \quad (2)$$

$$co_{it} = \beta_0 + \beta_1 fdi_{it} + \beta_2 open_{it} + \beta_3 industry_{it} + \beta_4 fine_{it} + \theta_{it} \quad (3)$$

$$rd_{it} = \alpha_0 + \alpha_1 fdi_{it} + \alpha_2 open_{it} + \alpha_3 industry_{it} + \alpha_4 fine_{it} + \tau_{it} \quad (4)$$

where i is province and t is year. α_i, β_i, $(i = 0, 1, \ldots, 4)$, and γ_i, $(i = 0, 1, \ldots, 5)$ are regression coefficients. ε_{it}, θ_{it}, τ_{it} are residual terms. *co* refers to carbon emission efficiency, *rd* is technological innovation, and *fdi* is foreign direct investment. *open*, *industry*, and *fine* are control variables, referring to the degree of openness, industrial structure, and fiscal expenditure, respectively. Equations (2)–(4) jointly verify the intermediate effect of technological innovation [62].

The step method proposed by Wen and Ye [63] is adopted to estimate the intermediate effect of FDI. Firstly, the significance of coefficient β_1 in Equation (3) is tested. If it is significant, subsequent tests will be conducted. Otherwise, the test will be terminated. Secondly, the significance of α_1 in Equation (4) and γ_1 in Equation (2) is tested. If both coefficients are significant, it proves the existence of the intermediate effect. If one of them is insignificant, the Sobel test will be conducted. When the null hypothesis of the Sobel test (H_0:$\alpha_2\gamma_3 = 0$) is rejected, the intermediate effect is supported. Finally, the significance of γ_2 in Equation (2) is checked. If it is significant, there is a partial intermediate effect; otherwise, there is a complete intermediate effect. The intermediate effect can be calculated by $\alpha_1 * \gamma_1$, and the direct effect can be estimated by γ_2. Therefore, the total effect is $\alpha_1 * \gamma_1 + \gamma_2$.

We first check whether there is serious multicollinearity. If the data are highly correlated, it will lead to distorted regression results and an inaccurate estimation result. Table 2 reports the results of the correlation analysis of the panel data. The maximum coefficient of correlation between *rd* and *fdi* is 0.614, and that of the correlation between *fine* and *fdi* is −0.566. Therefore, there are no serious problems of multicollinearity, and we can conduct a regression analysis.

Table 2. Results of the correlation analysis.

Variable	*co*	*fdi*	*open*	*industry*	*fine*	*rd*
co	1.000					
fdi	0.159	1.000				
open	0.146	0.451	1.000			
industry	0.012	0.129	−0.101	1.000		
fine	0.012	−0.566	−0.168	−0.274	1.000	
rd	0.117	0.614	0.309	0.021	−0.262	1.000

Table 3 presents what we get from descriptive statistical analysis. There are 480 observed values. The mean of all variables except for *open* is greater than the standard error. The mean value of the variable *open* is 0.053, and the standard error is 0.076. This means the value is smaller than the standard error, which shows that this variable is a little scattered. Since the sample size is greater than 30, the regression result will not be affected. This table gives a list of the maximum and minimum values of each variable. The largest value of these maximum values is the variable *rd*, but the smallest value of these maximum values is the variable *open*. The largest value of these minimum values is the variable *rd*, but the smallest value of these minimum values is the variable *fdi*. The variable *rd* has the largest gap between the maximum value and the minimum value, being 8.386, and the variable *co* has the smallest difference, standing at 0.207.

Table 3. Descriptive statistical analysis.

Variable	Obs	Mean	Std.Dev.	Min	Max
co	480	0.999	0.029	0.941	1.148
fdi	480	3.662	1.517	−0.778	6.793
open	480	0.053	0.076	0.002	0.442
industry	480	0.426	0.119	0.161	2.126
fine	480	0.199	0.101	0.071	0.846
rd	480	13.831	1.797	8.863	17.249

4. Results and Discussion

4.1. Analysis of the Impact of FDI and Technological Innovation on CEE

The impact of FDI and TI on CEE is analyzed first. When processing panel data, we should choose whether to adopt the fixed-effect model or the random effect model. The null hypothesis is that "unobservable random variables are not related to all explanatory variables". If the null hypothesis is accepted, the random effect model should be used. If the null hypothesis is rejected, the fixed-effect model should be used. According to Hausman test, we find that panel data is significant at the statistical level of 1%. Hence, the fixed-effect model should be used. Table 4 reports the regression results of the fixed-effect model.

Table 4. Impact of FDI and technological innovation on CEE.

Variable	*co*	*co*	*rd*
	(2)	(3)	(4)
rd	0.002 ***		
	(0.001)		
fdi	0.003 *	0.004 ***	1.054 ***
	(0.002)	(0.001)	(0.076)
open	0.096 ***	0.082 **	−8.398 ***
	(0.037)	(0.037)	(2.111)
industry	−0.009	−0.009	−0.019
	(0.008)	(0.008)	(0.485)
fine	0.026 ***	0.041 ***	8.003 ***
	(0.014)	(0.012)	(0.711)
_cons	0.958 ***	0.974 ***	8.823 ***
	(0.011)	(0.007)	(0.403)
R^2	0.024	0.031	0.351

Note: ***, **, and * indicate significant levels at 1%, 5%, and 10%, respectively; the data in parenthesis are standard errors.

FDI positively influences CEE. When it increases by 1%, CEE increases by 0.003%. Foreign investment brings advanced technologies, standards, and concepts on environmental protection to the host country. Furthermore, enterprises in the host country reduce carbon emissions and increase CEE thanks to the imitation effect and the demonstration effect. Foreign ideas of environmental protection are of great significance to improving CEE of the host country. Environmental protection not only concerns human health, but also plays a vital role in long-term social development. The spread of ideas on environmental protection allows the host country to recognize the role of the environment, pay attention to the improvement of environmental protection technology, and improve the intensity of environmental regulations. Meanwhile, the inflow of foreign investment improves the technical level of the whole industry, increases energy efficiency, saves energy, and reduces carbon emissions. Fang et al. [60] drew the opposite conclusion in their study of CE of 282 cities in China. They found that FDI negatively affected CEE. Therefore, each city should raise thresholds for environmental access when absorbing the inflow of FDI. In addition to reducing disorderly competition and energy rebound, foreign enterprises could enhance environmental governance and advance technological progress.

Technological innovation positively affects CEE. It is a process of increasing the technology level which brings TSE to enterprises, reduces CE, and increases CEE. Fang et al. [60] also found that the effect of technical development was positive, that is, the use of clean technology improved CEE. R&D expenditure on low-carbon industries could bring technological progress as well as the energy rebound effect. Technology investment may bring technological progress as well as the energy rebound effect. When the former was greater than the latter, R&D expenditure could improve CEE. Our results are also consistent with those of Rizwana et al. [64]. Rizwana et al. [64] studied the Belt and Road economies and found that technological innovation could save energy consumption costs and improve environmental quality.

FDI is conducive to enhancing TI of the host country. When it increases by 1%, technological innovation increases by 1.054%. Some FDI goes to OEM production, that is, producing, processing, and assembling products in China and finally exporting the products to other countries. In this process, advanced foreign technical standards and environmental protection standards are followed by local enterprises, bringing about technology spillover and improving CEE in China. Other FDI goes to R&D institutions in China. In order to gain more profits, multinationals attach great importance to developing technologies, which causes technology spillovers to Chinese enterprises. The technology spillover is realized through the demonstration effect of products and the flow effect of R&D personnel.

Among the control variables, *open* positively affects CEE. It is calculated as the ratio of international trade to GDP. The larger the ratio, the greater the degree of openness. When China is open enough, it is fully connected to the world. This means it can get advanced environmental protection technologies from developed countries, which is conducive to its energy conservation and CEE. *open* negatively affects R&D expenditure. The possible reason is that most of China's product exports win by quantity, with low technological content, so the exports do not produce strong technology spillover to domestic R&D. Industrial structure, which equals to the ratio of secondary industry output to the total output, negatively and insignificantly affects CEE. This is because the secondary industry consumes a large amount of energy and generates a lot of CE, which counts against CEE. This is consistent with the conclusion of Fang et al. [60]. According to Fang et al. [60], industrial structure was negatively related to CEE, because CE mainly came from the secondary industry. Fiscal expenditure positively affects CEE. After the reform and opening up is implemented, local governments in China have focused on developing the economy. They no longer sacrifice the environment for economic growth due to river pollution, ecological degradation, and human health threats caused by ignoring environmental protection. They are changing from an extensive development mode to emphasis on environmental protections. Financial expenditure on environmental protection has been increased, reducing CE and increasing CEE.

4.2. Robustness Test

In order to test whether the above regression results are consistent and stable when some parameters change, a robustness test was conducted. Three methods can be used for the robustness test: variable replacement, method replacement, and change of sample size. The Tobit regression method is used in this section, and the results are listed in Table 5. FDI positively affects CEE. When FDI increases by 1%, CEE rises by 0.003%. Technological innovation positively affects CEE. When technological innovation increases by 1%, CEE increases by 0.002%. FDI positively affects R&D expenditure. When FDI increases by 1%, R&D expenditure rises by 1.141%. Hence, the regression results are robust.

Table 5. Robustness test.

Variable	*co*	*co*	*rd*
	(2)	(3)	(4)
rd	0.002 **		
	(0.001)		
fdi	0.003 *	0.004 ***	1.141 ***
	(0.001)	(0.001)	(0.061)
open	0.073 **	0.063 **	−4.587 ***
	(0.031)	(0.031)	(1.519)
industry	−0.008	−0.008	−0.026
	(0.008)	(0.008)	(0.469)
fine	0.031 **	0.042 ***	7.207 ***
	(0.013)	(0.012)	(0.664)
_cons	0.961 ***	0.975 ***	8.467 ***
	(0.011)	(0.008)	(0.383)
rho	0.674	0.667	0.399
Wald	41.58 ***	37.57 ***	472.47 ***

Note: ***, **, and * indicate significant levels at 1%, 5%, and 10%, respectively; the data in parenthesis are standard errors.

4.3. Intermediate Effect Test

Based on Equations (2)–(4), the stepwise method is used to test the intermediate effect of technological innovation. Table 6 reports the test results with both fixed-effect regression and Tobit regression. First, the coefficient β_1 in Equation (3) indicates that the influence of FDI on CEE is significant. β_1 is 0.004 in both the fixed-effect regression and the Tobit regression, so subsequent tests can be performed. Secondly, the significance of coefficient α_1 in Equation (4) and coefficient γ_1 in Equation (2) is tested. It is found that both coefficients are significant in the fixed-effect regression and the Tobit regression, which proves the existence of an intermediate effect.

Table 6. Intermediate effect test.

Method	Sobel Test	Direct Effect	Indirect Effect	Total Effect	Percentage of Indirect
Fixed-effect regression	2.0998 **	0.0028	0.0017	0.0045	37.78%
Tobit regression	1.9884 **	0.0026	0.0018	0.0044	40.91%

Note: ** indicates the significant level at 5%.

In the fixed-effect regression, the direct effect of FDI on CEE is 0.0028, and the indirect effect is 0.0017, of which the indirect effect accounts for 37.78%. This shows that FDI directly affects CEE and indirectly affects the latter through TI. In the Tobit regression, the direct effect of FDI on CEE is 0.0026, and the indirect effect is 0.0018, of which the indirect effect accounts for 40.91%. This not only confirms that FDI affects CEE both directly and indirectly, but also indicates that the conclusion is robust. As for carbon emission efficiency, He et al. [65] also proved the existence of an intermediate effect. In this study, we used the stepwise method to analyze the relationships among FDI, technological innovation, and carbon efficiency, which is different from the research of He et al. [65]. He et al. [65] used the panel threshold model to examine the relationships among technological innovation, market forces, and carbon efficiency. The intermediate variable in this study is technological innovation, but it was market forces in the research of He et al. [65]. Although the intermediate variables are different, we all find that there is an intermediate effect in the research of CEE.

4.4. Sub Sample Regression

We find that some places have a relatively developed economy and more foreign investment of higher quality, which help improve the local technical level and raise the ratio of environmentally friendly enterprises, thus improving CEE and reducing environmental pollution. In other regions, the economy is relatively backward. To ensure economic growth, they sacrifice the ecological environment. Among the foreign enterprises introduced by them, there are a large number of non-environmentally friendly enterprises, which reduces CEE and increases regional environmental pollution. Therefore, regional economy has different effects on FDI, technological innovation, and CEE. China has a developed economy in eastern regions and a backward economy in central and western regions. In this section, the total sample data is divided by region according to geographical location for detailed regression. The results of the regression are listed in Table 7.

Table 7. Results of sub-sample regression.

Variable	East Regions			Central and West Regions		
	co	*co*	*rd*	*co*	*co*	*rd*
rd	0.003 ***			−0.007 ***		
	(0.001)			(0.001)		
fdi	0.001	−0.004	1.561 ***	−0.001	0.005 ***	0.837 ***
	(0.003)	(0.002)	(0.212)	(0.002)	(0.001)	(0.078)
open	0.006	0.032	−6.811 ***	−0.068	0.199	35.983 ***
	(0.033)	(0.034)	(2.714)	(0.203)	(0.211)	(9.931)
industry	0.002	0.001	0.401	−0.071 ***	−0.101 ***	−3.986 ***
	(0.006)	(0.007)	(0.578)	(0.025)	(0.026)	(1.253)
fine	0.068 ***	0.041 ***	7.198 ***	−0.012	0.044 ***	7.677 ***
	(0.019)	(0.019)	(1.521)	(0.017)	(0.015)	(0.742)
_cons	1.043 ***	1.018 ***	6.564 ***	0.935 ***	1.011 ***	10.271 ***
	(0.015)	(0.015)	(1.186)	(0.017)	(0.013)	(0.643)
R^2	0.152	0.066	0.402	0.264	0.161	0.577

Note: *** indicates significant levels at 1%; the data in parenthesis are standard errors.

FDI positively affects CEE in east China, with an insignificant impact coefficient. It negatively affects CEE in central and western regions, with an insignificant impact coefficient. Regarding the direction and degree of impact, we can see that the eastern region has attracted a great quantity of FDI inflows with high technological content, which improves local technological innovation, CEE, and the ecological environment. Central and western regions have received many non-environmentally friendly FDI enterprises, which has a slight positive and even negative effect on CEE. FDI positively affects R&D expenditure in all three of the regions, which indicates that there is TSE in China. Nevertheless, whether technology spillover improves local technology innovation and CEE depends on the absorption capacity of each region.

5. Conclusions and Policy Implications

This paper firstly employs the super efficiency DEA model to obtain the CEE of 30 provinces and cities in China, which reveals that there are regional imbalances. During the whole period, the highest mean value of CEE falls in the eastern region, followed by the central region and the western region. The efficiency shows a downward trend in the central region and the western region but an upward trend in the eastern region. On this basis, we test the relationship among FDI, TI, and CEE with regression models. The results reveal that both FDI and technological innovation positively and significantly affect CEE, and FDI positively and significantly affects technological innovation. The intermediate effect test confirms that FDI directly affects regional CEE and indirectly affects CEE through TI. Finally, we divide the total sample into the eastern region, the central region, and the western region and test the regional heterogeneity of FDI, technological innovation, and CEE. Based on research results, we propose several policy implications which are stated below.

Foreign investment in central and western regions is more likely to transfer pollution. It promotes local economic development but brings about serious environmental problems. The central and western regions should commit to low-carbon development and increase environmental standards to attract environmentally friendly foreign investment. Therefore, in the process of attracting FDI inflow, the central and western regions should perform comprehensive low-carbon planning, construction, and management and urge all parties to meet low-carbon requirements. The central and western regions should focus on economic growth and improve their absorptive capacity so that they can transform FDI technology spillovers into independent innovation and realize sustainable development.

Independent innovation is a key driving factor for high-quality economic development. Promoting independent innovation through FDI technology spillover is an important means for improving China's technology. In general, the government should increase investment in R&D and enhance its patent system to increase FDI spillovers. Considering regional differences, the eastern region, which embraces a high degree of innovation, should focus on independent innovation capability. It should enhance competitive advantages through absorbing high-technology FDI. The central and western regions should strengthen the attraction of foreign investment through improving the innovation market, cultivating technical talents, and expanding innovation subjects, so as to form the innovation catch-up effect. Moreover, some of the areas in the western region should not blindly carry out independent R&D activities. They should invest more in technological transformation, increase economic output based on effective innovation, and then improve carbon emission efficiency.

It is difficult to change the status quo that China's economy will depend on resources for a long time, and carbon emissions will continue to increase. Therefore, it is more realistic to reduce carbon emissions and increase carbon emission efficiency by using renewable energy. It is recommended that the government formulate industrial and macroeconomic policies to support renewable energy and green energy technologies, supervise related R&D activities, and provide financial support for them. Also, the government is advised to encourage domestic and foreign enterprises to cooperate in the development of renewable energy. The renewable energy industry is a high-risk industry characterized by high investment. Foreign enterprises face more risks and challenges due to transnational factors. For example, high financing costs and less experience in the renewable energy industry often hinder their overseas investment. To solve these problems, it is a better choice for domestic and foreign enterprises to jointly invest in this industry.

This study represents more in-depth research on the relationships among FDI, technological innovation, and CEE, but there are still deficiencies and limitations. We suggest the following future research directions. First, alternative estimation methods (considering structural breaks and non-linear/asymmetric) can be adopted to investigate the relationships among the three variables and explore whether future results support empirical research in different panels. Second, the analysis can be performed in the Environmental Kuznets Curve (EKC) framework. Different stages of open economy may have diverse effects on FDI and CEE, which is of great significance for China to formulate targeted energy policies. These analyses should be more fruitful and helpful.

Author Contributions: Methodology, Q.W.; formal analysis, S.L.; investigation, S.L.; resources, S.L.; data curation, Q.W.; writing—original draft preparation, Q.W.; writing—review and editing, Q.W. All authors have read and agreed to the published version of the manuscript.

Funding: This research received no external funding.

Data Availability Statement: All relevant data have been submitted within this manuscript.

Conflicts of Interest: The authors declare no conflict of interest.

References

1. Alam, W.; Ikram, F.; Kumar, P.; Haseeb, M.; Ali, N. Asymmetric Effects of Foreign Direct Investment on Economic Growth: Fresh Evidence from India Using NARDL Simulation. *Millennial Asia* **2022**, *in press*. [CrossRef]
2. He, J. Pollution haven hypothesis and environmental impacts of foreign direct investment: The case of industrial emission of sulfur dioxide (SO_2) in Chinese province. *Ecol. Econ.* **2006**, *60*, 228–245. [CrossRef]
3. Kamal, M.; Usman, M.; Jahanger, A.; Balsalobre-Lorente, D. Revisiting the Role of Fiscal Policy, Financial Development, and Foreign Direct Investment in Reducing Environmental Pollution during Globalization Mode: Evidence from Linear and Nonlinear Panel Data Approaches. *Energies* **2021**, *12*, 6968. [CrossRef]
4. Xing, Y.; Kolstad, C. Do lax environmental regulations attract foreign investment? *Environ. Resour. Econ.* **2002**, *21*, 1–22. [CrossRef]
5. Bae, J.H.; Li, D.D.; Rishi, M. Determinants of CO_2 emission for post-Soviet Union independent countries. *Clim. Policy* **2017**, *17*, 591–615. [CrossRef]
6. Arain, H.; Sharif, A.; Akbar, B.; Younis, M.Y. Dynamic connection between inward foreign direct investment, renewable energy, economic growth and carbon emission in China: Evidence from partial and multiple wavelet coherence. *Environ. Sci. Pollut. Res.* **2020**, *27*, 40456–40474. [CrossRef] [PubMed]
7. Zakarya, G.Y.; Mostefa, B.; Abbes, S.M.; Seghir, G.M. Factors affecting CO_2 emissions in the BRICS countries: A panel data analysis. *Procedia Econ. Financ.* **2015**, *26*, 114–125. [CrossRef]
8. Behera, S.R.; Dash, D.P. The effect of urbanization, energy consumption, and foreign direct investment on the carbon dioxide emission in the SSEA (south and south-east Asian) region. *Renew. Sustain. Energy Rev.* **2017**, *70*, 96–106. [CrossRef]
9. Liu, Y.; Hao, Y.; Gao, Y. The environmental consequences of domestic and foreign investment: Evidence from China. *Energy Policy* **2017**, *108*, 271–280. [CrossRef]
10. Ahmad, M.; Jabeen, G.; Wu, Y. Heterogeneity of pollution haven/halo hypothesis and Environmental Kuznets Curve hypothesis across development levels of Chinese provinces. *J. Clean. Prod.* **2020**, *285*, 124898. [CrossRef]
11. Zhu, H.; Duan, H.; Guo, Y.; Yu, K. The effects of FDI, economic growth and energy consumption on carbon emissions in ASEAN-5: Evidence from panel quantile regression. *Econ. Model.* **2016**, *58*, 237–248. [CrossRef]
12. Ma, W.; Nasriddinov, F.; Haseeb, M.; Ray, S.; Kamal, M.; Khalid, N.; Ur Rehman, M. Revisiting the impact of energy consumption, foreign direct investment, and geopolitical risk on CO_2 emissions: Comparing developed and developing countries. *Front. Environ. Sci.* **2022**, *10*, 985384. [CrossRef]
13. Yildrim, E. Energy use, CO_2 emission and foreign direct investment: Is there any inconsistence between causal relations? *Front. Energy* **2014**, *8*, 269–278. [CrossRef]
14. Khurram, S.; Liu, X.; Muddassar, S. Envisaging the asymmetrical association among FDI, ICT, and climate change: A case from developing country. *Carbon Manag.* **2021**, *12*, 123–137. [CrossRef]
15. Zia, U.R.; Wu, C.; Manzoor, A. An (a)symmetric analysis of the pollution haven hypothesis in the context of Pakistan: A non-linear approach. *Carbon Manag.* **2019**, *10*, 227–239. [CrossRef]
16. Wu, W.; Li, X.; Lu, Z.; Gozgor, G.; Wu, K. Energy subsidies and carbon emission efficiency in Chinese regions: The role of the FDI competition in local governments. *Energy Sources Part B Econ. Plan. Policy* **2022**, *17*, 2094035. [CrossRef]
17. Guo, Z.Q.; Liu, H.B. The impact of carbon tax policy on energy consumption and CO_2 emission in China. *Energy Sources Part B Econ. Plan. Policy* **2016**, *11*, 725–731. [CrossRef]
18. Macdougall, G.D.A. The benefits and costs of private investment from abroad: A theoretical approach. *Oxf. Bull. Econ. Stat.* **1960**, *22*, 189–211. [CrossRef]
19. Keller, W.; Yeaple, S.R. Multinational Enterprises, International Trade, and Productivity Growth: Firm-Level Evidence from the United States. Working Paper 9504. 2003. Available online: http://www.nber.org/papers/w9504 (accessed on 1 February 2003).
20. Nyeadi, J.D.; Adjasi, C. Foreign direct investment and firm innovation in selected sub-Saharan African Countries. *Cogent Bus. Manag.* **2020**, *7*, 1763650. [CrossRef]
21. Luo, Y.; Salman, M.; Lu, Z. Heterogeneous impacts of environmental regulations and foreign direct investment on green innovation across different regions in China. *Sci. Total Environ.* **2021**, *759*, 143744. [CrossRef]
22. Munteanu, A.-C. Knowledge spillovers of FDI. *Procedia Econ. Financ.* **2015**, *32*, 1093–1099. [CrossRef]
23. Fernandes, A.M.; Paunov, C. Foreign direct investment in services and manufacturing productivity: Evidence for Chile. *J. Dev. Econ.* **2012**, *97*, 305–321. [CrossRef]
24. Yang, Z.; Shao, S.; Yang, L.; Liu, J. Differentiated effects of diversified technological sources on energy-saving technological progress: Empirical evidence from China's industrial sectors. *Renew. Sustain. Energy Rev.* **2017**, *72*, 1379–1388. [CrossRef]
25. Negash, E.S.; Zhu, W.; Lu, Y.; Wang, Z. Does Chinese Inward Foreign Direct Investment Improve the Productivity of Domestic Firms? Horizontal Linkages and Absorptive Capacities: Firm-Level Evidence from Ethiopia. *Sustainability* **2020**, *12*, 3023. [CrossRef]

26. Marco, L.C.K.; Steve, Y.F.; Zhang, Z.; Leung, V.K.K. Determinants of innovative activities: Evidence from Europe and central Asia region. *Singap. Econ. Rev.* **2015**, *60*, 1550004. [CrossRef]
27. Kim, M.; Choi, M.J. R&D spillover effects on firms' export behavior: Evidence from South Korea. *Appl. Econ.* **2019**, *51*, 3066–3080. [CrossRef]
28. Feng, Y.; Wang, X.; Du, W.; Wu, H.; Wang, J. Effects of environmental regulation and FDI on urban innovation in China: A spatial Durbin econometric analysis. *J. Clean. Prod.* **2019**, *235*, 210–224. [CrossRef]
29. Hu, J.; Wang, Z.; Huang, Q.; Zhang, X. Environmental regulation intensity, foreign direct nvestment, and green technology spillover—An empirical study. *Sustainability* **2019**, *11*, 2718. [CrossRef]
30. Cole, M.A.; Elliott, R.J.R. FDI and the capital intensity of "dirty" sectors: A missing piece of the pollution haven puzzle. *Rev. Dev. Econ.* **2005**, *9*, 530–548. [CrossRef]
31. Anwar, S.; Nguyen, L.P. Is foreign direct investment productive? A case study of the regions of Vietnam. *J. Bus. Res.* **2014**, *67*, 1376–1387. [CrossRef]
32. Hu, J.; Wang, Z.; Lian, Y.; Huang, Q. Environmental regulation, foreign direct investment and green technological progress—Evidence from Chinese manufacturing industries. *Int. J. Environ. Res. Public Health* **2018**, *15*, 221. [CrossRef] [PubMed]
33. Itoo, H.H.; Ali, N. Analyzing the causal nexus between CO_2 emissions and its determinants in India: Evidences from ARDL and EKC approach. *Manag. Environ. Qual.* **2022**, *in press*. [CrossRef]
34. Wen, Y.; Shabbir, M.S.; Haseeb, M.; Kamal, M.; Anwar, A.; Khan, M.F.; Malik, S. The dynamic effect of information and communication technology and renewable energy on CO_2 emission: Fresh evidence from panel quantile regression. *Front. Environ. Sci.* **2022**, *10*, 953035. [CrossRef]
35. Huang, Y.; Haseeb, M.; Usman, M.; Ozturk, I. Dynamic association between ICT, renewable energy, economic complexity and ecological footprint: Is there any difference between E-7 (developing) and G-7 (developed) countries? *Technol. Soc.* **2022**, *68*, 101853. [CrossRef]
36. Jiang, Y.; Batool, Z.; Raza, S.M.F.; Haseeb, M.; Ali, S.; Abidin, S.Z.U. Analyzing the Asymmetric Effect of Renewable Energy Consumption on Environment in STIRPAT-Kaya-EKC Framework: A NARDL Approach for China. *Int. J. Environ. Res. Public Health* **2022**, *19*, 7100. [CrossRef] [PubMed]
37. Fisher-Vanden, K.; Jefferson, G.H.; Liu, H.; Tao, Q. What is driving China's decline in energy intensity? *Resour. Energy Econ.* **2004**, *26*, 77–97. [CrossRef]
38. Fernández, F.Y.; López, M.A.F.; Blanco, B.O. Innovation for sustainability: The impact of R and D spending on CO_2 emissions. *J. Clean. Prod.* **2017**, *172*, 3459–3467. [CrossRef]
39. Yu, Y.; Du, Y. Impact of technological innovation on CO_2 emissions and emissions trend prediction on 'New Normal' economy in China. *Atmos. Pollut. Res.* **2019**, *10*, 152–161. [CrossRef]
40. Abid, A.; Mehmood, U.; Tariq, S.; Haq, Z.U. The effect of technological innovation, FDI, and financial development on CO_2 emission: Evidence from the G8 countries. *Environ. Sci. Pollut. Res.* **2022**, *29*, 11654–11662. [CrossRef]
41. Lee, K.; Min, B. Green R&D for eco-innovation and its impact on carbon emissions and firm performance. *J. Clean. Prod.* **2015**, *108*, 534–542. [CrossRef]
42. Kong, L.; Hasanbeigi, A.; Price, L. Assessment of emerging energy-efficiency technologies for the pulp and paper industry: A technical review. *J. Clean. Prod.* **2016**, *122*, 5–28. [CrossRef]
43. Sgobbi, A.; Simoes, S.G.; Magagna, D.; Nijs, W. Assessing the impacts of technology improvements on the deployment of marine energy in Europe with an energy system perspective. *Renew. Energy* **2016**, *89*, 515–525. [CrossRef]
44. Zeeshan, K.; Zhu, S.; Yang, S. Environmental regulations an option: Asymmetry effect of environmental regulations on carbon emissions using non-linear ARDL. *Energy Sources Part A Recovery Util Environ. Eff.* **2019**, *41*, 137–155. [CrossRef]
45. Shaari, M.S.; Abdullah, D.N.C.; Alias, N.S.B.; Adnan, N.S.M. Positive and negative effects of research and development. *Int. J. Energy Econ. Policy* **2016**, *6*, 767–770.
46. Demir, C.; Cergibozan, R.; Ari, A. Environmental dimension of innovation: Time series evidence from Turkey. *Environ. Dev. Sustain.* **2020**, *22*, 2497–2516. [CrossRef]
47. Gu, W.; Zhao, X.; Yan, X.; Wang, C.; Li, Q. Energy technological progress, energy consumption, and CO_2 emissions: Empirical evidence from China. *J. Clean. Prod.* **2019**, *236*, 117666. [CrossRef]
48. Yii, K.J.; Geetha, C. The nexus between technology innovation and CO_2 emissions in Malaysia: Evidence from Granger causality test. *Energy Procedia* **2017**, *105*, 3118–3124. [CrossRef]
49. Dauda, L.; Long, X.; Mensah, C.N.; Salman, M. The effects of economic growth and innovation on CO_2 emissions in different regions. *Environ. Sci. Pollut. Res.* **2019**, *26*, 15028–15038. [CrossRef]
50. Erdoğan, S.; Yıldırım, S.; Yıldırım, D.Ç.; Gedikli, A. The effects of innovation on sectoral carbon emissions: Evidence from G20 countries. *J. Environ. Manag.* **2020**, *267*, 110637. [CrossRef]
51. Zhang, C.; Chen, P. Industrialization, urbanization, and carbon emission efficiency of Yangtze River Economic Belt—Empirical analysis based on stochastic frontier model. *Environ. Sci. Pollut. Res.* **2021**, *28*, 66914–66929. [CrossRef]
52. Wang, S.; Zeng, J.; Liu, X. Examining the multiple impacts of technological progress on CO_2 emissions in China: A panel quantile regression approach. *Renew. Sustain. Energy Rev.* **2019**, *103*, 140–150. [CrossRef]
53. Huang, J.; Chen, X.; Yu, K.; Cai, X. Effect of technological progress on carbon emissions: New evidence from a decomposition and spatiotemporal perspective in China. *J. Environ. Manag.* **2020**, *274*, 110953. [CrossRef] [PubMed]

54. Yan, D.; Lei, Y.; Li, L.; Song, W. Carbon emission efficiency and spatial clustering analyses in China's thermal power industry: Evidence from the provincial level. *J. Clean. Prod.* **2017**, *156*, 518–527. [CrossRef]
55. Santra, S. The effect of technological innovation on production-based energy and CO_2 emission productivity: Evidence from BRICS countries. *Afr. J. Sci. Technol. Innov. Dev.* **2017**, *9*, 503–512. [CrossRef]
56. Qader, M.R.; Khan, S.; Kamal, M.; Usman, M.; Haseeb, M. Forecasting carbon emissions due to electricity power generation in Bahrain. *Environ. Sci. Pollut. Res.* **2022**, *29*, 17346–17357. [CrossRef]
57. Charnes, A.; Cooper, W.W.; Rhodes, E. Measuring the Efficiency of Decision Making Units. *Eur. J. Oper. Res.* **1978**, *2*, 429–444. [CrossRef]
58. Banker, R.D.; Charnes, A.; Cooper, W.W. Some models for estimating technical and scale inefficiencies in dea. *Manag. Sci.* **1984**, *32*, 1613–1627. [CrossRef]
59. Andersen, P.; Petersen, N.C. A procedure for ranking efficient units in data envelopment analysis. *Manag. Sci.* **1993**, *39*, 1261–1264. [CrossRef]
60. Fang, G.; Gao, Z.; Tian, L.; Fu, M. What drives urban carbon emission efficiency?—Spatial analysis based on nighttime light data. *Appl. Energy* **2022**, *312*, 118772. [CrossRef]
61. Liu, Q.; Hao, J. Regional Differences and Influencing Factors of Carbon Emission Efficiency in the Yangtze River Economic Belt. *Sustainability* **2022**, *14*, 4814. [CrossRef]
62. Pei, Y.; Zhu, Y.; Wang, N. How do corruption and energy efficiency affect the carbon emission performance of China's industrial sectors? *Environ. Sci. Pollut. Res.* **2021**, *28*, 31403–31420. [CrossRef] [PubMed]
63. Wen, Z.L.; Ye, B.J. Analyses of mediating effects: The development of methods and models. *Adv. Psychol. Sci.* **2014**, *22*, 731–745. [CrossRef]
64. Rizwana, Y.; Cui, Z.; Wasi, U.H.S.; Muhammad, A.K.; Anwar, K. Exploring the role of biomass energy consumption, ecological footprint through FDI and technological innovation in B&R economies: A simultaneous equation approach. *Energy* **2022**, *244*, 122703. [CrossRef]
65. He, A.; Xue, Q.; Zhao, R.; Wang, D. Renewable energy technological innovation, market forces, and carbon emission efficiency. *Sci. Total Environ.* **2021**, *796*, 148908. [CrossRef] [PubMed]

energies

MDPI

Article

R&D Human Capital, Renewable Energy and CO_2 Emissions: Evidence from 26 Countries

Grzegorz Mentel [1,*], Waldemar Tarczyński [2], Hossein Azadi [3,4,5], Kalandar Abdurakmanov [6], Elina Zakirova [7] and Raufhon Salahodjaev [8,9,10]

1 Department of Quantitative Methods, The Faculty of Management, Rzeszow University of Technology, 35-959 Rzeszow, Poland
2 Department of Sustainable Finance and Capital Markets, Institute of Economics and Finance, University of Szczecin, 71-101 Szczecin, Poland
3 Department of Economics and Rural Development, Gembloux Agro-Bio Tech, University of Liège, 5030 Gembloux, Belgium
4 Faculty of Environmental Sciences, Czech University of Life Sciences Prague, 165 00 Prague, Czech Republic
5 Faculty of Environmental Science and Engineering, Babeș-Bolyai University, 400000 Cluj-Napoca, Romania
6 Tashkent State University of Economics, Tashkent 100066, Uzbekistan
7 Institute of Additional Education, Ural State University of Economics, 620144 Yekaterinburg, Russia
8 Center for Research in Social Sciences, AKFA University, Tashkent 111221, Uzbekistan
9 Department of Research, ERGO Analytics, Tashkent 100022, Uzbekistan
10 Department of Mathematical Method in Economics, Tashkent State University of Economics, Tashkent 100022, Uzbekistan
* Correspondence: gmentel@prz.edu.pl

Abstract: This study examines the long-term relationship between carbon emissions and a number of researchers engaged in Research and Development (R&D), economic development, foreign capital inflows, renewable energy and population growth in 26 countries between 1995 and 2015. Pedroni's panel cointegration test confirms the cointegrating relationship between the variables. Long-term elasticities are derived from FMOLS regression. Researchers in R&D and renewable energy are negatively and significantly related to carbon emissions. There is a positive and significant long-term relationship between GDPs per capita and CO_2 and between the FDI and CO_2. Dumitrescu and Hurlin's panel causality test revealed unidirectional causality running from economic development to carbon emissions and feedback hypotheses between the FDI and CO_2 and between renewable energy and CO_2.

Keywords: human capital; renewable energy; CO_2 emission

Citation: Mentel, G.; Tarczyński, W.; Azadi, H.; Abdurakmanov, K.; Zakirova, E.; Salahodjaev, R. R&D Human Capital, Renewable Energy and CO_2 Emissions: Evidence from 26 Countries. *Energies* **2022**, *15*, 9205. https://doi.org/10.3390/en15239205

Academic Editor: Attilio Converti

Received: 1 November 2022
Accepted: 27 November 2022
Published: 5 December 2022

1. Introduction

During the last decades, research on the link between innovation and pollution has expanded significantly [1–5]. This is both due to the rapid increase of CO_2 emissions by nearly 90% since 1970 and continuing enhancing environmental concerns and technological advancements, which can effectively mitigate the consequences of environmental pollution and climate change without destroying economic growth [6]. At the same time, policymakers aim to achieve economic growth without "without inflicting environmental harm" [7].

However, a separate strand of studies explores the effect of human capital on pollution. Led by the cognitive capitalism theory, some studies suggest that nations with greater cognitive capital tend to achieve more wealth led by innovation, effectively adopt modern technologies, develop strong institutions and demonstrate higher environmental awareness and pro-environmental behavior [8–10]. Lin et al. [11] stated that economies' technology development and implementation are highly dependent on the so-called "innovative human capital". This is basically research and development personnel, whose knowledge and skills serve as a unique resource of cutting-edge technologies and innovations, including those aimed at clean production, clean energy generation and environmental sustainability.

To the best of our knowledge, few research studies have considered innovative human capital as a pollution-reducing factor. Some scholars who explored the relationship employed education-based proxies of innovative human capital, such as tertiary education, which might have created certain issues, as tertiary education measures human capital stock rather than specifies innovative human capital.

Based on the study of Lin et al. [11], our paper employed Research and Development (R&D hereafter)-based measures of innovative human capital, which is the number of researchers involved in R&D.

Our study explores the relationship between innovative human capital and carbon emissions in 26 industrialized economies between 1995 and 2015. This paper's structure is as follows. We first provide an overview of related literature. Based on the literature review, we construct a model and design a model estimation framework, which is described in detail in the Data and Methods section. Following that methodological description, we provide and discuss the econometric results. The final section concludes our findings.

2. Literature Review

Numerous researchers have explored the link between carbon emissions and innovation, considering innovation and technological as the significant factor to mitigate climate change issues [12] and enhance environmental wellbeing without harming economic growth. Existing research can measure innovation by expenditure in energy R&D, expenditure in R&D, renewable energy consumption, the number of patent families, number of researchers engaged in R&D, industry–university–research cooperation and even green project financing (e.g., green bonds) [13], etc. Authors have explored the effect of public expenditure in energy R&D and carbon emissions per GDP. Authors have investigated the relationship between public energy R&D and per capita carbon emissions using a sample of 13 developed economies, including Canada, Denmark, Finland, France, Germany, Italy, Japan, Norway, the Netherlands, Spain, Sweden, the UK and the USA between 1980 and 2004 and employing the Granger causality approach. They distinguished two channels by which public expenditure in energy R&D may affect carbon emissions, which are carbon emissions and carbon intensity. Their findings suggest that spending on energy R&D improves the energy efficiency, yet it did not significantly relate to the carbon factor or carbon intensity. On the other hand, carbon trends are related to the formation of energy R&D budgets.

However, later research provided evidence on the significant effects of innovation and technological advances on various environmental indicators, including carbon emissions. Lee and Min [14] observed the negative effect of firm innovation (green R&D) on carbon dioxide emissions using data on Japanese manufacturing firms during 2001–2010. Irandoust [3] studied the relationship between renewable energy consumption, technological innovation, economic growth and carbon dioxide emissions in Nordic countries. Technological innovation is measured by real R&D spending on the energy sector. Their findings suggest that technological innovation has an significant impact on renewable energy consumption. Álvarez-Herránz et al. [15] explained that, despite the time lags required for energy innovation to reach their maximum effect, it reduces the energy intensity and pollution. Zhang et al. [16] observed that environmental innovation can reduce carbon emissions using data from 30 provinces of China between 2000 and 2013. Similarly, Ganda [17] observed a negative relationship between innovations—in particular, renewable energy consumption and R&D expenditure—and carbon emissions in OECD countries. Khan et al. [18] studied the relationship between environmental innovation, renewable energy consumption, carbon emissions, trade and income in G7 between 1990 and 2017. Their findings revealed the negative impact of exports, environmental innovation and renewable energy consumption on carbon dioxide emissions in the long run. Similar findings were also obtained by Nguyen et al. [19], who attributed technology and spending on innovation as the main drivers of lowering carbon emissions in 13 countries of G20. Wahab et al. [20] suggested that the adoption of new technologies aimed at cleaner production can reduce carbon emissions in G7 economies between 1996 and 2017. At the same time, some studies revealed that the innovation–pollution relationship is sector-dependent. For example, Zhang et al. [21] (2020) attempted to identify whether carbon emissions policies are related to the implementation of low-carbon technology innovations. Their findings suggest that carbon emissions policies improve technology innovation in the power and aviation sectors only and are not related to each other in the steel, chemical, building material, petrochemical, nonferrous metals and paper industries. Erdoğan et al. [4] suggested that policies aimed at lowering carbon emissions should be designed and implemented in each sector separately, as the effects of technology innovation on carbon emissions vary. Using data on G20 between 1991 and 2017, the authors findings revealed that innovations in the industrial sector discourage carbon emissions and, on the other hand, enhance carbon emissions in the construction sector. The latest research considers the role of technological innovation, measured by patent applications in the remittances–renewable energy—CO_2 relationship [22]. Their findings suggest to incorporate R&D with carbon policies to quickly attain low-carbon growth.

On the other hand, another strand of research considered human capital as a pollution-reducing factor [23,24]. Those studies explained that greater human capital measured either by intelligence or cognitive abilities implies greater commitment to environmentalism and pollution reduction. Considering cognitive capital and the number of researchers occupied in research and development as elements of national human capital, it indeed may decrease the pollution levels and encourage public and private commitment to cleaner and ecologically friendly technologies [25].

Hassan et al. [26] examined the effect of the index of per capita human capital and biocapacity on the ecological footprint, measured by the area of bio-productive lands, as they both may have an impact on human pro-environmental behavior. For this purpose, the authors employed cointegration and the Granger causality approach. Their findings suggested a neutral relationship between the variables. Yao et al. [27] used a sample of 20 OECD economies between 1870 and 2014 using parametric and nonparametric tests to explore the effect of human capital on carbon emissions. Their results implied a negative yet time-variant relationship between human capital and carbon dioxide emissions. However, human capital may be indirectly related to pollution reduction. While investigating the effect of fiscal decentralization on CO_2 emissions, Khan et al. [28] observed the significant role of human capital. Their study also highlighted the one-way causal effect running from eco-innovation to CO_2 emissions. In contrast, Wang and Wu [29] investigated if air pollution could be related to brain drain in China and India. The empirical results revealed that a higher concertation of pollutants in the air negatively impact the stock of technologically innovative human capital—a higher-educated population engaged in R&D in enterprises, research centers and universities. Eshchanov et al. [30] observed that cognitive abilities positively related to pro-environmental behavior and a favorable attitude toward renewable energy sources using household data of Uzbek households. Finally, Lin et al. [11] explored the relationship between innovative human capital, carbon dioxide emissions and economic growth in Chinese provinces during 2003–2017 by employing System GMM. Innovative human capital was measured by the number of patents of every one million R&D staff full-time equivalent. Their findings suggested that innovative human capital decreases carbon dioxide emissions, and its further development will enhance the environmental sustainability of China.

3. Data and Methods

The current study examines the long-term relationships between atmospheric pollution, research in R&D fields, economic development, foreign capital inflows, renewable energy and population growth. Our key independent variable is research in R&D, measured by the number of researchers engaged in R&D. Variables' descriptions and summary statistics are represented in Table 1. Our sample is restricted to the countries with full data series covering Belgium, Canada, China, the Czech Republic, France, Germany, Hungary, Ireland, Italy, Japan, South Korea, Latvia, Lithuania, Mexico, the Netherlands, Poland, Portugal, Romania, the Russian Federation, Singapore, the Slovak Republic, Slovenia, Spain, Turkey, the United Kingdom and the United States.

Table 1. Summary statistics.

Variable	Indicator	Source	Mean	Std. Dev.	Min	Max
CO_2	CO_2 emissions (metric tons per capita)	World Development Indicators (WDI)	8.61	3.99	2.54	21.29
RRD	Researchers in R&D (per million people)	World Development Indicators (WDI)	2685.72	1447.22	213.58	7013.49
FDI	Foreign direct investment, net inflows (% of GDP)	World Development Indicators (WDI)	5.48	9.13	−15.84	86.59
GDP pc	GDP per capita (constant 2010 US$)	World Development Indicators (WDI)	25,840.96	15,655.76	1332.41	65,432.75
RE	Renewable energy consumption (% of total final energy consumption)	World Development Indicators (WDI)	10.85	8.78	0.33	40.37
POPG	Population growth (annual %)	World Development Indicators (WDI)	0.39	0.87	−2.26	5.32

The dependent variable is carbon emissions measured as CO_2 emissions in metric tons per capita. The average CO_2 emission in our sample is 8.6 metric tons per capita (Table 1). Between 1995 and 2015, carbon dioxide emissions decreased rapidly. Economies significantly decreased their pollution levels between 2007 and 2015. Before 2005, the amount of CO_2 emissions fluctuated, yet with an ascending tendency.

The figures below represent a time series of the selected variables as a pretest measure to identify trends and breaks as suggested by [31].

Independent variable—the number of researchers engaged in R&D per million population. The average number of researchers in R&D in our sample is 2.7 thousand people (Table 1). Between 1996 and 2015, the number of researchers engaged in R&D grew rapidly (Figure 1). In 2015, two outstanding economies were South Korea and Singapore, where the number of researchers exceeded 7 thousand per million people, followed by Ireland and Japan (>5 thousand) and Germany (4.7 thousand). The bottom 5 were Latvia, Turkey, China, Romania and Mexico, where the number of researchers in R&D was less than 2 thousand (Figure 2).

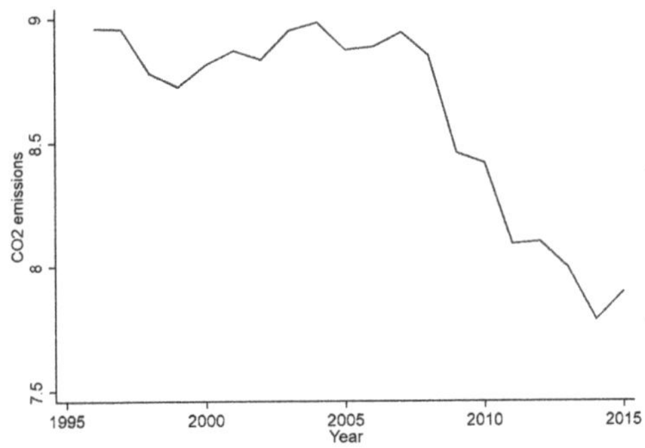

CO_2

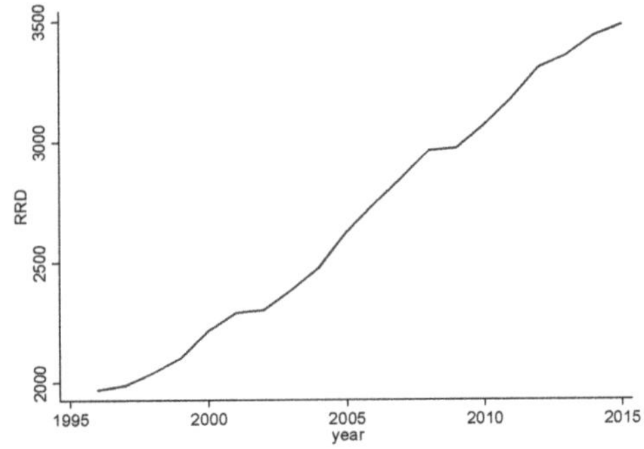

RRD

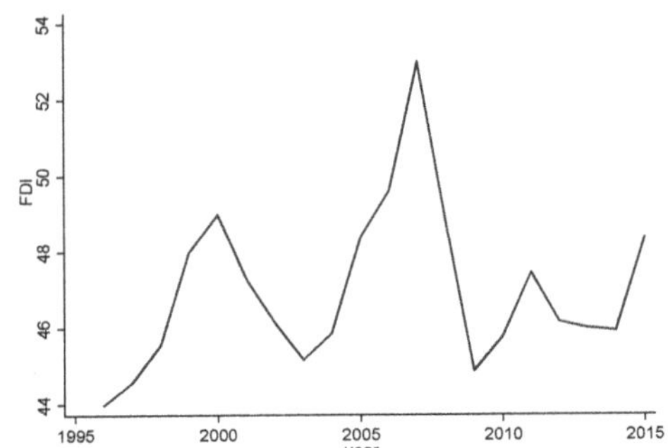

FDI

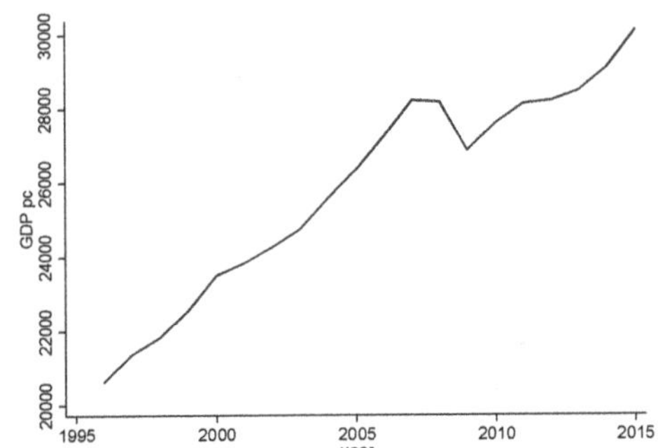

GDP pc

Figure 1. *Cont.*

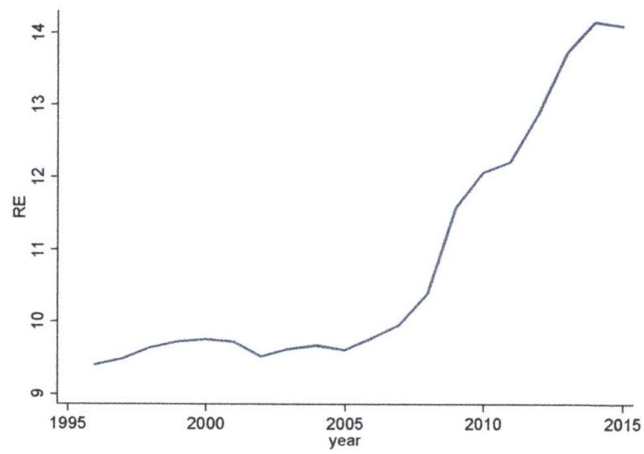

RE

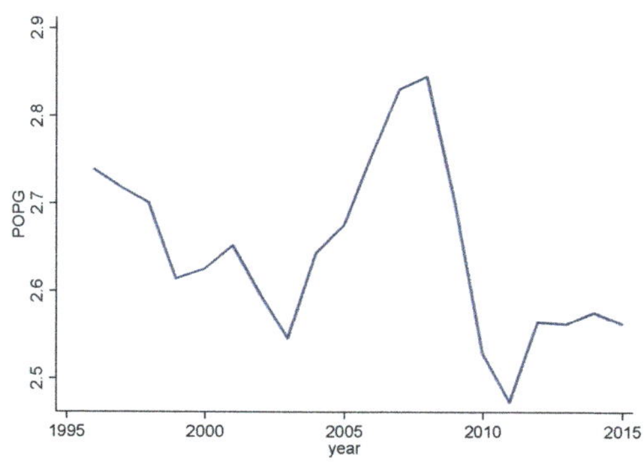

POPG

Figure 1. Pretest analysis: time series of the variables between 1995 and 2015.

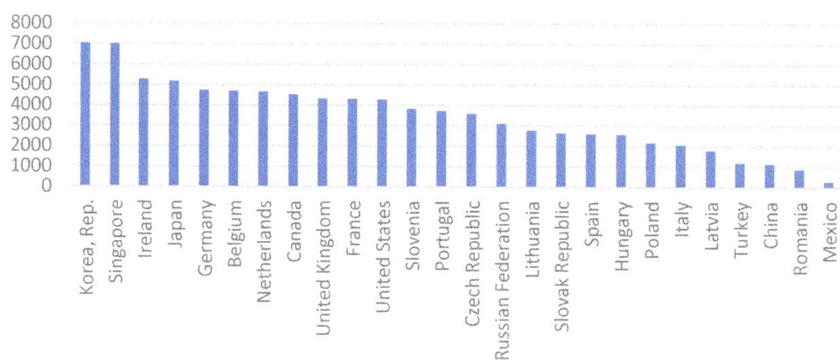

Figure 2. Researchers engaged in R&D in 2015 (per million population).

The control variables in the model include net inflows of the FDI as a GDP share, GDP per capita, share of renewable energy in the total final energy consumption and the annual rate of population growth. The FDI, GDP per capita and population growth experienced a significant fall down between 2007 and 2010, which may be a response to the global financial crisis. On the other hand, renewable energy consumption started a growing trend in 2005.

Our study describes CO_2 emissions as follows:

$$CO_{2i,t} = f(RRD_{i,t}, GDP\ pc_{i,t}, FDI_{i,t}, RE_{i,t}, POPG_{i,t}) \tag{1}$$

where CO_2 is the carbon dioxide emissions per capita, GDP pc is the GDP per capita, FDI—foreign direct investment as a share of the GDP, RE—renewable energy consumption as a % of the total final energy consumption and POPG is the population growth rate. i and t represent individual and temporal dimensions.

As one can note, some variables represent shares, while others are expressed in the number of people, dollars or tons. Such disparities may complicate further analyses and decrease the results quality due to data sharpness. We therefore transform Equation (1) by taking natural logs from each variable, so that each variable is now expressed in the same measurement unit.

$$\ln CO_{2i,t} = \beta_0 + \beta_1 \ln RRD_{i,t} + \beta_1 \ln GDP\ pc_{i,t} + \beta_2 \ln FDI_{i,t} + \beta_3 \ln RE_{i,t} + \beta_4 \ln POPG_{i,t} \quad (2)$$

The first step of our analyses assumes a stationarity check. For this purpose, we employ five panel unit root tests, which are the LLC test by [32], Breitung's test by [33], the IPS test by [34] and Fisher-type tests by Maddala & Wu [35]. We check each variable in two different forms—level and first difference—to test the null hypothesis of stationarity. In case our level model is nonstationary, then there may exist a cointegrating relationship [36].

Once all variables are stationary after first differencing, we may proceed with the cointegration analysis. Here, we adopt the panel cointegration test introduced by Pedroni [36,37] for heterogeneous panels. It tests the model for the cointegrating vector by employing seven parametric and nonparametric statistics. The test's output is grouped by group and panel estimates. The null hypothesis states no cointegration and may be rejected in the favor of the majority of the test statistics.

After confirming a cointegrating relationship between variables, we estimate the regression coefficients, which, in our case, are long-term elasticities, since natural logs are taken from each variable. We estimate our model with a Fully Modified Ordinary Least Squares (FMOLS) estimator that is usually applied if the model exhibits a cointegrating vector [32].

FMOLS coefficients provide us with the effect of selected variables on CO_2 and their significance, yet it does not account for the causality direction. We employ Dumitrescu and Hurlin's [33] panel causality test, which is based on Granger's technique. The test is designed for heterogeneous panel data and assumes a noncausal relationship. Dumitrescu and Hurlin's test described the regression equation as follows:

$$y_{i,t} = \alpha_i + \sum_{k=1}^{K} \gamma_{ik} y_{i,t-k} + \sum_{k=1}^{K} \beta_{ik} x_{i,t-k} + \varepsilon_{i,t} \quad (3)$$

where $x_{i,t}$ and $y_{i,t}$ are observations of two stationary variables in a strongly balanced panel dataset for individual i in a period.

The test runs separate regressions to check different causal directions. The test output results in a table providing Wald statistics and underlying z statistics. The final decision is made based on z-bar statistics [38].

4. Results

We first demonstrate the results of panel unit root tests (Table 2). We test each variable in the level- and first-difference forms with the LLC, Breitung, IPS and Fisher tests on both the ADF and PP. The dependent variable—log of carbon emissions—is nonstationary at the level under all the test statistics but stationary after first differencing and strongly significant. Researchers in R&D (RRD) demonstrate stationarity at the level form under the LLC test, while other test statistics demonstrate the opposite. First, differencing eliminates the unit root at $p < 0.01$. The level forms of GDP pc and RE are nonstationary but become stationary after the first differencing. On the other hand, the level forms of the FDI and population growth (POPG) are stationary both at the level and first-difference forms. Our results suggest that most of the variables are nonstationary at the level but stationary after detrending, which enables the further investigation of panel cointegration.

Table 2. Panel unit root test results.

Form	Variable	Test				
		LLC	Breitung	IPS	ADF Fisher	PP Fisher
Level	ln CO_2	1.2358 (0.8917)	2.5056 (0.9939)	4.0181 (1.0000)	45.9329 (0.7101)	34.7407 (0.9685)
First-difference	Δln CO_2	−8.1625 *** (0.0000)	−10.146 *** (0.0000)	−10.2910 *** (0.0000)	229.0940 *** (0.0000)	487.9605 *** (0.0000)
Level	ln RRD	−3.0719 *** (0.0011)	7.7004 (1.0000)	3.0491 (0.9989)	59.9960 (0.2085)	49.9982 (0.5530)
First-difference	Δln RRD	−8.6075 *** (0.0000)	−8.9015 *** (0.0000)	−9.8892 *** (0.0000)	208.9463 *** (0.0000)	351.0317 *** (0.0000)
Level	ln GDP pc	−2.5543 (0.0053)	8.9488 (1.0000)	−0.6820 (0.2476)	44.0777 (0.7746)	76.2989 (0.0157)
First-difference	Δln GDP pc	−7.8853 *** (0.0000)	−8.0403 *** (0.0000)	−6.7039 *** (0.0000)	175.9595 *** (0.0000)	204.6046 *** (0.0000)
Level	ln FDI	−6.1026 *** (0.0000)	−7.1488 *** (0.0000)	−5.9123 *** (0.0000)	133.3009 *** (0.0001)	158.4418 *** (0.0000)
First-difference	Δln FDI	−13.237 *** (0.0000)	−14.465 *** (0.0000)	−11.8957 *** (0.0000)	337.4060 *** (0.0000)	617.9227 *** (0.0000)
Level	ln RE	0.2612 (0.6030)	7.6633 (1.0000)	5.4611 (1.0000)	20.7966 (1.0000)	31.0526 (0.9907)
First-difference	Δln RE	−7.2622 *** (0.0000)	−9.5504 *** (0.0000)	−10.2925 *** (0.0000)	226.8364 *** (0.0000)	452.8732 *** (0.0000)
Level	ln POPG	−8.1323 *** (0.0000)	0.4863 (0.6866)	−1.5792 ** (0.0571)	161.9481 *** (0.0000)	106.7020 *** (0.0000)
First-difference	Δln POPG	−7.7530 *** (0.0000)	−8.4494 *** (0.0000)	−6.7115 *** (0.0000)	262.6809 *** (0.0000)	343.1562 *** (0.0000)

Note: *** and ** indicate significance at the 1% and 5% levels, respectively. The figures in parentheses are *p*-values.

Additionally, a panel unit root with a structural break is employed to check the stationarity of the variables, as structural breaks may mislead stationarity tests to accept the unit root while the opposite is true [39,40]. To be considered for a structural break, we employed Karavias and Tzavalis' [41] test for panel data. This methodology allows to test the unit root in the presence of one or two structural breaks in the intercepts of the individual series or in both intercepts and linear trends. The null hypothesis states the unit root in all panels and no structural breaks, while, alternatively, one assumes the stationarity of some panels and structural break(s). We assume one known structural break during the global subprime mortgage crisis between 2008 and 2009. We tested all variables for the unit root in the presence of a structural break in 2008 in both intercepts and trends. Table 3 demonstrates that the unit root are mostly contained in linear trends of ln CO_2, ln RRD, ln GDP pc and ln RE. Once the variables are detrended, the data is stationary.

Panel cointegration test results are depicted in Table 4. Out of seven test statistics, four confirm the hypothesis of a long-term relationship between variables at $p < 0.01$. In addition, Kao's ADF and Westerlund's variance ratio also demonstrate that the variables in our model are cointegrated. Our results suggest a long-term relationship between carbon emissions, researchers in R&D, GDP per capita, foreign capital investment, renewable energy share and population growth, which means that, in the long term, these variables can impact carbon dioxide emissions, yet the effect must be examined. For this purpose, we ran a fully modified OLS regression (Table 5). Since all variables are in logarithmic form, the coefficients represent long-term elasticities.

In our sample, the number of researchers in R&D is negatively related to carbon emissions, or, in other words, the more researchers are employed in R&D sector, the less are volumes of carbon dioxide emissions. As for other variables, the GDP per capita and FDI are positively related to CO_2 emissions. Shares of renewable energy, however, decrease carbon emissions in selected economies. Population growth is not related to carbon dioxide emissions.

Table 3. Karavias and Tzavalis' panel unit root test with structural breaks.

Variables	Intercept	Linear Trend
ln CO_2	−3.9318 ***	−0.8835
Δln CO_2	−17.7569 ***	−9.6421 ***
ln RRD	−7.7910 ***	−0.2121
Δln RRD	−18.4935 ***	−9.1178 ***
ln GDP pc	−10.2388 ***	2.1072
Δln GDP pc	−12.1393 ***	−5.0840 ***
ln FDI	−10.4964 ***	−6.0351 ***
Δln FDI	−5.0840 ***	−10.4964 ***
ln RE	−2.9500 ***	0.2949
Δln RE	−20.1432 ***	−12.5214 ***
ln POPG	−18.1378 ***	−7.8630 ***
Δln POPG	−25.8440 ***	−14.6723 ***

Note: *** indicates significance at the 1% levels, respectively.

Table 4. Pedroni's panel cointegration.

Test Statistic	Score
V-stat	−2.73 **
Panel rho-stat	1.221
Panel PP-stat	−8.909 ***
Panel ADF-stat	−0.9854
Group rho stat	2.954 ***
Group PP stat	−10.67 ***
Group ADF stat	0.8447
Kao's ADF	−12.9579 ***
Variance ratio	2.6777 ***

Note: *** and ** indicate significance at the 1% and 5% levels, respectively.

Table 5. Long-term elasticities.

	FMOLS
Δ ln RRD	−0.08 *** (−5.64)
Δ ln GDP pc	0.54 *** (35.75)
Δ ln FDI	0.05 *** (4.19)
Δ ln RE	−0.24 *** (−32.59)
Δ ln POPG	−0.02 0.09

Note: *** indicates significance at the 1% levels, respectively. Figures in parentheses are *t*-statistics.

Finally, it is necessary to understand the nature of the relationship, whether it is causal or linked through other channels. For this purpose, we conduct a panel causality test, introduced by Dumitrescu and Hurlin [42]. We test each independent variable with CO_2 for two hypotheses, which assume different causality directions. The first hypothesis tests the RRD–CO_2 nexus, i.e., causal relationship between the number of researchers in R&D and carbon dioxide emissions. According to Table 6, RRD and CO_2 are not causally related in our sample, although the variables are strongly and negatively related in the long term. Economic development, on the other hand, is causally related to carbon emissions. We observe a significant causal relationship running from the per capita GDP to carbon emissions at $p < 0.01$. A bidirectional causal relationship was also observed between the FDI and CO_2 (Figure 3). It means that foreign capital inflows may be directed to carbon-intensive sectors and thus increase atmospheric pollution. Similarly, there is a two-way causality between renewable energy and carbon dioxide emissions, meaning that higher shares of renewable energy consumption cause lower shares of carbon emissions.

Table 6. Dumitrescu and Hurlin's noncausality test.

Null Hypothesis: CO_2 Causalities	W-Stat	Zbar-Stat (p-Value)	Optimal Number of Lags (AIC)
RRD does not cause CO_2	1.0557	0.2010 (0.8407)	1
CO_2 does not cause RRD	0.9161	−0.3025 (0.7623)	1
GDP pc does not cause CO_2	2.4879	5.3649 *** (0.0000)	1
CO_2 does not cause GDP pc	1.0299	0.1077 (0.9143)	1
FDI does not cause CO_2	1.7056	2.5442 ** (0.0110)	1
CO_2 does not cause FDI	6.7268	4.9158 *** (0.0000)	4
RE does not cause CO_2	7.8991	7.0291 *** (0.0000)	4
CO_2 does not cause RE	4.2206	5.6614 *** (0.0000)	2
POPG does not cause CO_2	1.1809	0.6522 (0.5143)	1
CO_2 does not cause POPG	10.2530	11.2727 *** (0.0000)	4

Note *** and ** indicate significance at the 1% and 5% levels, respectively. Figures in parentheses are p-values.

Figure 3. Graphical results of the study.

Our results provide interesting insights on the relationship between scientific advancements of countries and atmospheric pollution. First, we found that researchers in R&D are positively related to carbon emissions in the long term, yet the causal link is not supported. Our findings are partly explained by Obydenkova and Salahodjaev [10], who find that higher human capital, combined with democracy, encourage environmental commitments. Secondly, we observed that economic development is significantly and positively related to carbon emissions. There is also a causal link running from economic development to carbon emissions. Similar findings were observed by [43,44], which generally stated that economic development initially encourages pollution. Further, there is a feedback effect between the FDI and carbon emissions. The relationship between the FDI and CO_2 is far from obvious, as it largely depends on the economy's development stage and environmental commitment [45]. Although inflows of foreign capital may enhance the development of green technologies [46], it can also be directed to support the most profitable pollution-extensive sectors [47]. Finally, we found the bidirectional and negative effects of renewable energy. Increasing shares of renewables consumed and policies aimed at green economy transition would thus decrease environmental harm. Similar findings were observed by [48].

5. Conclusions

This study tested the long-term relationships between carbon dioxide emissions, researchers engaged in R&D, GDP per capita, renewable energy and population growth in 26 economies between 1995 and 2015. The econometric estimates were based on Pedroni's cointegration test and Dumitrescu and Hurlin's panel causality test. Our results suggest a negative long-term relationship between the number of researchers in R&D, renewable energy and CO_2. A positive long-term relationship was observed between CO_2, economic growth and the FDI. A panel causality test revealed unidirectional causality running from GDPs per capita to carbon emissions. Bidirectional causality was observed between carbon emissions and the FDI and between renewable energy and carbon emissions.

Our study confirmed that innovations measured by researchers in R&D are negatively related to atmospheric pollution in the long term. Indeed, innovations result in the development of efficient and ecological solutions, aimed at improving the quality of life and wellbeing. Such logic is supported by a number of researchers [2,49], whose results concluded the pollution-reducing effect of innovations. Growing environmental concerns encourage both adopting pro-environmental policies and promoting research and development investments [50]. Thus, policymakers should encourage research and development both through investment and human capital. Besides innovation, pollution can be reduced by adopting renewable energy sources. On the other hand, the FDI exacerbates pollution. It might be useful to create incentives for the adoption of pro-environmental technologies and developing environmentally sustainable sectors, so that the direction of foreign capital would shift in favor of green projects.

Author Contributions: Conceptualization, K.A., E.Z., R.S., W.T. and G.M.; methodology, R.S. and G.M.; software, R.S.; validation, K.A., E.Z., W.T. and H.A.; formal analysis, R.S., G.M. and E.Z.; investigation, R.S.; and writing—original draft preparation, K.A., E.Z., R.S., W.T. and H.A. All authors have read and agreed to the published version of the manuscript.

Funding: This research received no external funding.

Data Availability Statement: Data are available from World Bank.

Conflicts of Interest: The authors declare no conflict of interest.

References

1. Calef, D.; Goble, R. The allure of technology: How France and California promoted electric and hybrid vehicles to reduce urban air pollution. *Policy Sci.* **2007**, *40*, 1–34. [CrossRef]
2. Carrión-Flores, C.E.; Innes, R. Environmental innovation and environmental performance. *J. Environ. Econ. Manag.* **2010**, *59*, 27–42. [CrossRef]
3. Irandoust, M. The renewable energy-growth nexus with carbon emissions and technological innovation: Evidence from the Nordic countries. *Ecol. Indic.* **2016**, *69*, 118–125. [CrossRef]
4. Zeng, S.; Zhou, Y. Foreign Direct Investment's Impact on China's Economic Growth, Technological Innovation and Pollution. *Int. J. Environ. Res. Public Health* **2021**, *18*, 2839. [CrossRef]
5. Erdoğan, S.; Yıldırım, S.; Yıldırım, D.; Gedikli, A. The effects of innovation on sectoral carbon emissions: Evidence from G20 countries. *J. Environ. Manag.* **2020**, *267*, 110637. [CrossRef]
6. Irga, P.J.; Pettit, T.J.; Torpy, F.R. The phytoremediation of indoor air pollution: A review on the technology devel-opment from the potted plant through to functional green wall biofilters. *Rev. Environ. Sci. Bio/Technol.* **2018**, *17*, 395–415. [CrossRef]
7. Özbay, R.D.; Athari, S.A.; Saliba, C.; Kirikkaleli, D. Towards Environmental Sustainability in China: Role of Glob-alization and Hydroelectricity Consumption. *Sustainability* **2022**, *14*, 4182. [CrossRef]

8. Nentjes, A.; Wiersma, D. Innovation and pollution control. *Int. J. Soc. Econ.* **1988**, *18*, 18–29. [CrossRef]
9. Salahodjaev, R. Is there a link between cognitive abilities and environmental awareness? Cross-national evidence. *Environ. Res.* **2018**, *166*, 86–90. [CrossRef]
10. Obydenkova, A.; Salahodjaev, R. Intelligence, democracy, and international environmental commitment. *Environ. Res.* **2016**, *147*, 82–88. [CrossRef]
11. Lin, X.; Zhao, Y.; Ahmad, M.; Ahmed, Z.; Rjoub, H.; Adebayo, T.S. Linking innovative human capital, economic growth, and CO_2 emissions: An empirical study based on Chinese provincial panel data. *Int. J. Environ. Res. Public Health* **2021**, *18*, 8503. [CrossRef] [PubMed]
12. Yan, L.; Wang, H.; Athari, S.A.; Atif, F. Driving green bond market through energy prices, gold prices and green energy stocks: Evidence from a non-linear approach. *Econ. Res.-Ekon. Istraživanja* **2022**, *35*, 6479–6499. [CrossRef]
13. Garrone, P.; Grilli, L. Is there a relationship between public expenditures in energy R&D and carbon emissions per GDP? An empirical investigation. *Energy Policy* **2010**, *38*, 5600–5613. [CrossRef]
14. Lee, K.-H.; Min, B. Green R&D for eco-innovation and its impact on carbon emissions and firm performance. *J. Clean. Prod.* **2015**, *108*, 534–542. [CrossRef]
15. Álvarez-Herránz, A.; Balsalobre, D.; Cantos, J.M.; Shahbaz, M. Energy Innovations-GHG Emissions Nexus: Fresh Empirical Evidence from OECD Countries. *Energy Policy* **2017**, *101*, 90–100. [CrossRef]
16. Zhang, Y.-J.; Peng, Y.-L.; Ma, C.-Q.; Shen, B. Can environmental innovation facilitate carbon emissions reduction? Evidence from China. *Energy Policy* **2017**, *100*, 18–28. [CrossRef]
17. Ganda, F. The impact of innovation and technology investments on carbon emissions in selected organisation for economic Co-operation and development countries. *J. Clean. Prod.* **2019**, *217*, 469–483. [CrossRef]
18. Khan, Z.; Ali, S.; Umar, M.; Kirikkaleli, D.; Jiao, Z. Consumption-based carbon emissions and International trade in G7 countries: The role of Environmental innovation and Renewable energy. *Sci. Total Environ.* **2020**, *730*, 138945. [CrossRef]
19. Nguyen, T.T.; Pham, T.A.T.; Tram, H.T.X. Role of information and communication technologies and innovation in driving carbon emissions and economic growth in selected G-20 countries. *J. Environ. Manag.* **2020**, *261*, 110162. [CrossRef]
20. Wahab, S.; Zhang, X.; Safi, A.; Wahab, Z.; Amin, M. Does energy productivity and technological innovation limit trade-adjusted carbon emissions? *Econ. Res.-Ekon. Istraživanja* **2021**, *34*, 1896–1912. [CrossRef]
21. Zhang, Y.J.; Shi, W.; Jiang, L. Does China's carbon emissions trading policy improve the technology innovation of relevant enterprises? *Bus. Strategy Environ.* **2020**, *29*, 872–885. [CrossRef]
22. Saliba, C.B.; Hassanein, F.R.; Athari, S.A.; Dördüncü, H.; Agyekum, E.B.; Adadi, P. The Dynamic Impact of Re-newable Energy and Economic Growth on CO2 Emissions in China: Do Remittances and Technological Innovations Matter? *Sustainability* **2022**, *14*, 14629. [CrossRef]
23. Obydenkova, A.; Nazarov, Z.; Salahodjaev, R. The process of deforestation in weak democracies and the role of Intelligence. *Environ. Res.* **2016**, *148*, 484–490. [CrossRef] [PubMed]
24. Omanbayev, B.; Salahodjaev, R.; Lynn, R. Are greenhouse gas emissions and cognitive skills related? Cross-country evidence. *Environ. Res.* **2018**, *160*, 322–330. [CrossRef]
25. Larkin, A.; Hystad, P. Towards Personal Exposures: How Technology Is Changing Air Pollution and Health Research. *Curr. Environ. Health Rep.* **2017**, *4*, 463–471. [CrossRef] [PubMed]
26. Hassan, S.T.; Baloch, M.A.; Mahmood, N.; Zhang, J. Linking economic growth and ecological footprint through human capital and biocapacity. *Sustain. Cities Soc.* **2019**, *47*, 101516. [CrossRef]
27. Yao, Y.; Ivanovski, K.; Inekwe, J.; Smyth, R. Human capital and CO_2 emissions in the long run. *Energy Econ.* **2020**, *91*, 104907. [CrossRef]
28. Khan, Z.; Ali, S.; Dong, K.; Li, R.Y.M. How does fiscal decentralization affect CO_2 emissions? The roles of institutions and human capital. *Energy Econ.* **2020**, *94*, 105060. [CrossRef]
29. Wang, F.; Wu, M. Does air pollution affect the accumulation of technological innovative human capital? Empirical evidence from China and India. *J. Clean. Prod.* **2020**, *285*, 124818. [CrossRef]
30. Eshchanov, B.; Abdurazzakova, D.; Yuldashev, O.; Salahodjaev, R.; Ahrorov, F.; Komilov, A.; Eshchanov, R. Is there a link between cognitive abilities and renewable energy adoption: Evidence from Uzbekistan using micro data? *Renew. Sustain. Energy Rev.* **2021**, *141*, 110819. [CrossRef]
31. Kirikkaleli, D.; Athari, S.A. Time-frequency co-movements between bank credit supply and economic growth in an emerging market: Does the bank ownership structure matter? *N. Am. J. Econ. Finance* **2020**, *54*, 101239. [CrossRef]
32. Levin, A.; Lin, C.-F.; Chu, C.-S.J. Unit root tests in panel data: Asymptotic and finite-sample properties. *J. Econ.* **2002**, *108*, 1–24. [CrossRef]
33. Breitung, J. Nonparametric tests for unit roots and cointegration. *J. Econ.* **2002**, *108*, 343–363. [CrossRef]
34. Im, K.S.; Pesaran, M.; Shin, Y. Testing for unit roots in heterogeneous panels. *J. Econ.* **2003**, *115*, 53–74. [CrossRef]
35. Maddala, G.S.; Wu, S. A Comparative Study of Unit Root Tests with Panel Data and a New Simple Test. *Oxf. Bull. Econ. Stat.* **1999**, *61*, 631–652. [CrossRef]
36. Pedroni, P. Fully modified OLS for heterogeneous cointegrated panels. In *Nonstationary Panels, Panel Cointegration, and Dynamic Panels*; Emerald Group Publishing Limited: Bradford, UK, 2001. [CrossRef]

37. Pedroni, P. Critical Values for Cointegration Tests in Heterogeneous Panels with Multiple Regressors. *Oxf. Bull. Econ. Stat.* **1999**, *61*, 653–670. [CrossRef]
38. Ahmed, N.; Sheikh, A.A.; Hamid, Z.; Senkus, P.; Borda, R.C.; Wysokińska-Senkus, A.; Glabiszewski, W. Exploring the Causal Relationship among Green Taxes, Energy Intensity, and Energy Consumption in Nordic Countries: Dumitrescu and Hurlin Causality Approach. *Energies* **2022**, *15*, 5199. [CrossRef]
39. Irani, F.; Athari, S.A.; Hadood, A.A.A. The impacts of country risk, global economic policy uncertainty, and macroeconomic factors on the Turkish tourism industry. *Int. J. Hosp. Tour. Adm.* **2022**, *23*, 1242–1265. [CrossRef]
40. Kondoz, M.; Kirikkaleli, D.; Athari, S.A. Time-frequency dependencies of financial and economic risks in South American countries. *Q. Rev. Econ. Finance* **2020**, *79*, 170–181. [CrossRef]
41. Karavias, Y.; Tzavalis, E. Testing for unit roots in short panels allowing for a structural break. *Comput. Stat. Data Anal.* **2014**, *76*, 391–407. [CrossRef]
42. Dumitrescu, E.-I.; Hurlin, C. Testing for Granger non-causality in heterogeneous panels. *Econ. Model.* **2012**, *29*, 1450–1460. [CrossRef]
43. Ozturk, I.; Acaravci, A. The long-run and causal analysis of energy, growth, openness and financial development on carbon emissions in Turkey. *Energy Econ.* **2013**, *36*, 262–267. [CrossRef]
44. Dogan, E.; Seker, F. The influence of real output, renewable and non-renewable energy, trade and financial devel-opment on carbon emissions in the top renewable energy countries. *Renew. Sustain. Energy Rev.* **2016**, *60*, 1074–1085. [CrossRef]
45. Salahodjaev, R.; Isaeva, A. Post-Soviet states and CO_2 emissions: The role of foreign direct investment. *Post-Communist Econ.* **2021**, *34*, 944–965. [CrossRef]
46. Hao, Y.; Liu, Y.-M. Has the development of FDI and foreign trade contributed to China's CO2 emissions? An empirical study with provincial panel data. *Nat. Hazards* **2014**, *76*, 1079–1091. [CrossRef]
47. Blanco, L.; Gonzalez, F.; Ruiz, I. The Impact of FDI on CO_2 Emissions in Latin America. *Oxf. Dev. Stud.* **2013**, *41*, 104–121. [CrossRef]
48. Baloch, M.A.; Mahmood, N.; Zhang, J.W. Effect of natural resources, renewable energy and economic development on CO_2 emissions in BRICS countries. *Sci. Total Environ.* **2019**, *678*, 632–638. [CrossRef]
49. Kumar, S.; Managi, S. Energy price-induced and exogenous technological change: Assessing the economic and en-vironmental outcomes. *Resour. Energy Econ.* **2009**, *31*, 334–353. [CrossRef]
50. Gu, G.; Wang, Z. Research on global carbon abatement driven by R&D investment in the context of INDCs. *Energy* **2018**, *148*, 662–675. [CrossRef]

energies

MDPI

Article

Energy Transition in Non-Euro Countries from Central and Eastern Europe: Evidence from Panel Vector Error Correction Model

Simona Andreea Apostu [1,2], Mirela Panait [2,3,*], Daniel Balsalobre-Lorente [4,5], Diogo Ferraz [6,7,8] and Irina Gabriela Rădulescu [9]

1 Department of Statistics and Econometrics, Bucharest University of Economic Studies, 010552 Bucharest, Romania
2 Institute of National Economy, 050771 Bucharest, Romania
3 Department of Cybernetics, Economic Informatics, Finance and Accounting, Petroleum-Gas University of Ploiesti, 100680 Ploiesti, Romania
4 Department of Political Economy and Public Finance, Economic and Business Statistics and Economic Policy, University of Castilla-La Mancha, 45071 Toledo, Spain
5 Department of Applied Economics, University of Alicante, 03690 San Vicente del Raspeig, Spain
6 Innovation Economics, Institute of Economics, University of Hohenheim, 70599 Stuttgart, Germany
7 Department of Economics, Institute of Social and Applied Sciences, Federal University of Ouro Preto (UFOP), Mariana 35420-000, Brazil
8 Department of Production Engineering, School of Engineering of Bauru, Campus Bauru, São Paulo State University (UNESP), Bauru 17033-360, Brazil
9 Department of Business Administration, Petroleum-Gas University of Ploiesti, 100680 Ploiesti, Romania
* Correspondence: mirela.matei@upg-ploiesti.ro

Citation: Apostu, S.A.; Panait, M.; Balsalobre-Lorente, D.; Ferraz, D.; Rădulescu, I.G. Energy Transition in Non-Euro Countries from Central and Eastern Europe: Evidence from Panel Vector Error Correction Model. *Energies* **2022**, *15*, 9118. https://doi.org/10.3390/en15239118

Academic Editors: Grzegorz Mentel and Sebastian Majewski

Received: 24 October 2022
Accepted: 28 November 2022
Published: 1 December 2022

Abstract: The countries of Central and Eastern Europe, from the non-euro area, have completed the process of economic transition before joining the European Union. Achieving a certain level of economic development and membership in the European Union have generated their involvement in a new transition process, namely the energy transition. Concerns about promoting the low carbon economy have become increasingly complex for those countries that are interested in the environmental impact of economic activity. This study aims to analyze the process of energy transition in the countries of Central and Eastern Europe on the basis of the causality relationship among specific variables for the period 1990–2018. The study is based on cross-sectional panel data and the panel vector error correction model (PVECM). The efforts made by these countries by joining the European Union have generated economic development, with positive effects being recorded on the protection of the environment, a fact due to the strict regulations adopted and rigorous implementation at the national level. Foreign capital had a positive impact on the transition to a low carbon economy because most of the FDI flows attracted by the non-euro countries in the CEE come from Western Europe, i.e., from EU member countries, located either among the founders or among the countries that joined during the first waves of union expansion. Membership in the European Union facilitates the energy transition process for the non-euro countries of Central and Eastern Europe, but the new geopolitical events generate the reconfiguration of the European strategy of considering the need to ensure energy security.

Keywords: renewable energy transition; non-euro area; panel data; Granger causality; VECM

1. Introduction

After the fall of communism, the countries of Central and Eastern Europe (CEE) were in a complex process of economic, environmental, social, and political metamorphosis, the efforts of the authorities being concentrated on the transition from the centralized economy to the market economy [1–4]. Some of them have succeeded and successfully completed the process of economic transition, with six of them joining the European Union (EU), as they had met the criteria imposed by European documents. The process of economic development is particularly complex, which is why not all CEE countries have joined the EU. In addition, failure to meet the convergence criteria set out in the Maastricht Treaty has led to the inclusion of CEE countries in the non-euro area [5–11]. Membership in the European Union generates challenges not only at the economic and social level but also at the level of environmental protection and promotion of sustainable development, with the energy transition towards a low carbon economy being one of the major objectives of the EU countries.

The opening of the national economies generated foreign capital inflows, differentiated by country, depending on the progress made in the transition to the market economy, the endowment with resources, or the extent of the privatization processes carried out by the public authorities [12–15]. In this way, foreign investors have set up private companies or taken over majority stakes in various companies, especially in the field of industry. Unfortunately, the interests of foreign companies have not always been compatible with the economic policy of the host countries, and the takeover of local companies has, in many cases, led to the deindustrialization of these economies [16,17] or the continuation and development of the activity, except for the lands they owned or for the equipment which was later sold for scrap iron [18].

Attractive sectors for foreign companies are the oil and gas industry as well as the energy industry, given the potential of the energy market. Market liberalization has generated not only the presence of foreign investors in the classical fields of the energy sector but also the emergence of companies involved in the production of renewable energy, which has led to a reduction in the concentration of this market [17,19,20]. Thus, the national industrial landscape in these countries was changed dramatically, and the decomposition and recomposition of industrial structures has been noticed by researchers [21,22]. Currently, these countries are in a process of energy transition, a fact generated by the European authorities' concerns of facilitating the transition to a low carbon economy to manage the challenges imposed by climate change [23–28]. Therefore, the countries of Central and Eastern Europe have completed the process of economic transition by joining the European Union and are currently in energy transition because reaching a certain level of development allows them to focus on concerns of a low carbon economy.

This study aims to analyze the process of energy transition in the countries of CEE that are not members of the euro area, namely Bulgaria, Croatia, Czechia, Hungary, Poland, and Romania. These countries were chosen by taking into account several considerations. All six countries were former communist countries which had a somewhat similar course in the process of transition to the market economy; their rates of economic development were different, and they joined the European Union in waves. In addition, in the period 1990–2020, these countries had similar economic and environmental paths, yet despite the progress made economically, institutionally, and politically, these countries do not meet the convergence criteria imposed by the Maastricht Treaty. As EU members, these countries assumed specific targets regarding sustainable development and are involved in the energy transition, a bold project that takes place through the promotion of the Energy Union launched in 2015. The CEE countries have an important renewable energy potential based on the fact that most of them achieved the targets set at the EU level for the share of renewable energy in consumption (20% until 2020). In order to meet the new targets established by the European Union, the member countries set up an integrated national energy and climate plan (NECP) for the period from 2021 to 2030. These plans have five pillars, namely greenhouse gas emissions reductions, energy efficiency, renewables, interconnections, and research and innovation.

Compared with other studies published in the international literature, the present research is differentiated because it is focused on a group of countries with a similar past economic development and, especially for the communist ones, the same concerns regarding the energy transition, considering the quality of members of the European Union (more precisely of the Energy Union). The energy transition process has certain specificities in these countries considering the deindustrialization process that these economies are going through after the fall of communism [17,18], the efforts to align with the standards promoted by the EU regarding sustainable development [27,29], impact of COVID-19 on economic activity [30] (Wang et al., 2022), the geopolitical context generated by the invasion of Ukraine by Russia, and the position of these countries taking into account the geographical location and dependence on energy resources in Russia. The need to ensure energy security will generate the reconfiguration of the energy transition process.

2. Literature Review

Given the economic, social, technological, and environmental challenges posed by the new energy transition, increasingly more studies are focusing on national, regional, and international efforts to move from fossil fuels to renewable energy [26,28,31,32]. Numerous researchers have focused on analyzing the impact of economic activity on environmental pollution, with multiple studies focusing on validating the Kuznets curve for different regions/countries/areas for different periods of time [33–43]. Lately, given the concerns of international political leaders regarding facilitating the energy transition, increasingly more scientific studies are using independent variables, such as conventional energy consumption, renewable energy consumption, energy intensity, energy efficiency, or energy innovation. Energy has, thus, become a common thread not only in human activity but also in scientific research, given the need for specialists and public authorities to find measures to help reduce the negative impact of energy production and consumption on the environment. The impact of the development of economic activity (under different aspects, such as the intensification of international trade, the change in the structure of foreign trade, urbanization, and the expansion of foreign capital) on energy consumption has been the subject of interesting studies that have been reported for certain groups of countries [44,45]. Therefore, the countries of CEE have completed the process of economic transition by joining the European Union and are currently in energy transition because reaching a specific level of development allows them to focus on concerns about low carbon economy [46].

In the literature, only a few studies have been identified that focus on the process of energy transition in which the countries of CEE are involved. Armeanu et al. (2019) developed research for eleven states from CEE over the period from 2000 to 2016 [47]. The results of the panel data estimations advocate for a non-linear relationship between renewable energy and economic growth and a long-run unidirectional causal relation from non-renewable energy to economic growth.

The study of Przychodzen and Przychodzen (2019) focused on 27 transition economies, from CEE and the Caucasus and Central Asia, for the period 1990–2014 [48]. A specific regression model was developed by Polish researchers in order to analyse the effects of different economic and political factors on renewable energy production. The reserchers concluded that renewable energy generation is positively influenced by factors such as higher economic growth, size of general government debt, rising level of unemployment, and implementation of the Kyoto protocol.

In order to analyse the relationship between economic growth and renewable energy consumption, Marinaș et al., 2019 focused their study on ten countries from CEE members of the European Union [49]. Specific statistical data were selected for the period 1990–2014, and the researchers used the auto-regressive and distributed lag (ARDL) modeling procedure. Despite some similarities among the selected countries, the results obtained revealed significant differences. For Romania and Bulgaria, gross domestic product and renewable energy consumption dynamics are independent, but for countries such as Lithuania, Hungary, and Slovenia, increasing renewable energy consumption generated economic growth.

The study of Simonescu (2021) focused on several new members of EU, such as Bulgaria, Slovakia, Slovenia, Czech Republic, Hungary, Romania and Poland [29]. The researcher used panel threshold and dynamic panel models, as well as vector error correction models, with data that was available for the period 1990–2019. The study used GHG emissions, GDP per capita, renewable energy consumption, foreign direct investment, gross inland energy consumption per capita, control of corruption index of economic freedom (corruption), human development index, and labour productivity. The author identified certain differences among the analyzed countries regarding the impact of different indicators on greenhouse gas emissions. An inverted N-shaped curve was detected between GDP and GHG, and a U-shaped renewable Kuznet curve was observed for selected countries, except for Poland.

Using parametric and semiparametric methods, the study of Butnaru et al., 2020 was developed for EU countries for the period 1960–2015 [50]. The research demonstrated the convergence of renewable energy consumption per capita for the selected countries, with fossil fuels being the most used energy in the short and medium run.

An interesting study run by Ćetković and Buzogány, 2019 focused on the position of the six countries from CEE (Hungary, Poland, Czech Republic, Bulgaria, Slovakia, and Romania) on EU energy-related legislation in the period of 2007–2018 [51]. The research is based on the position of the national officials in the Council of Ministers. No common regional positioning was detected for these countries. Even these countries are considered to be climate and energy policy laggards in the European Union, and the lack of regional coherence and the exclusive promotion of national interests has led to the adoption of bold goals for decarbonization at the EU level.

The interest on renewable energy transition has increased over time, and this is reflected in the large number of papers on this field. Filtering from the Web of Science platform regarding the studies on this theme, we found 15,026 papers published from 1981 until the present day. As shown in Figure 1, the number of published papers in the area illustrates a hyperbolical progression; there is a jump in the number of publications after 2013. Thus, there is a growing interest in the field, the main interest being on renewable energy.

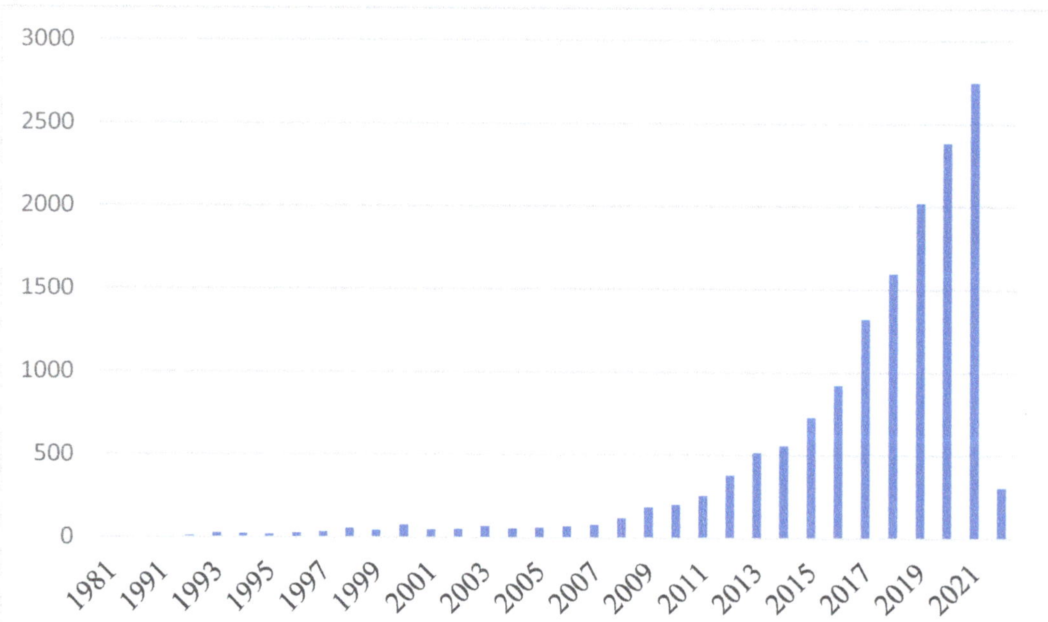

Figure 1. Dynamics on publications regarding renewable energy transition. Source: Authors' projection.

The studies conducted to analyze the impact of renewable resources on the energy transition process are increasingly more complex, both from the point of view of the statistical methods used as well as the variables used. Given the growing interdependencies among national economies as a result of globalization, increasingly more studies on the energy transition also take into account the impact generated by the intensification of trade and international capital flows. Foreign direct investment (FDI) has a significant impact on energy use both through the consumption it generates in the host countries as well as through the transfer of technology that transnational companies can make, which can determine a more rational use of fossil fuels and the promotion of renewable energy [52–55]. The intensification of international trade has generated an increase in national production and, thus, energy consumption, which is why many studies analyze the impact of trade on the use of conventional fuels and renewable energy for different periods and different levels (national, regional, or international) [56–60].

Taking in account the theoretical considerations presented by an in-depth literature review, carbon dioxide emission per capita is considered to be a proxy variable for renewable energy transition. Thus, the following research hypothesis has been defined in order to reach significant answers to our research aims:

Hypothesis 1. *GDP/capita, renewable energy consumption as percentage of total energy consumption, trade openness, FDI, and human development index cause carbon dioxide emission per capita in the CEE countries from the non-euro area over time.*

3. Data and Methodology

3.1. Methodology

To investigate the relationship among variables reflecting renewable energy transition over the time in the countries in the non-euro area, a panel VAR/VEC model was adopted. The vector autoregression (VAR) model was developed by Sims (1980) in order to analyze the dynamic response of the system as a result of shocks [61]. Its advantage does not depend on "incredible identification restrictions" inherent in structural models [62].

The VAR model represents a dynamic multivariate model aiming to treat a simultaneous set of variables equally, the endogenous variable being regressed on its own lags and the lags of all other variables considering a finite-order system.

In this study, the VAR (k) is modeled as:

$$y_t = \varphi + A_1 y_{t-1} + \ldots + A_i y_{t-k} + u_t \quad (1)$$

where y_t is a 6 × 1 vector including the variables of y_t, which are cointegrated as $y_t = [p_t, x_t, \ldots, b_t]$. A_i is a 6 × 6 parameter matrix, i = 1, ..., k. φ is an intercept vector. u_t is a 6 × 1 vector containing six error terms.

3.1.1. Testing Stationarity for Panel Data

Before beginning the estimation process, it is mandatory to pre-test the stationarity. For stationarity checking, the augmented Dickey–Fuller test (ADF) [63] can be used, but it registers a low power for rejecting the hypothesis of no stationarity, especially in the case of short-spanned data [64]. (Costantini and Martini, 2010). Recent panel unit root tests were introduced by Levin et al. (2002); LLC tests were introduced by Im et al. (2003); IPS tests by Breitung (2000); and BRT tests by Maddala and Wu (1999), Choi (2001), and Hadri (2000) [65–70].

The most popular tests used for checking stationarity are LLC and IPS; LLC considers homogeneity, and IPS considers heterogeneity of the autoregressive coefficients for all panel members, allowing for different orders of serial correlation through averaging the augmented Dickey–Fuller results [64].

In this study we have considered three unit root tests, namely LLC, ADF, and the Phillips–Perron test (PP) [71]. In order to investigate the existence of structural breaks, the robustness was checked both on single cross-sectional units and on the whole panel dataset.

3.1.2. Panel Cointegration

The cointegration analysis in the case of a single spatial series was significantly improved by the Pedroni panel cointegration technique (1999, 2000), allowing interdependence in the case of cross-sectional data with individual effects in the intercepts and slopes of the cointegrating equation [72,73]. According to Pedroni (1999), the time series panel regression can be written as follows [72]:

$$y_{i,t} = \alpha_t + \delta_i t + \beta_{1i} x_{1i,t} + \beta_{2i} x_{2i,t} + \ldots + \beta_{Mi} x_{Mi,t} + e_{i,t} \tag{2}$$

where t = 1, ... , T; I = 1, ... , N; and m = 1, ... , M. T represents the number of observations over time, N represents the number of individual cases in the panel, and M represents the number of regression variables. According to Pedroni, there are seven statistics for testing the cointegration in case of panel data. Four statistics consider the within-dimension cointegration, and three statistics consider the between-dimension cointegration [74]. Another test used in case of panel cointegration is that of Kao (1999), namely the panel cointegration test [75]. Other tests used in the case of residual-based panel cointegration were introduced by Westerlund (2005), Persyn and Westerlund (2008), and Westerlund and Edgerton (2008) [76–78]. The Westerlund (2005) test refers to the variance ratio statistics, and does not require corrections for the residual serial correlations [77]. The Persyn and Westerlund (2008) test presents an error correction based on the cointegration test [76]. Westerlund and Edgerton (2008) considered the presence of structural breaks within the panel [78].

3.1.3. Granger Causality

The implementation of the panel VAR/VEC model is the result of both the cointegration test [79] and the fact that VAR/VEC considers all the variables as a priori endogenous, controlling the interactions between dependent and independent variables [80]. In this regard, Granger (1988) presented the causal effect of one variable on another, known as Granger causality, and it exists when an independent variable conduces predictions of the dependent variable [80,81].

In order to identify whether a cointegration exists and whether a long-run relationship exists among variables, Johansen's VAR procedure [82] and Pedroni's heterogeneous panel cointegration test were used. Using VECM, the causality is tested considering the procedure of Engle–Granger causality [83]. In our study, a panel-VECM with 5 independent variables was proposed for examining the causality between the variables, which can be written as follows:

$$\begin{aligned} CDE_{it} = c_{1i} + \textstyle\sum_{i=1}^{k} \alpha_{1ik} FDI_{it-k} + \sum_{i=1}^{k} \beta_{1ik} \Delta GDP_{it-k} + \sum_{i=1}^{k} \gamma_{1ik} \Delta RE_{it-k} + \\ \textstyle\sum_{i=1}^{k} \delta_{1ik} \Delta TO_{it-k} + \sum_{i=1}^{k} \varphi_{1ik} \Delta HDI_{it-k} + \varepsilon_{it} \end{aligned} \tag{3}$$

$$\begin{aligned} FDI_{it} = c_{2i} + \textstyle\sum_{i=1}^{k} \alpha_{1ik} CDE_{it-k} + \sum_{i=1}^{k} \beta_{1ik} \Delta GDP_{it-k} + \sum_{i=1}^{k} \gamma_{1ik} \Delta RE_{it-k} + \\ \textstyle\sum_{i=1}^{k} \delta_{1ik} \Delta TO_{it-k} + \sum_{i=1}^{k} \varphi_{1ik} \Delta HDI_{it-k} + \varepsilon_{it} \end{aligned} \tag{4}$$

$$\begin{aligned} \Delta GDP_{it} = c_{3i} + \textstyle\sum_{i=1}^{k} \alpha_{1ik} FDI_{it-k} + \sum_{i=1}^{k} \beta_{1ik} \mathrm{CDE}_{it-k} + \sum_{i=1}^{k} \gamma_{1ik} \Delta RE_{it-k} + \\ \textstyle\sum_{i=1}^{k} \delta_{1ik} \Delta TO_{it-k} + \sum_{i=1}^{k} \varphi_{1ik} \Delta HDI_{it-k} + \varepsilon_{it} \end{aligned} \tag{5}$$

$$\begin{aligned} \Delta RE_{it} = c_{4i} + \textstyle\sum_{i=1}^{k} \alpha_{1ik} FDI_{it-k} + \sum_{i=1}^{k} \beta_{1ik} \mathrm{CDE}_{it-k} + \sum_{i=1}^{k} \gamma_{1ik} \Delta GDP_{it-k} + \\ \textstyle\sum_{i=1}^{k} \delta_{1ik} \Delta TO_{it-k} + \sum_{i=1}^{k} \varphi_{1ik} \Delta HDI_{it-k} + \varepsilon_{it} \end{aligned} \tag{6}$$

$$\begin{aligned} \Delta TO_{it} = c_{5i} + \textstyle\sum_{i=1}^{k} \alpha_{1ik} FDI_{it-k} + \sum_{i=1}^{k} \beta_{1ik} \mathrm{CDE}_{it-k} + \sum_{i=1}^{k} \gamma_{1ik} \Delta RE_{it-k} + \\ \textstyle\sum_{i=1}^{k} \delta_{1ik} \mathrm{CDE}_{it-k} + \sum_{i=1}^{k} \varphi_{1ik} \Delta HDI_{it-k} + \varepsilon_{it} \end{aligned} \tag{7}$$

$$\begin{aligned} \Delta HDI_{it} = c_{6i} + \textstyle\sum_{i=1}^{k} \alpha_{1ik} FDI_{it-k} + \sum_{i=1}^{k} \beta_{1ik} \mathrm{CDE}_{it-k} + \sum_{i=1}^{k} \gamma_{1ik} \Delta RE_{it-k} + \\ \textstyle\sum_{i=1}^{k} \delta_{1ik} \Delta TO_{it-k} + \sum_{i=1}^{k} \varphi_{1ik} \Delta CDE_{it-k} + \varepsilon_{it} \end{aligned} \tag{8}$$

where Δ represents the first difference; ECT_{t-1} represents the lagged ECT; k represents the lag length; and ε_{it}, υ_{it}, and ε_{it} represent the serially uncorrelated error terms. The direction of panel causations can be identified by testing the coefficients' significance of dependent variables in Equations (3)–(8) [74].

The panel data VECM methodology represents a mix between the traditional VAR approach and the panel-data approach. The VAR model considers all the variables in the system as endogenous, and panel data permits unobserved individual heterogeneity.

In the case where the variables of y_t are cointegrated, according to the cointegrating methodology of Johansen and Juselius (1992), a VECM can be estimated as [84]:

$$\Delta y_t = \prod y_{t-1} + \sum_{j=1}^{k-1} \Gamma_j \Delta y_{t-j} + \varepsilon_t \tag{9}$$

where Δy_t is a 6×1 vector that includes GDP, renewable energy consumption, trade openess, FDI, human development index, and carbon dioxide emission as $[\Delta p_t, \Delta x_t, \ldots, \Delta b_t]$. Γj represents the adjustment on short-run, and Πy_{t-1} represents the error correction term.

The error term ε_t is a vector of innovations that are independent and identically distributed [62]. The error correction term must be significant and negative to highlight the long-run causality [79]. Thus, error correction relates to the last period deviation from long-run equilibrium, influencing the short-run dynamics of the dependent variable [85]. To examine whether the variables are cointegrated, we used the likelihood ratio of maximal eigenvalue test and the trace test [86].

3.1.4. Panel DOLS Estimates

To estimate the regression equation, we considered FMOLS (fully modified OLS) [73], DOLS (dynamic OLS) [87], PMG (pooled mean group estimator) [88], GMM (generalized method of moments) or QML (quasi maximum likelihood). In the case of cointegration, the relationship on the long run can be estimated using the DOLS estimator [74,89,90].

3.2. *Data*

The variables reflecting renewable energy transition used annual data provided by the World Bank database over the time spanning from 1990 to 2018 (Table 1). The length of the period is dictated by the availability of data on energy consumption. Six countries are selected for the sample, representing the non-euro countries (Bulgaria, Czech Republic, Croatia, Hungary, Poland, and Romania). In order to draw an image of the status of renewable energy, most of the studies consider carbon dioxide emissions and renewable energy consumption as the core variables [91–95]. According to the extant literature, the economic development significantly influences the status of renewable energy. The level of economic development leading to renewable energy transition is usually reflected through investments [96–98], GDP [99–102], human development [103–105], and trade openness [106–108]. Therefore, the variables included in the analysis are carbon dioxide emission per capita (CDE), GDP/capita, renewable energy consumption as percentage of total energy consumption (RE), trade openness (TO), FDI, and human development index (HDI).

To examine the characteristics of the sample, the descriptive statistics were calculated and are presented in Table 2. Thus, the average CDE of the European countries in the sample in this study is 6.65 tonnes/capita, the lowest CDE is 3.12 tonnes/capita, and the highest CDE is 14.54 tonnes/capita with a standard deviation of 2.64 tonnes/capita. The average FDI is 4.61%, with the highest value of 54.24% and the lowest value of −40.33% and standard deviation of 8.41%. The medium GDP/capita is USD 15738.71, ranging from USD 4504.19 to USD 41143.09, with a standard deviation of USD 8156.46. The average renewable energy consumption as percentage of total energy consumption is 15.36%, with the lowest value being 1.92% and the highest value being 34.13%, with a standard deviation of 8.75%. HDI registers a mean of 0.78, ranging between 0.67 and 0.9, with a standard deviation of 0.06. Trade openness, the sum of trade as a percentage of GDP, varies between 43.72 and 168.24, with the mean value being 91.73 and a standard value of 32.84.

Table 1. Exhibition of the variables.

Variable	Description	Period	Source
CDE	Carbon dioxide emissions are the result of burning fossil fuels and the manufacturing of cement.	1990–2018	https://data.worldbank.org/indicator/EN.ATM.CO2E.PC
FDI	Net inflows represent the inward direct investment.	1990–2018	https://data.worldbank.org/indicator/BX.KLT.DINV.WD.GD.ZS
GDP	Gross domestic product (GDP) is reported to the size of the population.	1990–2018	https://data.worldbank.org/indicator/NY.GDP.PCAP.PP.CD
RE	Renewable energy consumption represents the share of renewable energy from the total consumption.	1990–2018	https://data.worldbank.org/indicator/EG.FEC.RNEW.ZS
HDI	HDI is a composite index of human development.	1990–2018	https://databank.worldbank.org/Human-development-index/id/363d401b
TO	Trade represents exports and imports as a share of GDP.	1990–2018	https://data.worldbank.org/indicator/NE.TRD.GNFS.ZS

Table 2. Summary Statistics of Dependent and Explanatory Variables.

Statistics	CDE	FDI	GDP	RE	HDI	TO
Mean	6.65	4.61	15738.71	15.37	0.78	91.73
Min.	3.12	−40.33	4504.19	1.92	0.90	43.72
Max.	14.54	54.24	41134.09	34.13	0.67	168.24
Std. Dev.	2.64	8.41	8156.46	8.75	0.06	32.84

In Figure 2, it can be observed that all countries register an ascending trend regarding FDI, RE, FI, GDP, and HDI and a slightly descending trend regarding CDE, from which it can be concluded that all countries have taken measures to achieve the transition to renewable energy. The six countries are homogenous regarding the variables analyzed, not registering large differences among their dynamics.

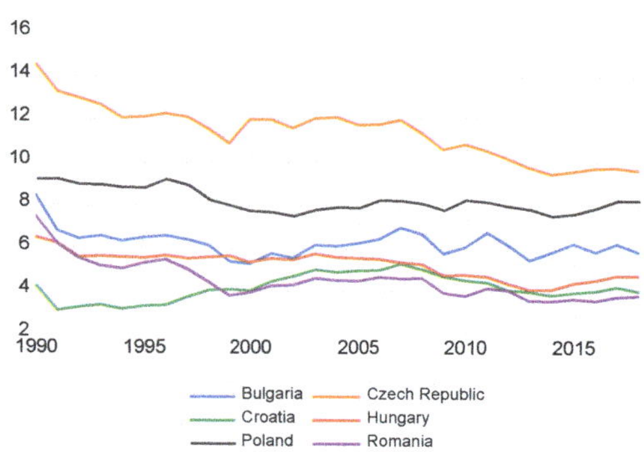

CDE

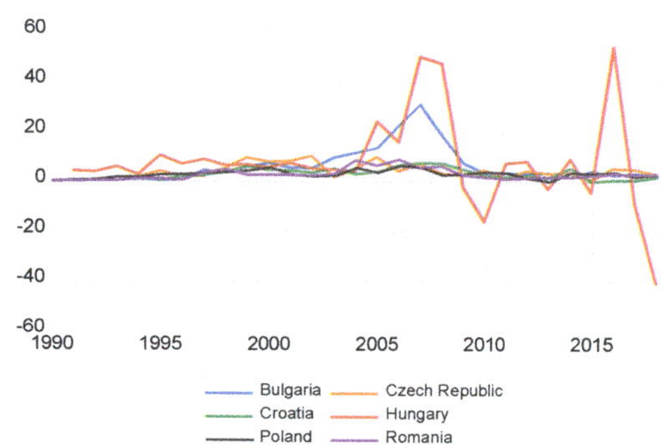

FDI

Figure 2. *Cont.*

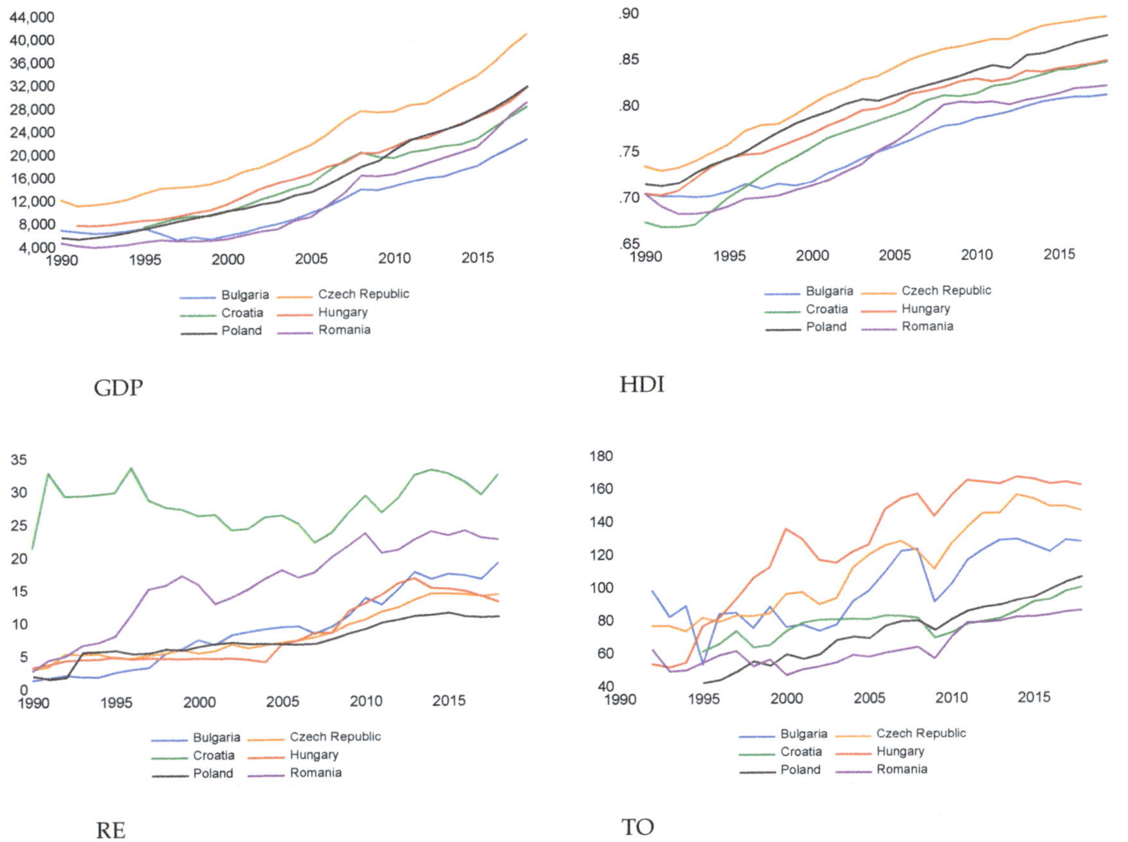

Figure 2. Trends regarding the variables reflecting renewable energy transition for the period 1990–2018. Source: Authors' projection, using Eview.

4. Empirical Results

In time series data, before beginning the analysis of cointegration and causality, the most important requirement is to check the stationarity [109]. In order to test the stationarity, we used Levin, ADT, and PP panel unit root tests for the full sample, the results being presented in Table 3. For four variables (GDP, RE, HDI, and TO), the null hypothesis of a unit root cannot be rejected, being nonstationary and integrated of order one. When structural breaks are considered, using Zivot and Andrews (1992) and Kwiatkowski et al. (1992) tests [110,111], we found that most of cross-sectional units are I(1) series and only few are I(0) in levels. Using the LM panel unit root test, we obtained stable results when considering series integrated of order one.

In order to test the existence of a long-run relationship, we used Pedroni's heterogeneous panel test and Johansen's tests [112]. The Johansen test results are presented in Table 4, indicating that in all countries except Czech Republic, the null hypothesis of no cointegration was rejected at the 10% significance level. In the case of renewable energy transition, Czech Republic does not exhibit a long-run relationship, while Bulgaria, Croatia, Hungary, Poland, and Romania do exhibit long-run relationships.

Table 3. Unit root tests for the full sample.

Variables	Levin—Lin and Chu		ADF—Fisher Chi-Square		PP—Fisher Chi-Square	
	Statistic	Prob.	Statistic	Prob.	Statistic	Prob.
			CDE			
Level	−3.33	0.0004 ***	28.26	0.0051 ***	26.85	0.0081 ***
			FDI			
Level	2.93	0.0017 ***	26.74	0.0084 ***	26.21	0.0100 ***
			GDP			
Level	12.74	1.0000	0.02	1.0000	0.00011	1.0000
First Difference	−15.99	0.0000 ***	191.88	0.0000 ***	224.66	0.0000 ***
			RE			
Level	3.82	0.9999	1.04	1.0000	0.83	1.0000
First Difference	−11.81	0.0000 ***	133.81	0.0000 ***	140.26	0.0000 ***
			HDI			
Level	12.49	1.0000	0.05	1.0000	0.01	1.0000
First Difference	−3.01	0.0013 ***	20.7	0.0500 **	34.51	0.0006 ***
			TO			
Level	4.27	1.0000	1.13	1.0000	0.71	1.0000
First Difference	−11.33	0.0000 ***	127.37	0.0000 ***	127.09	0.0000 ***

Note: ** indicates significance at the 5% level; *** indicates significance at the 1% level.

Table 4. Johansen's cointegration tests.

Country	H_0	Trace Statistics	Prob.	Country	H_0	Trace Statistics	Prob.
Bulgaria	None	60.3246	0.0960 *	Hungary	None	81.52	0.0008 ***
	At most 1	27.93	0.6262		At most 1	52.79	0.0039
Czech Republic	None	50.14	0.4076	Poland	None	81.22	0.0009 ***
	At most 1	30.46	0.4752		At most 1	34.58	0.2626
Croatia	None	67.60	0.0235 **	Romania	None	77.39	0.0024 ***
	At most 1	35.95	0.2081		At most 1	38.06	0.1407

Note: * represents significance at the 10% level; ** represents significance at the 5% level; *** represents significance at the 1% level [113].

In Table 5 are reported the results of the panel cointegration. Except for the Group ρ statistics for the full sample, all the statistics reject the null hypothesis of no cointegration for the sample. Hence, all six test statistics support a panel cointegration relationship among the variables.

Table 5. Heterogeneous panel cointegration results.

	Statistic	Prob.	Weighted Statistic	Prob.
Panel v	1.5383	0.0620 *	0.9761	0.1645
Panel ρ	−0.3419	0.3662	−0.0299	0.4881
Panel *pp*	−4.7129	0.0000 ***	−3.4082	0.0003 ***
Panel ADF	−4.7182	0.0000 ***	−0.5951	0.2759
Group ρ	0.8096	0.7909		
Group *pp*	−4.1510	0.0000 ***		
Group ADF	−4.1261	0.0000 ***		

Note: * indicates significance at the 10% level; *** indicates significance at the 1% level.

In the case of the Kao test, the cointegration is significant at the 1% level of significance (Table 6), confirming a panel cointegration relationship among the variables.

Thus, cointegration tests confirm the existence of a long-term relationships among CDE, GDP, FDI, HDI, RE, and TO in the six non-euro European countries. The *p*-value of 0.000 is less than 0.05; thus, the null hypothesis is rejected. This indicates that the long-run relationships exists among FDI, CDE, TO, GDP, CDE, and HDI. According to Pedroni's and Kao's residual cointegration tests, it is highlighted that variables are cointegrated in the long-term [114].

Table 6. Kao Test ADF.

	t-Statistic	Prob.
ADF	−4.9895	0.0000 ***
Residual Variance	0.1034	
Residual Variance	0.0982	

Note: *** indicates significance at the 1% level.

A panel vector error correction was conducted to see the convergence or the long-run causality. The cointegration equation and error correction revealed the long-run and short-run relationships among FDI, CDE, TO, GDP, CDE, and HDI.

In Table 7 is presented VECM, allowing us to identify short-term and long-term dynamic relationships among variables in the analysis. The variables with a negative sign and significant coefficient present a long-term relationship, and those with a negative sign but a non-significant coefficient present a short-term dynamic relationship. The results indicate a long-term relationship: the p-value is significant and amounts to 0.002%.

Table 7. The Long-Run and Short-Run Relationships.

Cointegrating Equation:			**CointEquation (1)**			
CDE(−1)			1.0000			
DGDP(-1)			−0.0048 (0.0032) [−1.4880]			
DHDI(−1)			−5.8374 (520.478) [−0.011]			
DRE(−1)			18.7188 (2.3329) [8.0239]			
DTO(−1)			1.5830 (0.4041) [3.9176]			
FDI(−1)			−1.1452 (0.2967) [−3.8605]			
C			−7.1369			
Error Correction:	D(CDE)	D(GDP)	D(HDI)	D(RE)	D(DTO)	D(FDI)
CointEq1	−0.0515 (0.0017) [−3.0841]	4.8829 (3.3478) [248.4671]	-1.64×10^{-6} (2.1×10^{-5}) [−0.0776]	−0.0402 (0.0072) [−5.6028]	−0.0824 (0.0477) [−1.7279]	0.1244 (0.0493) [2.5221]

Note: Standard errors in ( ) and t-statistics in [ ].

The error correction part represents the short-run relationship among variables. In the short run, when CDE lies above the long-term balance, the GDP and FDI will increase. A positive relationship between these variables in the short term indicates that the distribution of income and investments within a community group is unequal along with the increase in carbon dioxide emissions. As described earlier, in the long run, there is a negative relationship between CDE and GDP; in the short term, the relationship between the two variables is positive.

According to the result of PVECM above (Table 6), the cointegration equation of the variables is estimated as:

$$
\begin{aligned}
D(CDE) = {} & -0.00514751948054 \times CDE(-1) - 0.00481394966293 \times DGDP(-1) - 5.83742964978 \times DHDI(-1) + \\
& 18.7188391542 \times DRE(-1) + 1.5830023031 \times DTO(-1) - 1.14517261031 \times FDI(-1) - 7.13699144353) - \\
& 0.036777319216 \times D(CDE(-1)) - 0.304188452533 \times D(CDE(-2)) - 2.6766w8886627 \times 10^{-6} \times D(DGDP(-1)) - \\
& 3.61670141248 \times 10^{-5} \times D(DGDP(-2)) + 0.365300917573 \times D(DHDI(-1)) + 2.35080335289 \times D(DHDI(-2)) + \\
& 0.0549792019336 \times D(DRE(-1)) + 0.0182998305835 \times D(DRE(-2)) + 0.00767258447886 \times D(DTO(-1)) + \\
& 0.00238641318271 \times D(DTO(-2)) - 0.00221982795193 \times D(FDI(-1)) - 0.00344475574056 \times D(FDI(-2)) - \\
& 0.0529420314058.
\end{aligned}
\tag{10}
$$

Equation (10) above provides the empirical evidence with respect to the long-run relationships among FDI, CDE, TO, GDP, CDE, and HDI. In the long run, there is a negative relationship between CDE and GDP, HDI, and FDI, but a positive relationship among CDE, TO, and RE. In other words, increasing carbon dioxide emissions in the non-euro European countries, in the long run, encourages the increase of trade openness and renewable energy share, and the decrease of GDP, HDI, and FDI.

In Table 8, the error correction term (ECT) is seen to be negative and significant (−0.0051), indicating convergence, but with dampened fluctuations [115].

Table 8. The Summary of The Panel Vector Error Correction Model (PVECM) Result.

Exogenous Variable	Endogenous Variable D(CDE)	D(GDP)	D(HDI)	D(RE)	D(DTO)	D(FDI)
Coint Equation1	−0.0051 (0.0017) [−3.0841]	4.8829 (3.3478) [1.4585]	-1.64×10^{-6} (2.1×10^{-5}) [−0.0776]	−0.04016 (0.0072) [−5.6028]	−0.0824 (0.0477) [−1.7279]	0.1244 (0.0493) [2.5221]
D(CDE(−1))	−0.0368 (0.0976) [−0.3769]	248.4671 (195.680) [1.2698]	−0.0017 (0.0012) [−1.3588]	−0.1181 (0.4189) [−0.2819]	3.6178 (2.7857) [1.2987]	2.8008 (2.8837) [0.9713]
D(CDE(−2))	−0.3042 (0.0939) [−3.2395]	−333.9738 (188.348) [−1.7891]	−0.0028 (0.0012) [−2.3785]	0.6752 (0.4033) [1.6742]	−5.5992 (−2.6813) [−2.0882]	2.4260 (2.7757) [0.8740]
D(DGDP(−1))	2.68×10^{-6} (4.7×10^{-5}) [−00575]	−0.4135 (0.0933) [−4.4307]	5.82×10^{-8} (5.9×10^{-7}) [0.0988]	−0.0005 (0.0002) [−2.3940]	−0.0036 (0.0013) [−2.7187]	−0.0001 (0.0014) [0.0908]
D(DGDP(−2))	-3.62×10^{-5} (4.7×10^{-5}) 0.7695	−0,2302 (0.0943) [−2.4414]	4.45×10^{-8} (5.9×10^{-7}) [0.0749]	0.0003 (0.0002) [1.3632]	−0.0039 (0.0013) [−2.9463]	−0.0003 (0.0014) [2.5059]
D(DHDI(−1))	0.3653 (6.6272) [0.055]	880.0882 (13292.9) [0.0662]	−0.4109 (0.0838) [4.9025]	32.8285 (28.4614) [1.1534]	439.2954 (189.238) [2.3214]	253.3226 (195.897) [1.2931]
D(DHDI(−2))	2.3508 (6.2056) [0.3788]	14215.29 (12447.3) [1.1420]	−0.1535 (0.0785) [−1.9553]	9.0937 (26.6509) [0.3412]	128.2871 (177.200) [0.7239]	−90.3354 (183.435) [−0.4925]
D(DRE(−1))	0.0549 (0.0259) [2.1169]	−114.6060 (52.0921) [−2.20007]	-2.42×10^{-5} (0.0003) [−0.0737]	−0.1032 (0.1115) [−0.9255]	1.2879 (0.7416) [1.7367]	−1.4869 (0.7677) [−1.9369]
D(DRE(−2))	0.0183 (0.0154) [1.1895]	−53.2202 (30.8580) [−1.7247]	−0.0004 (0.0002) [−2.0982]	−0.0482 (0.0661) [−0.6990]	0.5235 (0.4393) [1.1916]	−0.6676 (0.4548) [−1.4681]
D(DTO(−1))	0.0077 (0.0036) [2.1605]	−5.6168 (7.1232) [−0.7885]	3.07×10^{-5} (4.5×10^{-5}) [−0.6829]	0.0509 (0.0153) [3.3385]	−0.6435 (0.1014) [6.3461]	−0.0157 (0.1049) [−0.1500]
D(DTO(−2))	0.0024 (0.0030) [0.7887]	−1.7762 (6.0692) [−0.2927]	1.89×10^{-5} (3.8×10^{-5}) [0.4928]	0.0412 (0.0129) [3.1681]	−0,2096 (0.0864) [−2.4260]	0.0892 (0.0894) [0.9971]
D(FDI(−1))	−0.0022 (0.0032) [−0.7048]	3.4302 (6.3176) [0.5429]	7.73×10^{-5} (4×10^{-5}) [1.9399]	−0.0322 (0.0135) [−2.3842]	−0.1827 (0.8994) [−2.0317]	−0.3090 (0.0931) [−3.3195]
D(FDI(−2))	−0.0034 (0.0037) [−0.9358]	3.8556 (7.3837) [0.5222]	8.2×10^{-5} (4.7×10^{-5}) [1.7619]	−0.0196 (0.0158) [−1.2378]	−0.3528 (0.1051) [−3.3562]	−0.4176 (0.1088) [−3.8379]
C	−0.0529 (0.0271) [−1.9551]	100.8955 (54.3167) [1.8575]	−0.0003 (0.0003) [−0.9836]	0.0348 (0.1163) [0.2992]	0.8974 (0.7733) [1.1606]	0.0652 (0.8005) [0.0814]

Regarding the causality relationship among the variables, the result of PVECM reveals that GDP and FDI have a negative and significant effect CDE at the one-year horizon and two-year horizon, and the other variables have a positive impact on CDE. The result of the PVECM, which explains the causality relationship among variables, can be seen in Table 8.

Using the impulse response function, the variables response to CDE is fluctuating. The reaction of each endogenous variable to the structural shocks occurring in the exogenous variables is shown in Figure 3 below.

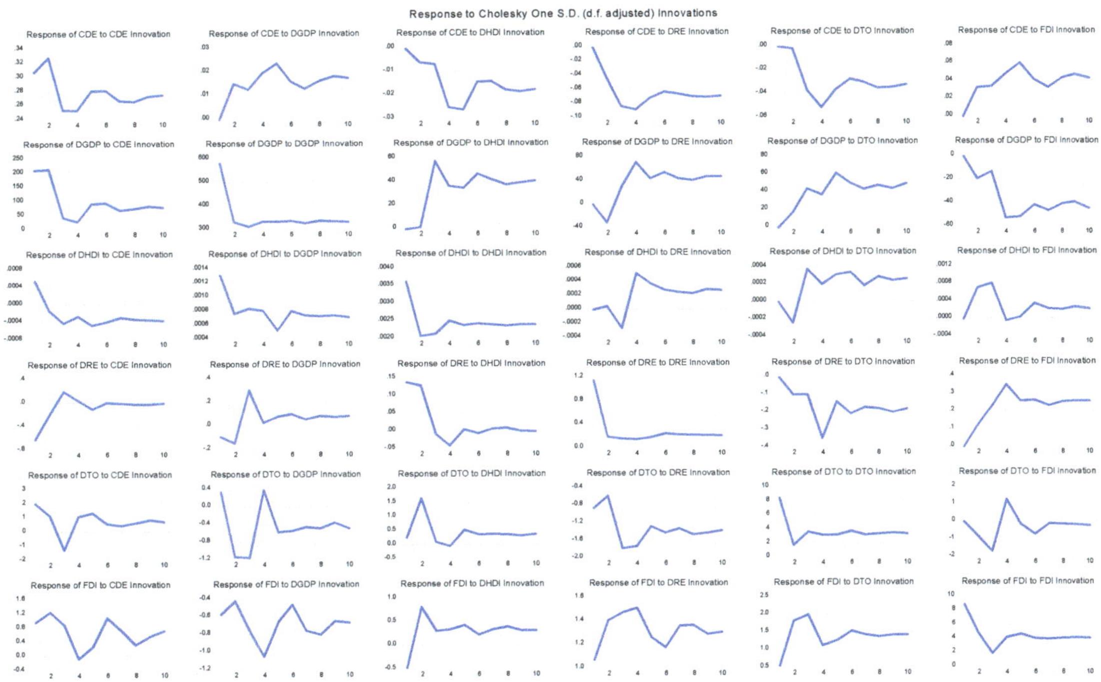

Figure 3. Impulse Response Function.

The Granger causality test is used to determine the causal relationship among variables, and the test results are presented in Table 9.

Table 9. The result of VAR Granger Causality/Block Exogeneity Wald Test.

Dependent Variable	Dependent Variable Excluded					
	D(CDE)	D(DGDP)	D(DHDI)	D(DRE)	D(DTO)	D(FDI)
D(CDE)		0.6442 (0.7246)	0.1457 (0.9297)	4.5015 (0.1053)	5.5941 (0.0610)	0.9698 (0.6157)
D(DGDP)	4.8957 (0.086)		1.3649 (0.5054)	5.0669 (0.0795)	0.7405 (0.6906)	0.3883 (0.8235)
D(DHDI)	7.3905 (0.0248)	0.0116 (0.9942)		6.776 (0.0338)	2.2274 (0.3283)	4.7151 (0.0947)
D(DRE)	2.9003 (0.2345)	11.2895 (0.0035)	1.3317 (9.5138)		12.5853 (0.0018)	5.7075 (0.0576)
D(DTO)	6.1459 (0.0463)	11.9063 (0.0026)	5.4011 (0.0672)	3.0412 (0.2186)		11.5568 (0.0031)
D(FDI)	1.6775 (0.4322)	0.0729 (0.9642)	2.4250 (0.2975)	3.8757 (0.1440)	2.3225 (0.3131)	

Note: The number in parenthesis ( ) is a probability value.

On the basis of the p-value, we conclude that there is statistical significance at the critical value level of 10%, which means that there is a bidirectional relationship or causality between TO and CDE in the long run [116] and unidirectional relationships from GDP to CDE and from HDI to CDE. Conversely, there is no causality running from CDE to GDP or HDI in the case of countries in non-euro area. This information is important because it tells us that the GDP affects the reduction of CDE [112].

The results regarding GDP and CDE reflect a negative and significant relationship, similar to those found in the literature [117–123]. Although the relationship expected in order to achieve renewable energy transition is negative, there are also studies in the literature in which this relation is positive, such as Tucker (1995), Chaabouni and Saidi (2017), and Cederborg and Snöbohm (2016) [124–126]. These studies suggested that a growing GDP leads to increased carbon dioxide emissions, as market economy mechanisms are not enough to lower the emissions. In this context, legal regulations are needed to avoid further environmental degradation because, as some theories claim, emissions start to decrease when a high enough GDP is reached.

The causality from trade openness and CO_2 emissions was found to be bilateral, confirmed also by Esty (2001), Mukhopadhyay (2007), Mukhopadhyay (2009), and Ertugrul et al. (2016) [57,127–129]. The improvement of globalization stimulates the dispersion of environmental technologies worldwide, promoting domestic environmental consciousness among firms and citizens [130]. However, there are studies in the literature, according to which the relationship between trade openness and Co_2 emissions is positive and insignificant [131,132].

The relationship between HDI and carbon dioxide emissions was found to be a unidirectional causality, as indicated in other studies by [133–138]. According to Ranis et al. (2000), there is no static link among these variables; a low HDI is not sustainable into the future, and policy reforms are necessary in order to help maintain at least this low level of human development [139].

FDI was negatively associated with carbon dioxide emissions, such as we found in studies by Tang and Tan (2015), Halicioglu (2009), Ahmed and Long (2012), Suri and Chapman (1998), Hossain (2011), Nahman and Antrobus (2005), Jorgenson (2007), Jorgenson (2009), and Ali et al. (2021) [54,56,140–144]. Although the relationship between FDI and carbon emissions is treated in various studies, the results are not conclusive. Thus, is necessary reinvestigating the association of these indicators for reliable empirical analysis [145].

5. Conclusions

The countries analyzed have made remarkable efforts to change the structures of production and the economic system, moving in 30 years from the centralized economy to the market economy, where their desire to join EU was essential for their development. The process of economic transition has been successfully completed for these countries with their accession to the European Union, but the challenges for them are not over. The necessity to promote the principles of sustainable development requires the entry of these countries into a new transition process, this time an energy transition that involves many changes in economic, social, technical, and environmental fields.

The main goal of this study was to analyze the causality relationship among variables reflecting the energy transition for six selected CEE countries from non-euro areas. The variables selected are carbon dioxide emission per capita (CDE), GDP/capita, renewable energy consumption as percentage of total energy consumption (RE), trade openness (TO), foreign direct investment (FDI), and human development index (HDI). For all selected countries, an ascending trend regarding FDI, RE, TO, GDP, and HDI and a slightly descending trend regarding CDE can be observed for the period of 1990–2018, showing that these economies, as members of European Union, have taken measures to achieve the transition to renewable energy. The six countries are homogenous regarding the variables analyzed, not registering large differences among their dynamics.

Using cross-sectional panel data and employing the panel vector error correction model (PVECM), the key conclusions of the study are as follows: (1) there is a long-run relationship among FDI, CDE, TO, GDP, CDE, and HDI; (2) in the long-run, the CDE is negatively and significantly related to GDP and FDI and positively related to TO and HDI; (3) there is a unidirectional causality running from GDP to CDE and from HDI to CDE and a bidirectional causality between TO and CDE.

The efforts made by these countries by joining the European Union have generated economic development, with positive effects also being recorded on the protection of the environment, a fact due to the strict regulations adopted and rigorous implementation at the national level. Foreign capital had a positive impact on the transition to a low carbon economy because most of the FDI flows attracted by the non-euro countries in the CEE come from Western Europe, i.e., from EU member countries, located either among the founders or among the countries that joined during the first waves of union expansion. As the development progresses, the negative impact of economic activity on the environment is also observed in these countries, with the liberalization of trade and capital movements generating the increase of production capacities. The openness of these countries' economies to foreign direct investment has generated massive capital inflows, especially from European Union countries, with geographical proximity, the existence of common European values, and previous business ties being the main factors that generated significant financial flows in Central and Eastern Europe. The transfer of technology that has accompanied capital flows has not always been up to date, with foreign investors often relocating obsolete technology to countries in Central and Eastern Europe that no longer meet the emission standards of their home countries.

The complexity of the energy transition process requires the involvement of all stakeholders; in addition to public authorities and local companies, an essential role can be played by foreign capital, which can bring high-performance technology and know-how to Central and Eastern European countries. In this way, implementation of energy innovation and an increase in the acceptance of renewable energy among consumers can be achieved. Raising living standards increases consumers' awareness of their role in the transition process and their involvement in the process of saving energy, avoiding the Jevons effect, and using green energy.

The energy transition must be carried out in conditions of energy security, and the military conflict in the area (Ukraine) generated a reconfiguration of the priorities of these countries regarding the use of different energy sources. The need to reduce gas dependence on Russia has led to a reconsideration of the use of coal in the energy mix as well as other energy sources. This proves the fragility of the ambitious objectives set by the European Union and implicitly the six countries analyzed as well as the importance of geostrategic competence in shaping the energy mix in the coming decades.

The study shows the importance of a unitary legal and institutional framework that facilitates both economic growth and the transition to a low carbon economy, where an important role is the foreign capital that comes from developed countries which brings to the host countries not only financial resources but also know-how, a certain organizational culture, and a new approach to business strategy under the banner of sustainable development.

The limitations of the research are given by the selection of the number of countries, by the indicators used as independent variables, and by the period chosen for the analysis. In the future, this research could be extended to all former communist countries in Europe and the analysis period extended to better capture the efforts made by these countries in both the process of economic transition and the process of energy transition. Other indicators, such as natural gas consumption per capita, can be taken as independent variables to capture the energy dependence of these countries on Russia—with which they had a special economic and political relationship during the communist period—and their reorientation towards increasing energy security, given the military context in the area.

Author Contributions: Conceptualization, S.A.A., M.P. and D.B.-L.; data curation, S.A.A. and D.B.-L.; formal analysis, S.A.A., M.P. and D.B.-L.; investigation, S.A.A., M.P. and D.F.; methodology, S.A.A. and D.B.-L.; project administration, S.A.A. and M.P.; resources, S.A.A., M.P., D.B.-L., D.F. and I.G.R.; software, S.A.A.; supervision, S.A.A., M.P., D.F. and I.G.R.; validation, S.A.A., M.P., D.B.-L., D.F. and I.G.R.; visualization, S.A.A., M.P., D.B.-L., D.F. and I.G.R.; writing—original draft, S.A.A. and M.P.; writing—review and editing, S.A.A., M.P., D.B.-L., D.F. and I.G.R. All authors have read and agreed to the published version of the manuscript.

Funding: This research received no external funding.

Data Availability Statement: The authors used public data for the study.

Acknowledgments: Diogo Ferraz acknowledges the Research Stay from the Deutscher Akademischer Austauschdienst (DAAD) for University Academics and Scientists, 2022, n. 91834685; and the National Council for Scientific and Technological Development–CNPq Process CNPq/MCTI/FNDCT N° 18/2021-Faixa A-Grupos Emergentes.

Conflicts of Interest: The authors declare no conflict of interest.

References

1. Vasile, V. Romania: A country under permanent public sector reform. In *Public Sector Shock*; Edward Elgar Publishing: Cheltenham, UK, 2013.
2. Zaman, G.; Vasile, V. Economic-financial and social vulnerabilities of Romania in the period 2013–2020. *Procedia Econ. Financ.* **2014**, *15*, 4–18. [CrossRef]
3. Cătălin, V.M. Changes of EU industrial policy and its effects on Romania's industrial paradigm. *Industrija* **2016**, *44*, 197–210. [CrossRef]
4. Iacob, L.Ș.V.; Dumbravă, Ș.G. The analysis of the main aspects regarding the industrial activity in Romania. *Rev. Română Stat.-Supl. Nr* **2020**, *3*, 40–54.
5. Varblane, U.; Vahter, P. An analysis of the economic convergence process in the transition countries. *Univ. Tartu Econ. Bus. Work. Pap.* **2005**, *2005*, 37. [CrossRef]
6. Manta, O. *Competitiveness, Entrepreneurship and Economic Convergence (CEEC)*; LAP LAMBERT Academic Publishing: Riga, Latvia, 2020.
7. Manta, O. The Eurozone in the context of national unity and the principles of global democratic society. *J. Lib. Conștiință* **2018**, *6*, 454–492.
8. Marelli, E.; Parisi, M.L.; Signorelli, M. Economic convergence in the EU and Eurozone. *J. Econ. Stud.* **2019**, *46*, 1332–1344. [CrossRef]
9. Głodowska, A.; Pera, B. On the relationship between economic integration, business environment and real convergence: The experience of the CEE countries. *Economies* **2019**, *7*, 54. [CrossRef]
10. Ioan, B.; Mozi, R.M.; Lucian, G.; Gheorghe, F.; Horia, T.; Ioan, B.; Mircea-Iosif, R. An empirical investigation on determinants of sustainable economic growth. Lessons from Central and Eastern European Countries. *J. Risk Financ. Manag.* **2020**, *13*, 146. [CrossRef]
11. Munteanu, I.; Grigorescu, A.; Condrea, E.; Pelinescu, E. Convergent insights for sustainable development and ethical cohesion: An empirical study on corporate governance in Romanian public entities. *Sustainability* **2020**, *12*, 2990. [CrossRef]
12. Matei, M. *Foreign Direct Investments. Functions and Evolutions*; Expert Publishing House: Bucharest, Romania, 2004.
13. Surugiu, M.R.; Surugiu, C. International trade, globalization and economic interdependence between European countries: Implications for businesses and marketing framework. *Procedia Econ. Financ.* **2015**, *32*, 131–138. [CrossRef]
14. Comes, C.A.; Bunduchi, E.; Vasile, V.; Stefan, D. The impact of foreign direct investments and remittances on economic growth: A case study in Central and Eastern Europe. *Sustainability* **2018**, *10*, 238. [CrossRef]
15. Çela, A.; Hysa, E.; Voica, M.C.; Panait, M.; Manta, O. Internationalization of Large Companies from Central and Eastern Europe or the Birth of New Stars. *Sustainability* **2021**, *14*, 261. [CrossRef]
16. Zaman, G.; Vasile, V.; Matei, M.; Croitoru, C.; Enescu, G. Some challenging (macro) economic aspects of FDI in Romania. *Rev. Romana Econ.* **2011**, *33*, 21–58.
17. Chivu, L.; Ciutacu, C.; Georgescu, G. *Deindustrialization and Reindustrialization in Romania*; Palgrave Macmillan: London, UK, 2017.
18. Ciutacu, C.; Chivu, L. Romania's Deindustrialisation. From the "Golden Age" to the "Iron Scrap Age". *Procedia Econ. Financ.* **2015**, *22*, 209–215. [CrossRef]
19. Zaman, G.; Vasile, V. Macroeconomic impact of FDI in Romania. *Procedia Econ. Financ.* **2012**, *3*, 3–11. [CrossRef]
20. Morina, F.; Ergün, U.; Hysa, E. Understanding Drivers of Renewable Energy Firm's Performance. *Environ. Res. Eng. Manag.* **2021**, *77*, 32–49. [CrossRef]
21. Platon, V.; Manea, G.; Antonescu, D. *Disparităţi Regionale şi Dezvoltarea Industrială în România*; Institutul de Economie a Industriei: Academia, Română, 1998.

22. Chivu, L.; Ciutacu, C.; Georgescu, G. *Descompunerea şi Recompunerea Structurilor Industriale din România*; Centrul de informare și Documentare Economică: București, Romania, 2016.
23. Vasile, V.; Balan, M. Impact of greenhouse effect gases on climatic changes. Measurement indicators and forecast models. *Ann. Univ. Apulensis Ser. Oeconom.* **2008**, *2*, 1–19.
24. Andrei, J.V.; Mieila, M.; Panait, M. The impact and determinants of the energy paradigm on economic growth in European Union. *PLoS ONE* **2017**, *12*, e0173282. [CrossRef]
25. Anghelache, C.; Anghel, M.G. Analysis of Romania's strategy of alignment with the EU environmental directives. *Rom. Stat. Rev. Suppl.* **2017**, *65*, 116–127.
26. Bucur, C.; Tudorică, B.G.; Oprea, S.V.; Nancu, D.; Duşmănescu, D.M. Insights Into Energy Indicators Analytics Towards European Green Energy Transition Using Statistics and Self-Organizing Maps. *IEEE Access* **2021**, *9*, 64427–64444. [CrossRef]
27. Popescu, G.H.; Mieila, M.; Nica, E.; Andrei, J.V. The emergence of the effects and determinants of the energy paradigm changes on European Union economy. *Renew. Sustain. Energy Rev.* **2018**, *81*, 768–774. [CrossRef]
28. Zaharia, M.; Pătrașcu, A.; Gogonea, M.R.; Tănăsescu, A.; Popescu, C. A cluster design on the influence of energy taxation in shaping the new EU-28 economic paradigm. *Energies* **2017**, *10*, 257. [CrossRef]
29. Simionescu, M. The nexus between economic development and pollution in the European Union new member states. The role of renewable energy consumption. *Renew. Energy* **2021**, *179*, 1767–1780. [CrossRef]
30. Wang, Q.; Li, S.; Li, R.; Jiang, F. Underestimated impact of the COVID-19 on carbon emission reduction in developing countries–a novel assessment based on scenario analysis. *Environ. Res.* **2023**, *204*, 111990. [CrossRef]
31. Mentel, G.; Tarczyński, W.; Dylewski, M.; Salahodjaev, R. Does Renewable Energy Sector Affect Industrialization-CO_2 Emissions Nexus in Europe and Central Asia? *Energies* **2022**, *15*, 5877. [CrossRef]
32. Morina, F.; Ergün, U.; Balomenou, C. The Impact of Renewable Energy Policies and Financial Development on CO_2 Emissions and Economic Growth in EU Countries. In *The Changing Financial Landscape*; Springer: Cham, Switzerland, 2021; pp. 97–107.
33. Adebayo, T.S.; Rjoub, H.; Akadiri, S.S.; Oladipupo, S.D.; Sharif, A.; Adeshola, I. The role of economic complexity in the environmental Kuznets curve of MINT economies: Evidence from method of moments quantile regression. *Environ. Sci. Pollut. Res.* **2021**, 1–13. [CrossRef]
34. Albulescu, C.T.; Boatca-Barabas, M.E.; Diaconescu, A. The asymmetric effect of environmental policy stringency on CO_2 emissions in OECD countries. *Environ. Sci. Pollut. Res.* **2022**, *29*, 27311–27327. [CrossRef]
35. Al-Mulali, U.; Ozturk, I.; Solarin, S.A. Investigating the environmental Kuznets curve hypothesis in seven regions: The role of renewable energy. *Ecol. Indic.* **2016**, *67*, 267–282. [CrossRef]
36. Khan SA, R.; Zaman, K.; Zhang, Y. The relationship between energy-resource depletion, climate change, health resources and the environmental Kuznets curve: Evidence from the panel of selected developed countries. *Renew. Sustain. Energy Rev.* **2016**, *62*, 468–477. [CrossRef]
37. Panait, M.; Voica, M.C.; Rădulescu, I. Approaches regarding environmental Kuznets curve in the European Union from the perspective of sustainable development. *Appl. Ecol. Environ. Res.* **2019**, *17*, 6801–6820. [CrossRef]
38. Anser, M.K.; Yousaf, Z.; Nassani, A.A.; Abro MM, Q.; Zaman, K. International tourism, social distribution, and environmental Kuznets curve: Evidence from a panel of G-7 countries. *Environ. Sci. Pollut. Res.* **2020**, *27*, 2707–2720. [CrossRef] [PubMed]
39. Sharif, A.; Afshan, S.; Chrea, S.; Amel, A.; Khan, S.A.R. The role of tourism, transportation and globalization in testing environmental Kuznets curve in Malaysia: New insights from quantile ARDL approach. *Environ. Sci. Pollut. Res.* **2020**, *27*, 25494–25509. [CrossRef] [PubMed]
40. Sharma, G.D.; Tiwari, A.K.; Erkut, B.; Mundi, H.S. Exploring the nexus between non-renewable and renewable energy consumptions and economic development: Evidence from panel estimations. *Renew. Sustain. Energy Rev.* **2021**, *146*, 111152. [CrossRef]
41. Wang, Q.; Zhang, F.; Li, R. Revisiting the environmental kuznets curve hypothesis in 208 counties: The roles of trade openness, human capital, renewable energy and natural resource rent. *Environ. Res.* **2023**, *216*, 114637. [CrossRef]
42. Li, R.; Yang, T.; Wang, Q. Does income inequality reshape the environmental Kuznets curve (EKC) hypothesis? A nonlinear panel data analysis. *Environ. Res.* **2023**, *216*, 114575.
43. Balsalobre-Lorente, D.; Ibáñez-Luzón, L.; Usman, M.; Shahbaz, M. The environmental Kuznets curve, based on the economic complexity, and the pollution haven hypothesis in PIIGS countries. *Renew. Energy* **2022**, *185*, 1441–1455. [CrossRef]
44. Wang, Q.; Wang, L.; Li, R. Trade protectionism jeopardizes carbon neutrality–Decoupling and breakpoints roles of trade openness. *Sustain. Prod. Consum.* **2023**, *35*, 201–215. [CrossRef]
45. Li, R.; Wang, Q.; Liu, Y.; Jiang, R. Per-capita carbon emissions in 147 countries: The effect of economic, energy, social, and trade structural changes. *Sustain. Prod. Consum.* **2021**, *27*, 1149–1164. [CrossRef]
46. Zaman, G.; Grădinaru, G.; Neagoe, I. The impact of the COVID-19 crisis on the manifestation of the rebound effect in energy consumption. *Rom. J. Econ.* **2020**, *50*, 29–46.
47. Armeanu, D.Ş.; Gherghina, Ş.C.; Pasmangiu, G. Exploring the causal nexus between energy consumption, environmental pollution and economic growth: Empirical evidence from central and Eastern Europe. *Energies* **2019**, *12*, 3704. [CrossRef]
48. Przychodzen, W.; Przychodzen, J. Determinants of renewable energy production in transition economies: A panel data approach. *Energy* **2020**, *191*, 116583. [CrossRef]

49. Marinaș, M.C.; Dinu, M.; Socol, A.G.; Socol, C. Renewable energy consumption and economic growth. Causality relationship in Central and Eastern European countries. *PLoS ONE* **2018**, *13*, e0202951. [CrossRef] [PubMed]
50. Butnaru, G.I.; Haller, A.P.; Clipa, R.I.; Ștefănică, M.; Ifrim, M. The nexus between convergence of conventional and renewable energy consumption in the present European Union states. Explorative study on parametric and semi-parametric methods. *Energies* **2020**, *13*, 5272. [CrossRef]
51. Ćetković, S.; Buzogány, A. The political economy of EU climate and energy policies in Central and Eastern Europe revisited: Shifting coalitions and prospects for clean energy transitions. *Politics Gov.* **2019**, *7*, 124. [CrossRef]
52. Lee, J.W. The contribution of foreign direct investment to clean energy use, carbon emissions and economic growth. *Energy Policy* **2013**, *55*, 483–489. [CrossRef]
53. Sbia, R.; Shahbaz, M.; Hamdi, H. A contribution of foreign direct investment, clean energy, trade openness, carbon emissions and economic growth to energy demand in UAE. *Econ. Model.* **2014**, *36*, 191–197. [CrossRef]
54. Tang, C.F.; Tan, B.W. The impact of energy consumption, income and foreign direct investment on carbon dioxide emissions in Vietnam. *Energy* **2015**, *79*, 447–454. [CrossRef]
55. Rahman, H.U.; Ghazali, A.; Bhatti, G.A.; Khan, S.U. Role of economic growth, financial development, trade, energy and FDI in environmental Kuznets curve for Lithuania: Evidence from ARDL bounds testing approach. *Eng. Econ.* **2020**, *31*, 39–49. [CrossRef]
56. Halicioglu, F. An econometric study of CO_2 emissions, energy consumption, income and foreign trade in Turkey. *Energy Policy* **2009**, *37*, 1156–1164. [CrossRef]
57. Mukhopadhyay, K. Trade and the Environment: Implications for Climate Change. *Decision* **2009**, *36*, 0304–0941.
58. Dogan, E.; Turkekul, B. CO_2 emissions, real output, energy consumption, trade, urbanization and financial development: Testing the EKC hypothesis for the USA. *Environ. Sci. Pollut. Res.* **2016**, *23*, 1203–1213. [CrossRef] [PubMed]
59. Jebli, M.B.; Youssef, S.B.; Ozturk, I. Testing environmental Kuznets curve hypothesis: The role of renewable and non-renewable energy consumption and trade in OECD countries. *Ecol. Indic.* **2016**, *60*, 824–831. [CrossRef]
60. Destek, M.A.; Sinha, A. Renewable, non-renewable energy consumption, economic growth, trade openness and ecological footprint: Evidence from organisation for economic Co-operation and development countries. *J. Clean. Prod.* **2020**, *242*, 118537. [CrossRef]
61. Sims, C.A. Macroeconomics and reality. *Econom. J. Econom. Soc.* **1980**, *48*, 1–48. [CrossRef]
62. Sun, L.; Ford, J.L.; Dickinson, D.G. Bank loans and the effects of monetary policy in China: VAR/VECM approach. *China Econ. Rev.* **2010**, *21*, 65–97. [CrossRef]
63. Dickey, D.A.; Fuller, W.A. Distribution of the estimators for autoregressive time series with a unit root. *J. Am. Stat. Assoc.* **1979**, *74*, 427–431.
64. Costantini, V.; Martini, C. The causality between energy consumption and economic growth: A multi-sectoral analysis using non-stationary cointegrated panel data. *Energy Econ.* **2010**, *32*, 591–603. [CrossRef]
65. Levin, A.; Lin, C.F.; Chu, C.S.J. Unit root tests in panel data: Asymptotic and finite-sample properties. *J. Econom.* **2002**, *108*, 1–24. [CrossRef]
66. Im, K.S.; Pesaran, M.H.; Shin, Y. Testing for unit roots in heterogeneous panels. *J. Econom.* **2003**, *115*, 53–74. [CrossRef]
67. Breitung, J. *The Local Power of Some Unit Root Tests for Panel Data*; Emerald Group Publishing Limited: Bingley, UK, 2001.
68. Maddala, G.S.; Wu, S. A comparative study of unit root tests with panel data and a new simple test. *Oxf. Bull. Econ. Stat.* **1999**, *61*, 631–652. [CrossRef]
69. Choi, I. Unit root tests for panel data. *J. Int. Money Financ.* **2001**, *20*, 249–272. [CrossRef]
70. Hadri, K. Testing for stationarity in heterogeneous panel data. *Econom. J.* **2000**, *3*, 148–161. [CrossRef]
71. Phillips, P.C.; Perron, P. Testing for a unit root in time series regression. *Biometrika* **1988**, *75*, 335–346. [CrossRef]
72. Pedroni, P. Critical values for cointegration tests in heterogeneous panels with multiple regressors. *Oxf. Bull. Econ. Stat.* **1999**, *61*, 653–670. [CrossRef]
73. Pedroni, P. Fully modified OLS for the heterogeneous cointegrated panels. *Adv. Econom.* **2000**, *15*, 93–130.
74. Mahmoodi, M. The relationship between economic growth, renewable energy, and CO_2 emissions: Evidence from panel data approach. *Int. J. Energy Econ. Policy* **2017**, *7*, 96.
75. Kao, C.; Chiang, M.H.; Chen, B. International R&D spillovers: An application of estimation and inference in panel cointegration. *Oxf. Bull. Econ. Stat.* **1999**, *61*, 691–709.
76. Persyn, D.; Westerlund, J. Error-correction–based cointegration tests for panel data. *STATA J.* **2008**, *8*, 232–241. [CrossRef]
77. Westerlund, J. New simple tests for panel cointegration. *Econom. Rev.* **2005**, *24*, 297–316. [CrossRef]
78. Westerlund, J.; Edgerton, D.L. A simple test for cointegration in dependent panels with structural breaks. *Oxf. Bull. Econ. Stat.* **2008**, *70*, 665–704. [CrossRef]
79. Nkalu, C.N.; Ugwu, S.C.; Asogwa, F.O.; Kuma, M.P.; Onyeke, Q.O. Financial development and energy consumption in Sub-Saharan Africa: Evidence from panel vector error correction model. *Sage Open* **2020**, *10*, 2158244020935432. [CrossRef]
80. Luetkepohl, H. Vector Autoregressive Models (Working Paper ECO 2011/30). European University Institute. 2011. Available online: https://cadmus.eui.eu/bitstream/handle/1814/19354/ECO_2011_30.pdf (accessed on 20 March 2022).
81. Granger, C. Some recent development in a concept of causality. *J. Econom.* **1988**, *39*, 199–211. [CrossRef]
82. Johansen, S. Statistical analysis of cointegration vectors. *J. Econ. Dyn. Control.* **1988**, *12*, 231–254. [CrossRef]

83. Engle, R.F.; Granger, C.W. Co-integration and error correction: Representation, estimation, and testing. *Econom. J. Econom. Soc.* **1987**, 251–276. [CrossRef]
84. Johansen, S.; Juselius, K. Testing structural hypotheses in a multivariate cointegration analysis of the PPP and the UIP for UK. *J. Econom.* **1992**, *53*, 211–244. [CrossRef]
85. Obi, P.; Sil, R.; Abuizam, R. Tourism stocks, implied volatility and hedging: A vector error correction study. *J. Account. Financ.* **2015**, *15*, 30–39.
86. Kuo, C.Y. Does the vector error correction model perform better than others in forecasting stock price? An application of residual income valuation theory. *Econ. Model.* **2016**, *52*, 772–789. [CrossRef]
87. Saikkonen, P. Asymptotically efficient estimation of cointegration regressions. *Econom. Theory* **1991**, *7*, 1–21. [CrossRef]
88. Pesaran, M.H.; Shin, Y.; Smith, R.P. Pooled mean group estimation of dynamic heterogeneous panels. *J. Am. Stat. Assoc.* **1999**, *94*, 621–634. [CrossRef]
89. Chiang, M.H.; Kao, C. *Nonstationary Panel Time Series Using NPT 1.3—A User Guide*; Center for Policy Research, Syracuse University: Syracuse, NY, USA, 2002.
90. Breitung, J.; Pesaran, M.H. Unit roots and cointegration in panels. In *The Econometrics of Panel Data*; Springer: Berlin/Heidelberg, Germany, 2008; pp. 279–322.
91. Leitão, N.C. Economic growth, carbon dioxide emissions, renewable energy and globalization. *Int. J. Energy Econ. Policy* **2014**, *4*, 391–399.
92. Long, X.; Naminse, E.Y.; Du, J.; Zhuang, J. Nonrenewable energy, renewable energy, carbon dioxide emissions and economic growth in China from 1952 to 2012. *Renew. Sustain. Energy Rev.* **2015**, *52*, 680–688. [CrossRef]
93. Adams, S.; Nsiah, C. Reducing carbon dioxide emissions; Does renewable energy matter? *Sci. Total Environ.* **2019**, *693*, 133288. [CrossRef] [PubMed]
94. Liu, X.; Zhang, S.; Bae, J. The impact of renewable energy and agriculture on carbon dioxide emissions: Investigating the environmental Kuznets curve in four selected ASEAN countries. *J. Clean. Prod.* **2017**, *164*, 1239–1247. [CrossRef]
95. Wang, Z.; Jebli, M.B.; Madaleno, M.; Doğan, B.; Shahzad, U. Does export product quality and renewable energy induce carbon dioxide emissions: Evidence from leading complex and renewable energy economies. *Renew. Energy* **2021**, *171*, 360–370. [CrossRef]
96. Wiser, R.H.; Pickle, S.J. Financing investments in renewable energy: The impacts of policy design. *Renew. Sustain. Energy Rev.* **1998**, *2*, 361–386. [CrossRef]
97. Bergmann, A.; Hanley, N.; Wright, R. Valuing the attributes of renewable energy investments. *Energy Policy* **2006**, *34*, 1004–1014. [CrossRef]
98. Ozorhon, B.; Batmaz, A.; Caglayan, S. Generating a framework to facilitate decision making in renewable energy investments. *Renew. Sustain. Energy Rev.* **2018**, *95*, 217–226. [CrossRef]
99. Chien, T.; Hu, J.L. Renewable energy: An efficient mechanism to improve GDP. *Energy Policy* **2008**, *36*, 3045–3052. [CrossRef]
100. Al-Mulali, U.; Fereidouni, H.G.; Lee, J.Y.; Sab, C.N.B.C. Examining the bi-directional long run relationship between renewable energy consumption and GDP growth. *Renew. Sustain. Energy Rev.* **2013**, *22*, 209–222. [CrossRef]
101. Amri, F. The relationship amongst energy consumption (renewable and non-renewable), and GDP in Algeria. *Renew. Sustain. Energy Rev.* **2017**, *76*, 62–71. [CrossRef]
102. Azam, A.; Rafiq, M.; Shafique, M.; Zhang, H.; Yuan, J. Analyzing the effect of natural gas, nuclear energy and renewable energy on GDP and carbon emissions: A multi-variate panel data analysis. *Energy* **2021**, *219*, 119592. [CrossRef]
103. Adekoya, O.B.; Olabode, J.K.; Rafi, S.K. Renewable energy consumption, carbon emissions and human development: Empirical comparison of the trajectories of world regions. *Renew. Energy* **2021**, *179*, 1836–1848. [CrossRef]
104. Wang, Z.; Bui, Q.; Zhang, B.; Nawarathna CL, K.; Mombeuil, C. The nexus between renewable energy consumption and human development in BRICS countries: The moderating role of public debt. *Renew. Energy* **2021**, *165*, 381–390. [CrossRef]
105. Hashemizadeh, A.; Ju, Y. Optimizing renewable energy portfolios with a human development approach by fuzzy interval goal programming. *Sustain. Cities Soc.* **2021**, *75*, 103396. [CrossRef]
106. Sebri, M.; Ben-Salha, O. On the causal dynamics between economic growth, renewable energy consumption, CO_2 emissions and trade openness: Fresh evidence from BRICS countries. *Renew. Sustain. Energy Rev.* **2014**, *39*, 14–23. [CrossRef]
107. Zeren, F.; Akkuş, H. The relationship between renewable energy consumption and trade openness: New evidence from emerging economies. *Renew. Energy* **2020**, *147*, 322–329. [CrossRef]
108. Wang, Q.; Zhang, F. Free trade and renewable energy: A cross-income levels empirical investigation using two trade openness measures. *Renew. Energy* **2021**, *168*, 1027–1039. [CrossRef]
109. Agrawal, G. Foreign direct investment and economic growth in BRICS economies: A panel data analysis. *J. Econ. Bus. Manag.* **2015**, *3*, 421–424. [CrossRef]
110. Zivot, E.; Andrews, D.W.K. Further evidence on the great crash, the oil-price shock, and the unit-root hypothesis. *J. Bus. Econ. Stat.* **2002**, *20*, 25–44. [CrossRef]
111. Kwiatkowski, D.; Phillips, P.C.; Schmidt, P.; Shin, Y. Testing the null hypothesis of stationarity against the alternative of a unit root: How sure are we that economic time series have a unit root? *J. Econom.* **1992**, *54*, 159–178. [CrossRef]
112. Mahadevan, R.; Asafu-Adjaye, J. Energy consumption, economic growth and prices: A reassessment using panel VECM for developed and developing countries. *Energy Policy* **2007**, *35*, 2481–2490. [CrossRef]

113. MacKinnon, J.G.; Haug, A.A.; Michelis, L. Numerical distribution functions of likelihood ratio tests for cointegration. *J. Appl. Econom.* **1999**, *14*, 563–577. [CrossRef]
114. Amri, K. Is there causality relationship between economic growth and income inequality?: Panel data evidence from Indonesia. *Eurasian J. Econ. Financ.* **2018**, *6*, 8–20. [CrossRef]
115. Narayan, P.K.; Smyth, R. What determines migration flows from low-income to high-income countries? An empirical investigation of Fiji–Us migration 1972–2001. *Contemp. Econ. Policy* **2006**, *24*, 332–342. [CrossRef]
116. Alam, M.; Rabbani, M.R.; Tausif, M.R.; Abey, J. Banks' performance and economic growth in India: A panel cointegration analysis. *Economies* **2021**, *9*, 38. [CrossRef]
117. Kurtovic, S.; Siljkovic, B.; Milanovic, M. Long-term impact of foreign direct investment on reduction of unemployment: Panel data analysis of the Western Balkans countries. *J. Appl. Econ. Bus. Res.* **2015**, *5*, 112–129.
118. Ahmed, K.; Long, W. Environmental Kuznets curve and Pakistan: An empirical analysis. *Procedia Econ. Financ.* **2012**, *1*, 4–13. [CrossRef]
119. Asumadu-Sarkodie, S.; Owusu, P.A. Carbon dioxide emissions, GDP, energy use and population growth: A multivariate and causality analysis for Ghana, 1971–2013. *Environ. Sci. Pollut. Res.* **2016**, *23*, 13508–13520. [CrossRef]
120. Chang, C.-C. A multivariate causality test of carbon dioxide emissions, energy consumption and economic growth in China. *Appl. Energy* **2010**, *87*, 3533–3537. [CrossRef]
121. Grossman, G.; Krueger, A. *Environmental Impacts of a North American Free Trade Agreement*; National Bureau of Economic Research: Cambridge, MA, USA, 1991; p. w3914.
122. Panayotou, T. *Empirical Tests and Policy Analysis of Environmental Degradation at Different Stages of Economic Development*; International Labour Organization: Geneva, Switzerland, 1993.
123. Mohiuddin, O.; Asumadu-Sarkodie, S.; Obaidullah, M. The relationship between carbon dioxide emissions, energy consumption, and GDP: A recent evidence from Pakistan. *Cogent Eng.* **2016**, *3*, 1210491. [CrossRef]
124. Tucker, M. Carbon dioxide emissions and global GDP. *Ecol. Econ.* **1995**, *15*, 215–223. [CrossRef]
125. Chaabouni, S.; Saidi, K. The dynamic links between carbon dioxide (CO_2) emissions, health spending and GDP growth: A case study for 51 countries. *Environ. Res.* **2017**, *158*, 137–144. [CrossRef] [PubMed]
126. Cederborg, J.; Snöbohm, S. Is There a Relationship between Economic Growth and Carbon Dioxide Emissions? 2016. Available online: https://www.diva-portal.org/smash/get/diva2:1076315/FULLTEXT01.pdf (accessed on 20 March 2022).
127. Esty, D.C. Bridging the trade-environment divide. *J. Econ. Perspect.* **2001**, *15*, 113–130. [CrossRef]
128. Mukhopadhyay, K. *Trade and Environment in Thailand: An Emerging Economy*; Serials Publications: New Delhi, India, 2007.
129. Ertugrul, H.M.; Cetin, M.; Seker, F.; Dogan, E. The impact of trade openness on global carbon dioxide emissions: Evidence from the top ten emitters among developing countries. *Ecol. Indic.* **2016**, *67*, 543–555. [CrossRef]
130. Dauda, L.; Long, X.; Mensah, C.N.; Salman, M.; Boamah, K.B.; Ampon-Wireko, S.; Dogbe, C.S.K. Innovation, trade openness and CO_2 emissions in selected countries in Africa. *J. Clean. Prod.* **2021**, *281*, 125143. [CrossRef]
131. Mahmood, H.; Maalel, N.; Zarrad, O. Trade openness and CO_2 emissions: Evidence from Tunisia. *Sustainability* **2019**, *11*, 3295. [CrossRef]
132. Shahbaz, M.; Tiwari, A.K.; Nasir, M. The effects of financial development, economic growth, coal consumption and trade openness on CO_2 emissions in South Africa. *Energy Policy* **2013**, *61*, 1452–1459. [CrossRef]
133. *Greenhouse Gas Emissions*; United States Environmental Protection Agency: Washington, DC, USA, 2019.
134. Churchill, S.A.; Inekwe, J.; Ivanovski, K.; Smyth, R. The environmental Kuznets curve in the OECD: 1870–2014. *Energy Econ.* **2018**, *75*, 389–399. [CrossRef]
135. Zakarya, G.Y.; Mostefa, B.; Abbes, S.M.; Seghir, G.M. Factors affecting CO_2 emissions in the BRICS countries: A panel data analysis. *Procedia Econ. Financ.* **2015**, *26*, 114–125. [CrossRef]
136. Le, H.P.; Ozturk, I. The impacts of globalization, financial development, government expenditures, and institutional quality on CO_2 emissions in the presence of environmental Kuznets curve. *Environ. Sci. Pollut. Res.* **2020**, *27*, 22680–22697. [CrossRef] [PubMed]
137. Alotaibi, A.A.; Alajlan, N. Using Quantile Regression to Analyze the Relationship between Socioeconomic Indicators and Carbon Dioxide Emissions in G20 Countries. *Sustainability* **2021**, *13*, 7011. [CrossRef]
138. Ribeiro, H.V.; Rybski, D.; Kropp, J.P. Effects of changing population or density on urban carbon dioxide emissions. *Nat. Commun.* **2019**, *10*, 1–9. [CrossRef] [PubMed]
139. Neumayer, E. The human development index and sustainability—A constructive proposal. *Ecol. Econ.* **2001**, *39*, 101–114. [CrossRef]
140. Suri, V.; Chapman, D. Economic growth, trade and energy: Implications for the environmental Kuznets curve. *Ecol. Econ.* **1998**, *25*, 195–208. [CrossRef]
141. Hossain, M.S. Panel estimation for CO_2 emissions, energy consumption, economic growth, trade openness and urbanization of newly industrialized countries. *Energy Policy* **2011**, *39*, 6991–6999. [CrossRef]
142. Nahman, A.; Antrobus, G. Trade and the environmental Kuznets curve: Is southern Africa a pollution haven? *S. Afr. J. Econ.* **2005**, *73*, 803–814. [CrossRef]

143. Jorgenson, A.K. Does foreign investment harm the air we breathe and the water we drink? A cross-national study of carbon dioxide emissions and organic water pollution in less-developed countries, 1975 to 2000. *Organ. Environ.* **2007**, *20*, 137–156. [CrossRef]
144. Jorgenson, A.K. The transnational organization of production, the scale of degradation, and ecoefficiency: A study of carbon dioxide emissions in less-developed countries. *Hum. Ecol. Rev.* **2009**, *16*, 64–74.
145. Bakhsh, K.; Rose, S.; Ali, M.F.; Ahmad, N.; Shahbaz, M. Economic growth, CO_2 emissions, renewable waste and FDI relation in Pakistan: New evidences from 3SLS. *J. Environ. Manag.* **2017**, *196*, 627–632. [CrossRef]

energies

Article

The Influence of Investors' Mood on the Stock Prices: Evidence from Energy Firms in Warsaw Stock Exchange, Poland

Waldemar Tarczyński [1], Urszula Mentel [2], Grzegorz Mentel [3,*] and Umer Shahzad [4]

[1] Department of Sustainable Finance and Capital Markets, Institute of Economics and Finance, University of Szczecin, 71-101 Szczecin, Poland; waldemar.tarczynski@usz.edu.pl
[2] Department of Projects Management and Security Policy, Faculty of Management, Rzeszow University of Technology, 35-959 Rzeszow, Poland; u.mentel@prz.edu.pl
[3] Department of Quantitative Methods, Faculty of Management, Rzeszow University of Technology, 35-959 Rzeszow, Poland
[4] School of Statistics and Applied Mathematics, Anhui University of Finance and Economics, Bengbu 233030, China; umer@aufe.edu.cn
* Correspondence: gmentel@prz.edu.pl; Tel.: +48-608591330

Abstract: The subject of this publication is an analysis of the sentiment of stock exchange investors in terms of making investment decisions in the energy sector of the Polish stock exchange. The investment mood is considered in the context of the possible impact of weather factors on investment decisions. Possible effects are verified in relation to the rates of return and the volume of trading of energy sector entities. The analysis is carried out both in terms of co-integration analyses as well as in econometric terms, in the cross-section of classic OLS models or causality analysis using VAR vector autoregression models. The main purpose of the issues discussed is the problem of indicating (illustrating) the presence or absence of mutual relations between weather factors and the stock market in terms of the methods considered.

Keywords: energy market; stock market; mood; emotions; weather; causality

Citation: Tarczyński, W.; Mentel, U.; Mentel, G.; Shahzad, U. The Influence of Investors' Mood on the Stock Prices: Evidence from Energy Firms in Warsaw Stock Exchange, Poland. *Energies* **2021**, *14*, 7396. https://doi.org/10.3390/en14217396

Academic Editor: Periklis Gogas

Received: 11 October 2021
Accepted: 3 November 2021
Published: 5 November 2021

1. Introduction

The traditional theory of finance assumes that market participants are rational, and the goal they all pursue is profit maximization, which translates into high market efficiency. However, since the 1980s, many studies have suggested the existence of behavioral aspects of the market that are difficult to explain on the basis of traditional financial theory. The examples include the research conducted by Rajnish Mehr and Edward C. Prescott [1]. They noticed that there was a 6% gap between the return on the S&P 500 index and the risk-free interest rates, which is difficult to explain in a rational way. Of course, there are many more examples of this type. Consequently, this leads to certain market disturbances caused by the lack of complete market information. As suggested by Daniel Kahneman, Jack L. Knetsch and Richard Thaler [2], people are concerned about phenomena, situations that are not described by traditional financial theory.

As a result, contemporary research is mainly related to the analysis of the impact of non-economic factors on financial markets and the mechanisms governing them. Consequently, this translates into an analysis of the behavior of investors as basic market participants. An important element of the analyses carried out in this area is an attempt to determine the possible impact of weather factors on the stock exchanges. According to psychologists, the weather has a significant impact on people's moods and, therefore, on their decision-making processes. This state of affairs may cause changes in, for example, share prices. According to many [3,4], climate change has a significant psychological impact on people, prompting them to behave in certain ways. However, in order for the factors under consideration to influence the decisions made, decision makers must be exposed to their influence [5].

Weather factors and natural biorhythms can be described as non-economic variables in terms of traditional equity pricing models. Differences in these variables have been shown to have a significant impact on people's moods [6]. This is important because, as argued by Loewenstein G. [7], feelings experienced during decision making "often direct behaviors in different directions from those dictated by balancing the long-term costs and benefits of different actions". As capital valuation relates essentially to the risk of future cash flow of capital and net cash flow participation rights of capital, an increasing number of behavioral finance scientists are investigating whether and to what extent mood swings, mainly caused by weather changes, affect or may affect the stock market.

Based on studies that have found a relationship between weather and sentiment, investors tend to price stocks higher (optimistically) when they are in a good mood due to the weather and lower (pessimistic) for negative weather influences. Saunders E.M. [8], who analyzed the relationship between cloudiness in New York and returns on the US stock market, moved towards this direction. In his research, he indicated the importance of cloud cover, and, thus, also insolation, for the percentage changes in share prices. Several years after these studies, Hirshleifer D. and Shumway T. [9], extending their scope, additionally analyzed the impact of rainfall and snow on returns in the Irish Datastream market index. Others, in turn, such as Floros C. [10], checked in their analyses the relationship between the temperature factor and market returns. The possible impact of weather on the variability of Korean KOSPI200 options was analyzed by Shim H. et al. [11] concluding that the volatility tends to increase on windless days. Subsequent research extended the study of the relationship between weather and stock prices to include different weather phenomena (i.e., temperature, precipitation, humidity, wind speed) and different countries [12–15]. These studies additionally confirm the theses that changes in mood caused by a number of weather phenomena are related to changes in stock prices.

There is really a lot of research indicating the significant impact of weather factors on stock market investments or those of slightly less importance in this respect, and it is impossible to list them all here. Nevertheless, a number of them document a strong link between weather patterns and stock index returns, providing indirect evidence of the impact of investor sentiment on asset prices.

Investigating the impact of weather patterns on the actual perception of investors is important, for instance, to establish the credibility of the weather effect from existing evidence. Therefore, in order to assess the significance of the possible impact of weather factors, research in this area should be continued.

Thus, the motivation of the considerations presented in the paper is not so much influenced by the traditional financial theory, but by the emotional reactions caused by the change in weather conditions affecting the performance of the quotations of the energy sector companies. Its content consists of considerations on weather factors as causative elements in the context of the investment sentiment analysis. Therefore, the research hypothesis is based on the assumption that in the case of quotations measured by the rate of return and the trading volume, the above-mentioned meteorological factors are an important causal element. The entire analysis is carried out both in terms of co-integration as well as formal causality in the Granger sense.

2. Emotions–Weather–Market

Psychological literature considers how emotions and moods influence people's decision making. People who are in a good mood seem to be more optimistic about their own choices. A very strong effect in this regard is that people in a positive mood have many kinds of positive assessments, such as satisfaction with life, with past events, with people or even consumer products [16,17]. There is a mood-compatible effect in which people with a bad or good mood tend to find that negative or positive material, respectively, is more accessible or more pronounced [18,19]. More importantly, it is stated that mood most strongly influences relatively abstract judgments about which no specific information is available [20,21].

Research also shows that in a good mood, we are more likely to use simplified heuristics to aid our decision-making process [22,23]. There is an ongoing debate as to whether this type of use of heuristics reflects possible cognitive deficiencies associated with good mood, or is it more the effective use of measures to simplify complex data.

Some of the studies conducted so far indicate that a bad mood induces investors to undertake analytical activities to a large extent, while a good mood refers to less critical methods of information processing [24–26]. As reported, for example, by Bless H. et al. [27] good moods result in greater reliance on information about the category and, thus, simplify stereotypes. Thus, in their view, good moods make people rely more on "pre-existing structures of knowledge", which does not necessarily result in a general decline in motivation or the ability to think effectively. Positive moods also have their advantages. In a good mood we often tend to create unusual associations, we become better at creative problem solving, and we show much greater mental flexibility. Moreover, being in a good mood, we tend to work out more detailed tasks with neutral/positive (at least not negative) stimulus material [23].

Emotions influence assessments of both the favorable outlook for the future [28,29] and risk assessments [30,31]. The direction of the influence of mood on the perception of risk is a complex process depending primarily on the task itself and the situation in which we are.

An important thread of emotions or moods (the so-called affective states theory) is that they provide individuals with information about the environment [24,32]. A significant amount of research confirms the informative role of the effect, as in [24,33] and [20]. The feelings-based decision-making procedure was named by Slovic P. et al. [31] as "Affect Heuristics".

We very often attribute our feelings to the wrong source, leading to wrong judgments. An example of this is that people are happier on sunny days and are less positive on cloudy days. The influence of sunlight on their perception of happiness is much weaker when we ask about the weather [34]. Presumably, this is due to the fact that they attribute the good mood to the sun and not to long-term considerations.

Psychology has long documented the relationship between weather and behavior, focusing its research mainly on the effect of insolation. The vast majority of analyses suggest that we feel better when we are directly exposed to sunlight. Therefore, we are more optimistic on sunny days and more likely to buy stocks. Thus, a positive correlation is very often seen between insolation and rates of return. Moreover, information (as in the weather forecast) that a particular day will be sunny should not trigger an immediate and full positive response from stock prices. Only the exposure to sunlight itself should cause the price movement of financial instruments. However, it should be remembered that insolation in one specific place is not representative of weather conditions in the entire economy. It is also important that sunlight is a transient variable. The amount of the expected insolation today is not strongly correlated with the amount that will occur in a few days, weeks, or months.

As was shown much earlier, there is evidence that the sun influences markets. The confirmation of this can be found in the research of Saunders E.M. Jr. [8], which was mentioned in the introduction. In his research, apart from showing a negative dependence between cloud cover and returns from the New York Stock Exchange, he also proves that this correlation is resistant to various stock index selections and regression specifications.

As one can see, bad weather can complicate the market situation, communication or other activities. Hence, it seems reasonable to analyze the possible impact of other weather factors on the stock market as not only the right amount of sun can cause significant changes in the stock markets. There are a number of studies showing the importance of other weather determinants, such as temperature or atmospheric pressure (Table 1). Therefore, further research in this area may turn out to be extremely interesting and possible observations will give a new look at some dependencies.

Table 1. Review of literature research on the emotions–weather–stock market relationship.

Author (Authors)	Conclusions
Howarth, E. and Hoffman, M.S [6]	They verified the impact of eight weather variables on ten mood measures. As a result of the research, they indicated that meteorological factors such as humidity, temperature and sunshine have a significant influence on the mood. In their opinion, the duration of sunlight was associated with higher scores in terms of optimism, while high levels of humidity with lower scores in terms of concentration, and a possible increase in temperature with lower scores in terms of anxiety and skepticism.
Loughran, T. and Schultz, P. [35]	They showed that cloudy days have little effect on a company's trading volume apart from extreme weather conditions, which can be attributed to other factors that may be unrelated to mood.
Saunders, E.M. [8]	He was the first to investigate the relationship between New York City weather and returns on the New York Stock Exchange (NYSE). As a result of the analyses, he drew the attention of economists to the possible impact of weather changes on the returns on the stock exchange. It showed a significant correlation between cloud cover and percentage changes in stock prices.
Kamstra, M.J., Kramer, L.A., and Levi, M.D. [36]	In their opinion, in autumn and winter, market profits are on average lower than in spring and summer. They characterized it as the beginning of a seasonal affective disorder, i.e., a depression associated with a decline in daylight. This phenomenon is particularly strong in the Scandinavian countries. According to them, due to the lack of sun, people become more depressed, which lowers the general good mood and the willingness to invest. If investors had realized this sooner, they could have prevented irrational decisions.
Keller, M.C., Fredrickson, B.L., Ybarra, O., Côté, S., Johnson, K., Mikels, J., Conway, A., Wager, T. [5]	They found that a pleasant temperature and atmospheric pressure are associated with better mood. Trading activity correlates strongly with the skills, personalities and moods of traders.
Hirshleifer, D. and Shumway, T. [9]	They investigated whether the sun can lead to a good mood, which would additionally translate into positive plot twists. In their research, they proved that there was a significant positive correlation between insolation and the rates of return. They found that investors could benefit from knowledge about their mood at any given time. Thanks to this, they can avoid mistakes caused by an unsuitable mood.
Kang, S.H., Jiang, Z., Lee, Y. and Yoon, S.M. [37]	They proved that weather conditions had an impact on type A funds and not on type B funds. In the period after market opening, only type B funds are heavily influenced by weather conditions, while in the profitability of both type A and type B funds, the weather has led to volatility; Shanghai Stock Exchange, 1996–2007.
Dowling, M. and Brian, M.L. [38]	A positive correlation was found between the results of the moisture analysis and the performance indicators in relation to slightly different opinions in this regard in the literature. The authors found the reasons for this different result of the analysis in the unusual weather conditions prevailing in Ireland; Irish Stock Exchange; 1988–2001.

Table 1. *Cont.*

Author (Authors)	Conclusions
Symeonidis, L., Daskalakis, G. and Markellos, R.N. [39]	According to the authors, cloudy days and extending and reducing night hours have a negative impact on the volatility of the stock markets. They state that the performance of the S&P index tends to have a negative effect on cloudy days; 26 international exchanges, 1982–1997.
Goetzman, W.N. and Zhu, N. [40]	On rainy days, market participants tend to finish trading earlier, according to the team. They note the effect of cloudy days on both liquidity and volatility; 2005.
Loughran, T. I Schultz, P. [35]	They addressed the issue of low volume. They saw the reasons for this condition in the difficulties in getting to work on days with snowstorms; 2004.
Hong, H., Kubik, J. and Stein, J. [41]	They argued that investors are more sociable and open to communication with each other on sunny days, which may explain the high volatility currently observed in the stock market; 2004
Vlady, S., Tufan, E. and Hamarat, B. [42]	They showed that there was a significant change in profitability on the Australian Stock Exchange on rainy days; Australia 1992–2006.
Kang, S.H., Jiang, Z. and Yoon, S.M. [37]	They stated that the Hong Kong Stock Exchange was not vulnerable to extreme weather conditions; Hong Kong Stock Exchange 1999–2008.
Worthington, A. [43]	He examined the effects of meteorological conditions such as evaporation, relative humidity, high and low temperatures, hours of sunshine, and the direction and speed of a severe storm on the Australian stock price index and found no impact on market earnings; 1958–2005.
Keef, S.P. and Roush, M.L. [44]	They observed that off-season temperatures had a stronger and negative impact on stocks compared to normal temperature parameters. On the other hand, the speed of the storm and the number of cloudy days did not have such an effect; 1992–2003
Floros, C. [10]	The impact of daily temperatures on the stock markets in Austria, Belgium, France, Greece and the UK was studied at different times and a negative correlation was identified between daily temperature parameters and the yields of the stock markets in Austria, Belgium and France. In turn, in the case of Greece and the UK, this correlation was positive but not statistically strong. In another study, Floros (2011) showed that weather conditions had a negative impact on market profitability, using data from the Lisbon Stock Exchange (PSI-20) in 1995–2007. The researcher also found that safety indicators were positive in January, mainly due to the low temperatures, which lead to aggressive risk taking; Austria, Belgium, France, Greece and UK Equity Markets, 2008.
Chang, T., Nieh, C.C., Yang, M.J. and Yang, T.Y. [14]	They examined the effect of weather conditions on earnings, against factors such as temperature, humidity and cloudy days. According to the research results, the following had a significant influence in this case: temperature and cloudy days; Taiwan Stock Exchange, July–October 2006.
Wang, Y., Lin, C.T. and Lin, J.D. [45]	They did not identify the impact of rainy days on market profits but showed that sunny days and temperature had a significant impact; Taiwan Stock Exchange, 2001–2007.
Tuna, G. [46]	He examined the effect of humidity and cloudy days on the profitability of the Istanbul Stock Index and found no effect; Istanbul Stock Exchange, 1987–2006.

Table 1. *Cont.*

Author (Authors)	Conclusions
Silva, P. and Almeida, L. [47]	Thanks to their analysis, a correlation between low temperature and high efficiency was identified; Portuguese Stock Exchange, 2000–2009.
Zadorozhna, O. [48]	Correlations between profitability indices, individual safety indices, commercial value, and weather conditions (storm, cloudy days, pressure, rain and humidity) at different times were examined. While correlations were identified in some countries, they were found to be low; 13 countries of Central and Eastern Europe, 2009

3. Materials and Methods

The research sample in these analyses are energy companies listed on the Warsaw Stock Exchange in 2015–2021. Possible causes of weather factors were analyzed in relation to the rates of return and the value of their trading volumes. The meteorological data were taken from the Institute of Meteorology and Water Management and concerned such weather factors as temperature, daily rainfall, sunshine duration, rainfall duration, average daily cloud cover, average daily wind speed, average relative humidity and average daily pressure at sea level. The location of the weather station was consistent with the seat of a given company listed. The research was limited to the Polish capital market due to the fact that there are no potential analyses of this type for the countries of Central and Eastern Europe. The cases of world studies concern either Asian markets, the American market or European countries with strong economies and well-functioning financial markets.

The analysis of the possible impact of meteorological factors on the stock market of energy companies should begin with the study of the phenomenon of co-integration. In this case, the correlation study did not bring satisfactory results as it did not talk about the long-term interdependence of the series. Thus, the correlation coefficients were not a suitable measure to rate the effect in question. Importantly, the co-integration effect may also occur when a low correlation is identified.

The most common approaches used to test the phenomenon are the Engle–Granger method [49] and the Johansen method [50]. The authors of the first proposed a relatively simple approach to the estimation of the degree of co-integration, namely the use of least squares regression and its application to the studied series. Then they proposed to perform a stationarity test (unit root test) for the residuals of the estimated regression model. However, both presented approaches are applicable in the case of non-stationary time series.

Therefore, initially, the stationarity of time series is analyzed using the most commonly used tests in this area, i.e., ADF (Augmented Dickey–Fuller test) and KPSS (Kwiatkowski, Phillips, Schmidt and Shine test) [51]. In the former, the null hypothesis is that the series is non-stationary. Thus, the alternative hypothesis is its negation. In the case of the KPSS test, we deal with the reverse system of hypotheses.

The use of these tests is also recommended by Hamulczuk, Grudkowska, Gędek, Klimkowski and Stańko [52] as the so-called confirming analysis. According to this, one deals with "strong" stationarity when it is found by both tests. In another variant, we talk about the lack of stationarity of the time series.

The possible finding of stationarity (or lack thereof) for dependent and independent variables generates two approaches to the causality analysis. In the case of a pair of doubly stationary variables, it is used in the classic OLS regression analysis (ordinary least squares). This approach allows the study of two-dimensional relationships between variables. Thus, this method is limited to model estimation

$$y_t = \alpha_0 + \alpha_1 x_{ti} + \xi_t, \tag{1}$$

where x_{ti} is the independent variable (regressor) for the i-th measurement.

In turn, in the variant when one of the variables analyzed or both are non-stationary, the best approach is VAR modeling (vector autoregressive model). In this case, assuming the same delays for both variables k, a test of the total significance of the delays of a given variable is applied in the equation explaining the second variable

$$y_t = \alpha_{10} + \sum_{j=1}^{k} \alpha_{1j} y_{t-j} + \sum_{j=1}^{k} \beta_{1j} x_{t-j} + \varepsilon_{1t} \tag{2}$$

$$x_t = \alpha_{20} + \sum_{j=1}^{k} \alpha_{2j} y_{t-j} + \sum_{j=1}^{k} \beta_{2j} x_{t-j} + \varepsilon_{2t} \tag{3}$$

The Akaike (AIC), Schwartz–Bayesian (BIC) or Hannan–Quinn (HQC) criteria are most often used when selecting lags. In this case, it is worth emphasizing that a one-way analysis showing the impact of meteorological variables on the stock market is substantively justified. The analysis in the reverse direction is pointless in terms of possible inference.

4. Results

The aforementioned analysis of the occurrence of the unit root in the context of the dependent variables (rate of return, trading volume) is illustrated by the results of Table 2.

Table 2. Results of stationarity tests of the analyzed stock market time series.

Instrument	ADF Test				KPSS Test	
	Delay	Test Statistic	p	Autocorrelation of First-Order Residuals	Test Statistic	Critical Value α = 5% and α = 1%
Będzin						
rate of return	0	−45.1744	0.0001	−0.015	0.267245	0.462/0.743
trading volume	0	−26.9194	1.220×10^{-44}	−0.048	1.551350	
Enea						
rate of return	0	−35.7359	1.209×10^{-24}	0.001	0.033697	0.462/0.743
trading volume	0	−29.3397	4.378×10^{-41}	−0.047	1.593400	
Energa						
rate of return	0	−38.5467	4.595×10^{-16}	0.000	0.167231	0.462/0.743
trading volume	0	−24.3335	2.797×10^{-46}	−0.077	15.80410	
Kogeneracja						
rate of return	0	−43.3854	0.0001	0.000	0.244913	0.462/0.743
trading volume	0	−32.4472	4.448×10^{-46}	−0.014	0.622624	
ML System						
rate of return	0	−23.4542	1.643×10^{-39}	0.008	0.227393	0.462/0.743
trading volume	0	−11.0043	5.512×10^{-21}	−0.132	17.56750	
PGE						
rate of return	0	−35.2625	4.451×10^{-26}	0.000	0.033137	0.462/0.743
trading volume	0	−32.7535	2.832×10^{-33}	−0.020	0.711201	
Polenergia						
rate of return	0	−37.1847	3.350×10^{-20}	0.002	0.512327	0.462/0.743
trading volume	0	−32.4881	5.693×10^{-34}	−0.017	0.564129	
Tauron						
rate of return	0	−34.277	5.311×10^{-29}	0.002	0.197904	0.462/0.743
trading volume	0	−22.4762	5.342×10^{-46}	−0.066	6.307180	
ZE PAK						
rate of return	0	−39.7838	1.823×10^{-12}	0.000	0.220253	0.462/0.743
trading volume	0	−31.2886	5.690×10^{-37}	−0.020	1.216120	

The analysis of the values of individual test statistics allows the conclusion that the rate of return of the instruments considered is a stationary value in the "strong" sense. On the other hand, slightly different results are obtained when verifying the time series in the form of the trading volume. They are not homogeneous for both the tests. The KPSS test clearly excludes the stationary effect in this case. With a critical value of 0.743 for $\alpha = 0.01$ in three cases, Kogeneracja, PGE and Polenergia, one could assume that the series is stationary. However, if we take the classical significance level of 0.05 (critical value 0.462), the occurrence of the unit root is common. Therefore, it is assumed that the trading volume is a non-stationary series. A possible increase in the number of delays does not significantly improve the value of the KPSS test statistics.

A similar analysis for weather variables requires a breakdown into individual weather stations (depending on the location of a given listed company) across the analyzed meteorological factors (Table 3).

Table 3. Results of stationarity tests of the analyzed weather time series.

Instrument	ADF Test				KPSS Test	
	Delay	Test Statistic	p	Autocorrelation of First-Order Residuals	Test Statistic	Critical Value $\alpha = 5\%$ and $\alpha = 1\%$
Gdańsk						
average daily temperature	0	−6.65853	3.412×10^{-09}	−0.072	1.770400	0.462/0.743
	4	−4.07270	0.001076	−0.005	0.383428	
daily sum of precipitation	0	−34.7505	1.292×10^{-27}	−0.003	0.163824	
insolation	0	−19.4848	5.023×10^{-43}	−0.153	0.640190	
	1	−13.3531	4.477×10^{-30}	−0.030	0.401188	
duration of rainfall	0	−29.8175	3.587×10^{-40}	−0.013	0.224766	
mean daily overall cloudiness	0	−21.8798	1.193×10^{-45}	−0.030	0.328337	
average daily wind speed	0	−25.6320	9.492×10^{-46}	−0.008	0.175068	
average daily relative humidity	0	−19.9723	1.030×10^{-43}	−0.080	1.547060	
	6	−8.10506	2.454×10^{-18}	−0.003	0.428725	
mean daily sea level pressure	0	−36.5630	4.060×10^{-22}	−0.001	0.082570	
Katowice						
average daily temperature	0	−7.13979	1.901×10^{-10}	−0.014	1.205010	0.462/0.743
	2	−5.85133	2.707×10^{-07}	−0.019	0.426140	
daily sum of precipitation	0	−29.7677	2.896×10^{-40}	0.011	0.251548	
insolation	0	−21.4800	2.414×10^{-45}	−0.090	0.562847	
	1	−15.8271	1.435×10^{-37}	−0.027	0.368371	
duration of rainfall	0	−29.1066	1.678×10^{-41}	−0.011	0.124021	
mean daily overall cloudiness	0	−23.4955	2.661×10^{-46}	−0.001	0.395364	
average daily wind speed	0	−23.9630	2.561×10^{-46}	−0.033	0.334759	
average daily relative humidity	0	−18.0161	1.659×10^{-40}	−0.091	0.953594	
	2	−11.5877	2.313×10^{-24}	−0.016	0.436156	

Table 3. *Cont.*

Instrument	ADF Test				KPSS Test	
	Delay	Test Statistic	p	Autocorrelation of First-Order Residuals	Test Statistic	Critical Value $\alpha = 5\%$ and $\alpha = 1\%$
mean daily sea level pressure	0	−37.7039	1.303×10^{-18}	−0.000	0.058875	
				Koło		
average daily temperature	0	−8.67207	9.310×10^{-15}	−0.072	1.414050	0.462/0.743
	3	−5.65398	7.779×10^{-07}	−0.013	0.394922	
daily sum of precipitation	0	−34.2158	3.527×10^{-29}	−0.010	0.382024	
insolation	0	−19.3793	7.278×10^{-43}	−0.111	1.743280	
	6	−6.88035	6.935×10^{-10}	−0.007	0.439224	
duration of rainfall *	-	-	-	-	-	
mean daily overall cloudiness *	-	-	-	-	-	
average daily wind speed	0	−24.3227	2.808×10^{-46}	0.004	2.583170	
average daily relative humidity	0	−20.5403	2.046×10^{-44}	−0.088	0.723716	
	2	−13.4426	2.324×10^{-30}	−0.006	0.360950	
mean daily sea level pressure	0	−24.4785	3.020×10^{-46}	0.018	0.237739	
				Poznań		
average daily temperature	0	−6.65828	3.418×10^{-09}	−0.072	1.768160	0.462/0.743
	4	−4.07820	0.001053	−0.005	0.382957	
daily sum of precipitation	0	−34.7403	1.222×10^{-27}	−0.003	0.163108	
insolation	0	−19.4890	4.976×10^{-43}	−0.153	0.637882	
	1	−13.3568	4.357×10^{-30}	−0.031	0.399831	
duration of rainfall	0	−30.5269	1.066×10^{-38}	−0.010	0.264330	
mean daily overall cloudiness	0	−23.0241	3.292×10^{-46}	−0.023	0.436696	
average daily wind speed	0	−25.5923	9.047×10^{-46}	−0.008	0.174762	
average daily relative humidity	0	−19.9831	1.002×10^{-43}	−0.080	1.534470	
	6	−8.11923	2.229×10^{-13}	−0.003	0.425760	
mean daily sea level pressure	0	−36.5527	3.827×10^{-22}	−0.001	0.082906	

Table 3. *Cont.*

Instrument	ADF Test				KPSS Test	
	Delay	Test Statistic	p	Autocorrelation of First-Order Residuals	Test Statistic	Critical Value $\alpha = 5\%$ and $\alpha = 1\%$
Rzeszów						
average daily temperature	0	−4.90820	3.714×10^{-05}	−0.086	2.201240	0.462/0.743
	5	−2.44556	0.1293	−0.003	0.410210	
daily sum of precipitation	0	−20.6487	1.856×10^{-39}	0.006	0.367610	
insolation *	-	-	-	-	-	
duration of rainfall	0	−19.7419	7.614×10^{-39}	−0.001	0.484120	
	1	−19.7419	7.614×10^{-39}	−0.001	0.393551	
mean daily overall cloudiness	0	−13.9563	6.826×10^{-29}	0.005	0.465584	
	1	−13.9563	6.826×10^{-29}	0.005	0.305371	
average daily wind speed	0	−13.9629	6.574×10^{-29}	−0.026	0.793649	
	2	−13.9629	6.574×10^{-29}	−0.026	0.416820	
average daily relative humidity	0	−10.5698	9.411×10^{-20}	−0.057	1.241470	
	3	−6.3716	1.453×10^{-08}	−0.010	0.442910	
mean daily sea level pressure	0	−10.5538	1.045×10^{-19}	0.123	0.253807	
Warszawa						
average daily temperature	0	−6.76346	1.839×10^{-09}	−0.011	1.511100	0.462/0.743
	3	−4.81845	4.731×10^{-05}	−0.011	0.405907	
daily sum of precipitation	0	−32.9775	1.137×10^{-32}	−0.015	0.181474	
insolation	0	−19.1301	1.774×10^{-42}	−0.108	1.491150	
	5	−8.64524	5.879×10^{-15}	−0.003	0.422109	
duration of rainfall	0	−28.8845	6.807×10^{-42}	−0.031	0.364404	
mean daily overall cloudiness	0	−21.0338	6.075×10^{-45}	−0.042	0.449521	
average daily wind speed	0	−26.2856	2.938×10^{-45}	0.016	0.207873	
average daily relative humidity	0	−15.1308	7.267×10^{-34}	−0.096	0.971395	
	2	−12.3351	8.475×10^{-27}	−0.022	0.560530	
mean daily sea level pressure	0	−9.91117	6.393×10^{-19}	−0.010	0.408821	
	1	−16.5259	2.494×10^{-37}	0.127	0.604866	

Table 3. *Cont.*

Instrument	ADF Test				KPSS Test	
	Delay	Test Statistic	p	Autocorrelation of First-Order Residuals	Test Statistic	Critical Value $\alpha = 5\%$ and $\alpha = 1\%$
				Wrocław		
average daily temperature	0	−7.19858	1.324×10^{-10}	−0.053	1.269010	0.462/0.743
	2	−5.60812	9.898×10^{-07}	−0.017	0.449098	
daily sum of precipitation	0	−34.2791	5.387×10^{-29}	−0.001	0.244142	
insolation	0	−20.6760	1.440×10^{-44}	−0.083	0.819525	
	2	−12.5505	1.695×10^{-27}	−0.017	0.406633	
duration of rainfall	0	−31.9303	2.108×10^{-35}	−0.007	0.146045	
mean daily overall cloudiness	0	−24.2092	2.697×10^{-46}	−0.004	0.726690	
	2	−16.4118	3.403×10^{-39}	−0.005	0.423752	
average daily wind speed	0	−23.7511	2.549×10^{-46}	0.005	0.425851	
average daily relative humidity	0	−17.8176	4.064×10^{-40}	−0.079	1.371260	
	4	−9.09078	2.496×10^{-16}	−0.007	0.434634	
mean daily sea level pressure	0	−16.6966	1.012×10^{-37}	0.119	0.880180	
	2	−15.7554	2.294×10^{-37}	0.001	0.407097	

* no data for a given weather station.

The data presented in the table above confirm to a large extent the non-stationarity of weather time series. Of course, there are cases of stationarity but they mainly relate to variables such as the daily sum of precipitation or the duration of rainfall. This is in some ways due to the nature of the data. One could assume some kind of "dichotomy" here. In ranks of this type, there are many observations when we deal with the lack of a phenomenon, hence, many values are just null.

Therefore, by making a distinction between stationary and non-stationary variables, one can perform the causality analysis in accordance with the principles presented a little earlier. Thus, in the case of a pair of double-stationary variables (explained and explanatory), the analysis of the influence of the weather factor on the market variable is carried out using classical econometric modeling (OLS). The results of this type of analysis are presented in Table 4.

Modeling, by means of regression equations, presented in Table 3 does not indicate the existence of a cause and effect relationship between the rate of return and the weather instrument of a stationary nature. However, it should be remembered that this type of analysis indicates linear relationships but nonlinear relationships may take place here as well. Since the presented analysis is based on a direct relationship at the same moment, it would be worth considering whether it is better to carry out similar tests but in relation to the shifted values of a given weather factor. After all, signals from a given meteorological regressor may have consequences on the next day, and, thus, affect the behavior of rates of return with a certain delay. There is little logical justification for examining a larger scale of delays than one. Hence, Table 5 presents the results of OLS estimation for the rates of return of the companies in question depending on the one-period lag of the stationary weather factor.

Table 4. OLS estimation results for the rate of return depending on the stationary weather factor along with testing the properties of the residual component.

	Testing the Significance of the Parameters		**Testing the Correctness of the Model Form**		**Normality Test**		**Autocorrelation Analysis**		**Heteroscedasticity Analysis**	
	***t*-Student Test**		**Nonlinearity Test for Squares**		**Doornik–Hansen Test**		**Durbin–Watson Test**		**Breusch–Pagan Test**	
Independent Variable	***t* Statistic**	***p***	**TR^2 Statistic**	***p***	**chi-Square Statistic (2)**	***p***	**d Statistic**	***p***	**LM Statistic**	***p***
Energa—Gdańsk										
daily sum of precipitation	0.4155	0.6778	0.538936	0.46287	989.230	0.0000	1.99345	0.44892	0.489294	0.48424
duration of rainfall	0.1203	0.9043	0.031045	0.86014	987.966	0.0000	1.99303	0.45175	0.581807	0.44560
mean daily overall cloudiness	0.2338	0.8152	0.160692	0.68852	988.727	0.0000	1.99356	0.44737	0.083981	0.77197
average daily wind speed	−0.2360	0.8135	0.401483	0.52632	987.547	0.0000	1.99280	0.44222	0.484559	0.48636
mean daily sea level pressure	1.0570	0.2908	0.048854	0.82507	990.031	0.0000	1.99468	0.46791	0.237134	0.62628
Tauron—Katowice										
daily sum of precipitation	−1.1410	0.2542	0.703611	0.40157	464.169	0.0000	1.76016	$1.7 \cdot 10^{-06}$	4.685670	0.03042
duration of rainfall	−0.0890	0.9291	0.062104	0.80320	464.985	0.0000	1.75960	$1.5 \cdot 10^{-06}$	0.437466	0.50835
mean daily overall cloudiness	−0.4410	0.6593	0.095458	0.75735	465.626	0.0000	1.75837	$1.3 \cdot 10^{-06}$	0.325955	0.56805
average daily wind speed	−0.5350	0.5927	0.186623	0.66574	460.212	0.0000	1.76062	$1.7 \cdot 10^{-06}$	1.531245	0.21593
mean daily sea level pressure	−0.1077	0.9143	0.293223	0.58816	465.123	0.0000	1.75907	$1.5 \cdot 10^{-06}$	0.126153	0.72245
ZE Pątnów—Koło										
daily sum of precipitation	−0.4699	0.6385	6.173630	0.01297	1288.18	0.0000	2.05608	0.86340	0.112137	0.73772
average daily wind speed	−0.4356	0.6632	0.097398	0.75498	1289.83	0.0000	2.05659	0.86580	1.244608	0.26459
mean daily sea level pressure	−0.0828	0.9340	1.345360	0.24609	1289.82	0.0000	2.05640	0.86053	3.586737	0.05824
Będzin—Poznań										
daily sum of precipitation	−0.7680	0.4426	0.218907	0.63987	1463.51	0.0000	2.30515	1.00000	7.552060	0.00599
duration of rainfall	0.7455	0.4561	1.032730	0.30952	1460.58	0.0000	2.30857	1.00000	16.61387	0.00005
mean daily overall cloudiness	0.7873	0.4312	0.112266	0.73758	1459.29	0.0000	2.30795	1.00000	9.245279	0.00236
average daily wind speed	0.0740	0.9410	0.070044	0.79127	1461.21	0.0000	2.30783	1.00000	8.849761	0.00293
mean daily sea level pressure	−0.2565	0.7976	0.468392	0.49373	1461.11	0.0000	2.30805	1.00000	0.245686	0.62013
Enea—Poznań										
daily sum of precipitation	1.1510	0.2498	1.031410	0.30983	423.939	0.0000	1.84101	0.00103	3.028158	0.08183
duration of rainfall	0.1389	0.8896	0.940882	0.33205	421.976	0.0000	1.84188	0.00108	3.063442	0.08007
mean daily overall cloudiness	−0.7836	0.4334	0.881376	0.34783	420.630	0.0000	1.84282	0.00112	1.085435	0.29749
average daily wind speed	0.4623	0.6440	0.242990	0.62206	427.272	0.0000	1.84247	0.00109	1.143693	0.28487
mean daily sea level pressure	1.6950	0.0903	1.948870	0.16271	407.771	0.0000	1.84513	0.00127	3.561509	0.05913
ML System—Rzeszów										
daily sum of precipitation	0.2710	0.7865	0.318978	0.57222	564.198	0.0000	1.87726	0.06118	0.683641	0.40834
mean daily sea level pressure	1.0790	0.2811	2.873300	0.09006	560.000	0.0000	1.88292	0.06773	2.398701	0.12144
PGE—Warszawa										
daily sum of precipitation	0.4121	0.6803	0.946749	0.33055	1061.63	0.0000	1.81504	0.00017	7.425154	0.00643
duration of rainfall	−1.2760	0.2022	0.173579	0.67695	1068.89	0.0000	1.81437	0.00016	1.945605	0.16306
mean daily overall cloudiness	−0.2345	0.8147	1.143880	0.28483	1067.41	0.0000	1.81533	0.00016	2.888960	0.08919
average daily wind speed	1.2950	0.1955	0.207160	0.64899	1081.99	0.0000	1.81814	0.00021	2.970178	0.08481

Table 4. *Cont.*

Independent Variable	Testing the Significance of the Parameters — t-Student Test: t Statistic	p	Testing the Correctness of the Model Form — Nonlinearity Test for Squares: TR^2 Statistic	p	Normality Test — Doornik–Hansen Test: chi-Square Statistic (2)	p	Autocorrelation Analysis — Durbin–Watson Test: d Statistic	p	Heteroscedasticity Analysis — Breusch–Pagan Test: LM Statistic	p
					Polenergia—Warszawa					
daily sum of precipitation	0.7020	0.4828	1.732770	0.18806	1362.74	0.0000	1.92028	0.06106	0.001298	0.97126
duration of rainfall	−0.9413	0.3467	2.126270	0.14479	1359.18	0.0000	1.92113	0.06271	4.954596	0.02602
mean daily overall cloudiness	0.9738	0.3303	0.302700	0.58219	1368.22	0.0000	1.92125	0.06305	0.080192	0.77704
average daily wind speed	0.3699	0.7115	4.804840	0.02838	1364.89	0.0000	1.92187	0.06516	1.637421	0.20068
					Kogeneracja—Wrocław					
daily sum of precipitation	2.2120	0.0271	0.443257	0.50556	1177.49	0.0000	2.22979	0.99999	1.820133	0.17729
duration of rainfall	0.6508	0.5153	0.005482	0.94098	1179.43	0.0000	2.22825	0.99999	1.784164	0.18164
average daily wind speed	0.0730	0.9418	1.069210	0.30112	1179.77	0.0000	2.22818	0.99999	8.293205	0.00398

Table 5. OLS estimation results for the rate of return depending on the one-period stationary lag of the weather factor together with testing the properties of the residual component.

Independent Variable	Testing the Significance of the Parameters — t-Student Test: t Statistic	p	Testing the Correctness of the Model Form — Nonlinearity Test for Squares: TR^2 Statistic	p	Normality Test — Doornik–Hansen Test: chi-Square Statistic (2)	p	Autocorrelation Analysis — Durbin–Watson Test: d Statistic	p	Heteroscedasticity Analysis — Breusch–Pagan Test: LM Statistic	p
					Energa—Gdańsk					
daily sum of precipitation	−0.2857	0.7751	0.449465	0.50259	984.144	0.0000	1.99164	0.43788	0.221894	0.63760
duration of rainfall	−0.5902	0.5551	0.211027	0.64596	980.543	0.0000	1.99148	0.43221	1.036973	0.30853
mean daily overall cloudiness	1.6430	0.1006	0.002522	0.95994	990.410	0.0000	1.99091	0.43538	7.889667	0.00497
average daily wind speed	0.8886	0.3744	0.012163	0.91218	976.356	0.0000	1.99326	0.44422	0.566568	0.45163
mean daily sea level pressure	0.4858	0.6272	0.034527	0.85259	986.594	0.0000	1.99338	0.44953	0.660328	0.41644
					Tauron—Katowice					
daily sum of precipitation	−0.8023	0.4225	0.753009	0.38553	462.422	0.0000	1.76025	$1.7 \cdot 10^{-06}$	12.86008	0.00034
duration of rainfall	−0.1507	0.8802	2.076000	0.14963	463.182	0.0000	1.75928	$1.5 \cdot 10^{-06}$	7.571485	0.00593
mean daily overall cloudiness	1.2590	0.2083	2.394450	0.12177	464.987	0.0000	1.75920	$1.4 \cdot 10^{-06}$	8.975215	0.00274
average daily wind speed	−1.2620	0.2070	0.001042	0.97425	454.976	0.0000	1.75933	$1.4 \cdot 10^{-06}$	1.963433	0.16115
mean daily sea level pressure	2.2650	0.0236	0.305251	0.58061	451.535	0.0000	1.76494	$2.6 \cdot 10^{-06}$	3.708909	0.05412

Table 5. *Cont.*

	Testing the Significance of the Parameters		Testing the Correctness of the Model Form		Normality Test		Autocorrelation Analysis		Heteroscedasticity Analysis	
	t-Student Test		Nonlinearity Test for Squares		Doornik–Hansen Test		Durbin–Watson Test		Breusch–Pagan Test	
Independent Variable	***t* Statistic**	***p***	**TR^2 Statistic**	***p***	**chi-Square Statistic (2)**	***p***	**d Statistic**	***p***	**LM Statistic**	***p***
ZE Pątnów—Koło										
daily sum of precipitation	0.2622	0.7932	5.117830	0.02368	1291.79	0.0000	2.05552	0.85831	5.324408	0.02103
average daily wind speed	−0.0807	0.9357	1.055140	0.30433	1288.13	0.0000	2.05631	0.86044	0.603302	0.43732
mean daily sea level pressure	0.3064	0.7593	0.046081	0.83003	1286.21	0.0000	2.05603	0.86063	0.322038	0.57039
Będzin—Poznań										
daily sum of precipitation	1.706	0.0883	5.49766	0.01904	1444.07	0.00000	2.30599	1.00000	1.987446	0.15861
duration of rainfall	0.5572	0.5774	0.252543	0.61529	1450.39	0.0000	2.30707	1.00000	0.030353	0.86169
mean daily overall cloudiness	0.0598	0.9523	0.361593	0.54762	1458.44	0.0000	2.30771	1.00000	17.95416	0.00002
average daily wind speed	0.0740	0.9410	0.070045	0.79127	1461.21	0.0000	2.30783	1.00000	8.849761	0.00293
mean daily sea level pressure	−0.2531	0.8002	0.470542	0.49274	1458.93	0.0000	1458.93	1.00000	0.110635	0.73942
Enea—Poznań										
daily sum of precipitation	1.3080	0.1910	0.887971	0.34603	424.912	0.0000	1.84457	0.00130	0.478429	0.48914
duration of rainfall	0.9757	0.3294	0.676544	0.41078	424.901	0.0000	1.84248	0.00114	1.699012	0.19242
mean daily overall cloudiness	−0.5893	0.5558	3.017120	0.08239	421.705	0.0000	1.84184	0.00106	0.770240	0.38014
average daily wind speed	1.7110	0.0872	0.116634	0.73271	436.200	0.0000	1.84156	0.00106	3.902850	0.04820
mean daily sea level pressure	−0.2861	0.7748	2.381920	0.12275	422.420	0.0000	1.84127	0.00101	1.544707	0.21392
ML System—Rzeszów										
daily sum of precipitation	1.3390	0.1810	0.029227	0.86426	570.439	0.0000	1.87622	0.06036	1.226202	0.26815
mean daily sea level pressure	0.7083	0.4791	0.205789	0.65009	556.432	0.0000	1.87715	0.05981	13.06857	0.00030
PGE—Warszawa										
daily sum of precipitation	−0.5808	0.5614	1.302680	0.25372	1065.26	0.0000	1.81608	0.00018	0.347363	0.55561
duration of rainfall	0.9768	0.3288	2.234830	0.13493	1073.48	0.0000	1.81185	0.00013	3.504961	0.06118
mean daily overall cloudiness	1.0510	0.2935	2.340300	0.12606	1063.41	0.0000	1.81333	0.00015	0.961013	0.32693
average daily wind speed	1.1170	0.2641	3.389190	0.06563	1082.99	0.0000	1.81813	0.00020	0.491088	0.48344
Polenergia—Warszawa										
daily sum of precipitation	−0.3721	0.7099	0.730131	0.39284	1361.23	0.0000	1.92104	0.06799	0.055163	0.81431
duration of rainfall	−0.4245	0.6712	0.097101	0.75534	1357.61	0.0000	1.92176	0.06443	2.794717	0.09458
mean daily overall cloudiness	0.0946	0.9801	0.228107	0.63293	1360.39	0.0000	1.92137	0.06474	0.076028	0.78276
average daily wind speed	1.4880	0.1370	2.246850	0.13389	1353.90	0.0000	1.92221	0.06802	1.848813	0.17392

Table 5. *Cont.*

	Testing the Significance of the Parameters		Testing the Correctness of the Model Form		Normality Test		Autocorrelation Analysis		Heteroscedasticity Analysis	
	t-Student Test		Nonlinearity Test for Squares		Doornik–Hansen Test		Durbin–Watson Test		Breusch–Pagan Test	
Independent Variable	*t* Statistic	*p*	TR^2 Statistic	*p*	chi-Square Statistic (2)	*p*	d Statistic	*p*	LM Statistic	*p*
					Kogeneracja—Wrocław					
daily sum of precipitation	0.2983	0.7655	1.817510	0.17761	177.484	0.0000	2.22857	0.99999	1.888471	0.16938
duration of rainfall	−0.7427	0.4578	1.598320	0.20614	1176.80	0.0000	2.22607	0.99999	4.402569	0.03589
average daily wind speed	0.9691	0.3327	0.000476	0.98259	1171.36	0.0000	2.22919	0.99999	23.58994	0.00001

The regression analysis illustrated in the table above does not show any clear relationships between the delayed weather factor and the rate of return.

The analysis of causality for pairs of non-stationary variables or those in which at least one variable is non-stationary is much more interesting. Essentially, it is the analysis of the dependent variable in terms of trading volume but not limited to this. Vector autoregression analysis (VAR) with co-integration tests has already been described in the methodological section. The results of the research are presented in Table 6.

Table 6. VAR causality test results for trading volume and weather factors with co-integration analysis.

Independent Variable	VAR			Johansen's Test						Engle–Granger Test	
	Delay	F Statistic	*p*	Row of the Matrix	Eigenvalue	λ_{trace} Test	*p*	λ_{max} Test	*p*	*t* Statistic	*p*
					Energa—Gdańsk						
average daily temperature	5	2.9141	0.0127	0	0.04954	92.606	0.000	75.966	0.000	−8.00159	5.465×10^{-08}
daily sum of precipitation	5	0.7631	0.5764	0	0.36116	904.30	0.000	669.47	0.000	−24.7666	6.223×10^{-51}
insolation	5	1.1266	0.3441	0	0.05908	151.25	0.000	91.042	0.000	−8.11206	2.590×10^{-12}
duration of rainfall	4	1.6614	0.1565	0	0.45144	1157.7	0.000	897.68	0.000	−28.9309	7.866×10^{-48}
mean daily overall cloudiness	4	0.4648	0.7616	0	0.45005	1055.2	0.000	893.91	0.000	−28.8953	6.965×10^{-48}
average daily wind speed	4	2.8127	0.0242	0	0.45061	1109.2	0.000	895.43	0.000	−28.4388	1.598×10^{-48}
average daily relative humidity	5	1.9973	0.0763	0	0.06720	168.75	0.000	104.00	0.000	−8.08309	3.152×10^{-12}
mean daily sea level pressure	4	0.1414	0.9668	0	0.44992	1207.1	0.000	893.54	0.000	−28.9597	8.688×10^{-48}

Table 6. *Cont.*

Independent Variable	VAR				Johansen's Test					Engle–Granger Test	
	Delay	F Statistic	p	Row of the Matrix	Eigenvalue	λ_{trace} Test	p	λ_{max} Test	p	t Statistic	p
Tauron—Katowice											
average daily temperature	5	0.3876	0.8576	0	0.05513	105.11	0.000	84.771	0.000	−9.02435	4.371×10^{-15}
daily sum of precipitation	4	0.5833	0.6748	0	0.41019	1072.5	0.000	789.30	0.000	−26.1031	1.242×10^{-50}
insolation	5	1.2505	0.2830	0	0.06217	174.31	0.000	95.963	0.000	−9.01031	4.835×10^{-15}
duration of rainfall	4	0.2780	0.8923	0	0.40887	1067.7	0.000	785.94	0.000	−26.1105	1.252×10^{-50}
mean daily overall cloudiness	4	0.0112	0.9998	0	0.40869	979.99	0.000	785.49	0.000	−26.1055	1.245×10^{-50}
average daily wind speed	4	0.4384	0.7809	0	0.40928	995.57	0.000	786.99	0.000	−26.0606	1.186×10^{-50}
average daily relative humidity	5	1.3608	0.2363	0	0.05960	161.69	0.000	91.863	0.000	−9.01169	4.787×10^{-15}
mean daily sea level pressure	4	1.1416	0.3352	0	0.40902	1108.3	0.000	786.34	0.000	−26.0966	1.233×10^{-50}
ZE Pątnów-Adamów-Konin—Koło											
average daily temperature	5	1.0597	0.3810	0	0.10145	184.83	0.000	159.82	0.000	−12.7955	2.035×10^{-27}
daily sum of precipitation	7	0.4583	0.8649	0	0.25201	608.13	0.000	432.93	0.000	−20.8275	5.928×10^{-48}
insolation	5	0.5252	0.7574	0	0.10039	223.14	0.000	158.06	0.000	−12.7453	2.973×10^{-27}
duration of rainfall *	-	-	-	-	-	-	-	-	-	-	-
mean daily overall cloudiness *	-	-	-	-	-	-	-	-	-	-	-
average daily wind speed	6	1.1919	0.3077	0	0.27398	628.87	0.000	477.71	0.000	−18.0341	1.315×10^{-42}
average daily relative humidity	4	0.3609	0.8365	0	0.11399	306.15	0.000	180.94	0.000	−12.9202	7.944×10^{-27}
mean daily sea level pressure	8	0.1075	0.9998	0	0.18964	420.92	0.000	312.90	0.000	−18.0277	1.362×10^{-42}
Będzin—Poznań											
average daily temperature	3	1.7209	0.1607	0	0.15566	278.99	0.000	253.13	0.000	−15.3990	1.359×10^{-35}
daily sum of precipitation	6	18.777	0.0000	0	0.32158	775.85	0.000	578.88	0.000	−22.5578	5.839×10^{-50}
insolation	2	0.3416	0.7107	0	0.19807	499.23	0.000	330.44	0.000	−16.5536	7.275×10^{-39}
duration of rainfall	6	3.3540	0.0028	0	0.32266	777.70	0.000	581.26	0.000	−22.5154	6.348×10^{-50}
mean daily overall cloudiness	6	1.4174	0.2042	0	0.32704	725.07	0.000	590.93	0.000	−22.5244	6.235×10^{-50}
average daily wind speed	6	1.0896	0.3663	0	0.32428	745.37	0.000	584.83	0.000	−22.4460	7.304×10^{-50}
average daily relative humidity	3	0.7523	0.5210	0	0.15539	400.10	0.000	252.65	0.000	−15.3476	1.927×10^{-35}
mean daily sea level pressure	6	0.3789	0.8928	0	0.32463	809.98	0.000	585.60	0.000	−22.5294	6.174×10^{-50}

Table 6. *Cont.*

Independent Variable	VAR				Johansen's Test					Engle–Granger Test	
	Delay	F Statistic	p	Row of the Matrix	Eigenvalue	λ_{trace} Test	p	λ_{max} Test	p	t Statistic	p
Enea—Poznań											
average daily temperature	5	2.5835	0.0246	0	0.08286	148.72	0.000	129.31	0.000	−10.2803	4.150×10^{-19}
daily sum of precipitation	5	0.5230	0.7590	0	0.36300	890.40	0.000	673.77	0.000	−25.3005	6.814×10^{-51}
insolation	5	0.8618	0.5061	0	0.08343	195.65	0.000	130.24	0.000	−10.1827	8.637×10^{-19}
duration of rainfall	5	0.6469	0.6639	0	0.36332	872.54	0.000	674.51	0.000	−24.6077	6.353×10^{-51}
mean daily overall cloudiness	9	0.7969	0.6318	0	0.23890	475.30	0.000	406.48	0.000	−18.8800	1.762×10^{-44}
average daily wind speed	5	2.0417	0.0702	0	0.36093	874.31	0.000	668.92	0.000	−25.2936	6.795×10^{-51}
average daily relative humidity	3	1.7005	0.1650	0	0.14339	324.14	0.000	231.70	0.000	−13.6977	2.411×10^{-30}
mean daily sea level pressure	5	0.5034	0.7739	0	0.36113	856.48	0.000	669.40	0.000	−18.7718	2.979×10^{-44}
ML System—Rzeszów											
average daily temperature	3	0.7697	0.5113	0	0.05273	45.701	0.000	33.751	0.000	−5.44588	1.857×10^{-05}
daily sum of precipitation	9	1.8289	0.0527	0	0.22283	204.51	0.000	155.04	0.000	−10.8954	3.939×10^{-21}
insolation *	-	-	-	-	-	-	-	-	-	-	-
duration of rainfall	3	1.3027	0.2726	0	0.19452	166.58	0.000	134.77	0.000	−5.40644	2.248×10^{-05}
mean daily overall cloudiness	3	0.3462	0.7919	0	0.13308	121.10	0.000	88.972	0.000	−5.39844	2.912×10^{-06}
average daily wind speed	3	0.3950	0.7566	0	0.14536	130.14	0.000	97.855	0.000	−5.45341	1.790×10^{-05}
average daily relative humidity	3	0.5936	0.6194	0	0.07885	83.565	0.000	51.165	0.000	−5.40631	2.249×10^{-05}
mean daily sea level pressure	2	0.3052	0.7371	0	0.58743	674.03	0.000	551.57	0.000	−20.7258	8.382×10^{-48}
PGE—Warszawa											
average daily temperature	4	1.9605	0.0981	0	0.13958	247.93	0.000	224.89	0.000	−13.4403	1.620×10^{-29}
daily sum of precipitation	5	0.6327	0.6748	0	0.35467	879.10	0.000	654.37	0.000	−25.8567	9.735×10^{-51}
insolation	3	0.5616	0.6404	0	0.17672	406.56	0.000	291.11	0.000	−15.4866	7.512×10^{-29}
duration of rainfall	4	1.8345	0.1197	0	0.13899	370.14	0.000	223.88	0.000	−13.3520	3.125×10^{-29}
mean daily overall cloudiness	5	1.7493	0.1203	0	0.35566	784.62	0.000	656.66	0.000	−25.7137	8.657×10^{-51}
average daily wind speed	5	0.7229	0.6063	0	0.35473	875.77	0.000	654.49	0.000	−25.6844	8.470×10^{-51}
average daily relative humidity	3	1.8621	0.1341	0	0.17679	386.58	0.000	291.24	0.000	−15.4671	8.564×10^{-36}
mean daily sea level pressure	3	0.8564	0.4632	0	0.17839	528.05	0.000	294.15	0.000	−15.4749	1.469×10^{-36}

Table 6. *Cont.*

Independent Variable	VAR				Johansen's Test					Engle–Granger Test	
	Delay	F Statistic	p	Row of the Matrix	Eigenvalue	λ_{trace} Test	p	λ_{max} Test	p	t Statistic	p
Polenergia—Warszawa											
average daily temperature	3	1.0736	0.3592	0	0.18866	343.14	0.000	312.77	0.000	−16.2404	5.264×10^{-38}
daily sum of precipitation	2	0.6598	0.5171	0	0.25947	887.13	0.000	449.67	0.000	−18.6107	6.605×10^{-44}
insolation	3	0.1714	0.9157	0	0.18862	427.96	0.000	312.70	0.000	−16.2300	5.629×10^{-38}
duration of rainfall	2	0.1277	0.8801	0	0.25479	805.61	0.000	440.26	0.000	−18.6316	5.952×10^{-44}
mean daily overall cloudiness	3	0.4706	0.7028	0	0.18898	497.84	0.000	313.35	0.000	−16.2540	4.826×10^{-38}
average daily wind speed	2	1.2637	0.2829	0	0.26709	891.80	0.000	465.17	0.000	−18.7068	4.098×10^{-44}
average daily relative humidity	3	0.2849	0.8363	0	0.18863	407.94	0.000	312.71	0.000	−16.2288	5.671×10^{-38}
mean daily sea level pressure	2	1.7225	0.1790	0	0.25732	742.44	0.000	445.33	0.000	−18.6071	6.726×10^{-44}
Kogeneracja—Wrocław											
average daily temperature	3	1.4784	0.2186	0	0.19417	354.10	0.000	322.95	0.000	−15.6308	2.850×10^{-36}
daily sum of precipitation	1	0.6244	0.4295	0	0.44553	1667.2	0.000	883.44	0.000	−22.8552	3.387×10^{-50}
insolation *	4	0.9464	0.4361	0	0.14203	337.36	0.000	229.01	0.000	−14.4136	1.324×10^{-32}
duration of rainfall	1	0.0479	0.8268	0	0.41688	1578.1	0.000	807.96	0.000	−22.8612	3.353×10^{-50}
mean daily overall cloudiness	1	0.1080	0.7425	0	0.41325	1293.9	0.000	798.68	0.000	−22.8629	3.343×10^{-50}
average daily wind speed	4	3.3118	0.0104	0	0.13993	440.17	0.000	225.36	0.000	−14.3291	2.427×10^{-50}
average daily relative humidity	3	1.1463	0.3291	0	0.19548	447.90	0.000	325.40	0.000	−15.6591	2.359×10^{-36}
mean daily sea level pressure	2	0.8535	0.4261	0	0.26064	753.60	0.000	452.06	0.000	−18.8891	1.687×10^{-36}

* no data for a given weather station.

In the causal analysis, using the VAR test, some dependencies of the trading volume on weather factors are noticed, but it is difficult to clearly identify the essential determinants in this respect. The most positive indications are from the variable wind speed. However, this is not a spectacular result because it only occurred in three cases out of nine analyzed entities. Other significant variables in this respect are average temperature, a daily sum of precipitation, duration of precipitation and, in a single case, humidity. Although the aforementioned relations are not very clear, they illustrate, to some extent, the impact of weather variables on the trading volume.

In order to complete the analysis, a causality study between the rate of return, stationary factor and non-stationary meteorological components should also be carried out (Table 7). The hitherto analysis of the rate of return concerned the regression analysis where a few factors were characterized by the lack of the unit root (Tables 4 and 5). Therefore, the remaining weather factors for which the series stationary was excluded were omitted. The table below supplements the considerations so far.

Table 7. VAR causality test results for the rate of return and non-stationary weather factors along with co-integration analysis.

Independent Variable	VAR				Johansen's Test					Engle–Granger Test	
	Delay	F Statistic	p	Row of the Matrix	Eigenvalue	λ_{trace} Test	p	λ_{max} Test	p	t Statistic	p
Energa—Gdańsk											
average daily temperature	3	0.0343	0.9915	0	0.37114	1135.1	0.000	693.46	0.000	−19.5541	8.061×10^{-46}
insolation	3	0.6701	0.5704	0	0.43516	1297.0	0.000	853.97	0.000	−19.5467	8.325×10^{-46}
average daily relative humidity	4	2.0181	0.0895	0	0.36893	1027.4	0.000	687.74	0.000	−18.2126	5.092×10^{-43}
Tauron—Katowice											
average daily temperature	3	0.2039	0.8937	0	0.36995	1132.4	0.000	690.61	0.000	−18.7265	3.721×10^{-44}
insolation	3	2.4909	0.0587	0	0.25564	591.49	0.000	441.66	0.000	−18.7269	3.713×10^{-44}
average daily relative humidity	2	1.9430	0.0500	0	0.33044	859.74	0.000	598.90	0.000	−15.8048	8.959×10^{-37}
ZE Pątnów-Adamów-Konin—Koło											
average daily temperature	3	1.1687	0.3204	0	0.38497	1175.0	0.000	726.70	0.000	−19.5512	8.163×10^{-46}
insolation	4	1.1284	0.3414	0	0.38592	1068.6	0.000	728.52	0.000	−17.5784	1.626×10^{-41}
average daily relative humidity	4	1.1366	0.3376	0	0.36625	1021.1	0.000	681.43	0.000	−17.6007	1.433×10^{-41}
Będzin—Poznań											
average daily temperature	3	0.5154	0.6717	0	0.37201	1327.6	0.000	695.53	0.000	−23.3028	1.719×10^{-50}
insolation	3	0.3157	0.8140	0	0.43482	1487.6	0.000	853.06	0.000	−23.3071	1.710×10^{-50}
average daily relative humidity	4	0.9332	0.4437	0	0.36905	1153.2	0.000	688.04	0.000	−20.9834	3.539×10^{-48}
Enea—Poznań											
average daily temperature	2	0.4365	0.6464	0	0.44470	1464.4	0.000	880.01	0.000	−22.5165	6.334×10^{-50}
insolation	4	0.5869	0.6721	0	0.37956	1062.7	0.000	713.12	0.000	−18.0414	1.265×10^{-42}
average daily relative humidity	3	1.3326	0.2621	0	0.41224	1231.0	0.000	794.50	0.000	−19.7667	3.261×10^{-46}
ML System—Rzeszów											
average daily temperature	4	0.5883	0.6713	0	0.34336	381.81	0.000	261.20	0.000	−10.2085	7.120×10^{-19}
duration of rainfall	5	1.4409	0.2076	0	0.39570	408.82	0.000	312.28	0.000	−9.29655	6.103×10^{-16}
mean daily overall cloudiness	4	2.0564	0.0851	0	0.37888	418.96	0.000	295.73	0.000	−10.1933	7.980×10^{-19}
average daily wind speed	4	0.6256	0.6444	0	0.37334	412.30	0.000	290.22	0.000	−10.2076	7.167×10^{-19}
average daily relative humidity	5	1.2802	0.2707	0	0.36187	374.94	0.000	278.52	0.000	−9.3072	5.648×10^{-16}
mean daily sea level pressure	2	0.5462	0.5794	0	0.47732	697.93	0.000	404.19	0.000	−15.7550	1.246×10^{-36}

Table 7. *Cont.*

Independent Variable	VAR			Johansen's Test						Engle–Granger Test	
	Delay	F Statistic	p	Row of the Matrix	Eigenvalue	λ_{trace} Test	p	λ_{max} Test	p	t Statistic	p
PGE—Warszawa											
average daily temperature	3	0.4157	0.7418	0	0.36547	1118.3	0.000	680.04	0.000	−19.3625	1.875×10^{-45}
insolation	3	1.6534	0.1752	0	0.25511	556.27	0.000	440.61	0.000	−19.3561	1.929×10^{-45}
average daily relative humidity	3	1.1135	0.3424	0	0.39355	1186.7	0.000	747.71	0.000	−19.3471	2.009×10^{-45}
mean daily sea level pressure	3	0.7133	0.5440	0	0.37141	1130.0	0.000	694.10	0.000	−19.3177	2.293×10^{-45}
Polenergia—Warszawa											
average daily temperature	2	1.9365	0.1446	0	0.44713	1509.1	0.000	886.59	0.000	−22.7157	4.333×10^{-50}
insolation	4	0.8714	0.4804	0	0.36359	1048.1	0.000	675.16	0.000	−17.8496	3.586×10^{-42}
average daily relative humidity	2	2.6845	0.0686	0	0.50785	1688.2	0.000	1060.6	0.000	−22.7127	4.356×10^{-50}
mean daily sea level pressure	3	0.0136	0.9978	0	0.37191	1138.2	0.000	695.27	0.000	−20.5336	1.650×10^{-47}
Kogeneracja—Wrocław											
average daily temperature	2	0.0877	0.9160	0	0.44752	1540.7	0.000	887.63	0.000	−22.2746	1.051×10^{-49}
duration of rainfall	5	0.6319	0.6754	0	0.33407	920.97	0.000	607.02	0.000	−17.1305	2.176×10^{-40}
mean daily overall cloudiness	3	0.3249	0.8073	0	0.44509	1310.0	0.000	880.47	0.000	−19.7930	5.163×10^{-47}
average daily relative humidity	5	1.2834	0.2684	0	0.31770	885.81	0.000	570.76	0.000	−17.1103	2.453×10^{-40}
mean daily sea level pressure	2	0.5638	0.5692	0	0.44485	1530.9	0.000	880.42	0.000	−22.2951	1.005×10^{-49}

The occurrence of simultaneous covariance between the random components of individual equations included in the VAR model allows the creation of the so-called structural models [53]. Structural VAR models enable the construction of the impulse response function (IRF), which allows the description of the time dependence. It illustrates the time distribution of changes in the magnitude of one variable in response to disturbances in the other's residuals. Most often, the IRF is presented graphically. The analysis of the impulse response function relates to three components: the direction of the impulse's interaction, the strength of the impulse as well as the time distribution and the rate of decay [52]. An example of the IRF function is shown in Figure 1. In the case under consideration, the structure of the graph is as follows: the value of the rate of returns caused by a change in a given weather factor is placed on the graph's ordinate, and the abscissa shows the time horizon of this impulse expressed in days. It is clear that changes in some weather variables trigger reactions in the percentage price changes (in this case, the Tauron company) in the first few periods. The decay of the pulses in both cases occurs fairly quickly, so the main reactions take place in the first few periods. The situation is similar in the case of impulse responses in relation to the trading volume (Energa case—Figure 2).

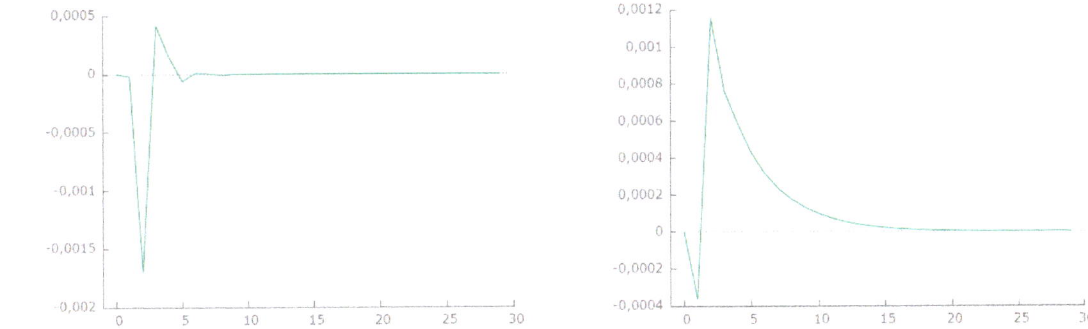

(**a**) (**b**)

Figure 1. Examples of functions of the rate of return reaction for Tauron, Katowice to an impulse: (**a**) insolation; (**b**) average daily relative humidity.

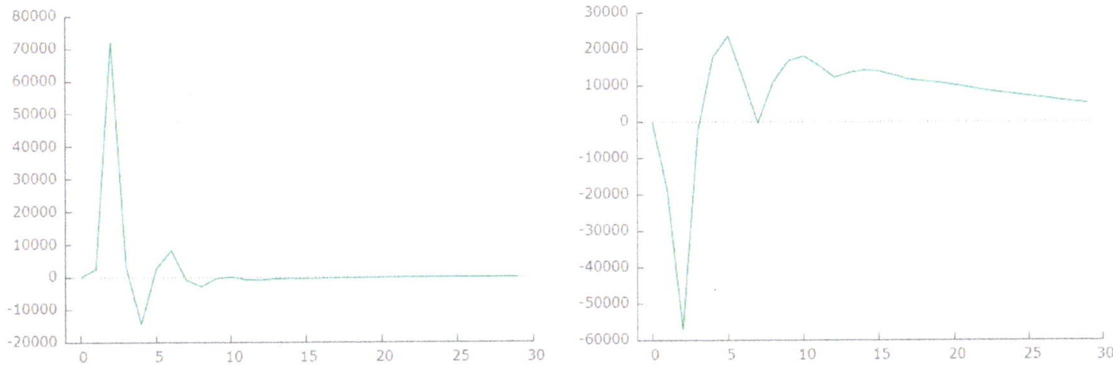

(**a**) (**b**)

Figure 2. Examples of the response functions of the trading volume for Energa, Gdańsk to the impulse: (**a**) average daily wind speed; (**b**) average daily relative humidity.

In the case of the VAR analysis for the relation rate of return ← the meteorological variable with the highest frequency of impact is relative humidity (33%). The variables insolation or cloudiness level have unitary indications in this regard. Thus, it is hard to define a tendency here. In this respect, it seems reasonable to make a summary list, which is a kind of compilation of the indications obtained so far in the section of the applied methodology. The results are presented in Table 8.

Table 8. The frequency of occurrence of the weather factor in the case of modeling the rate of return and the volume of trading in the cross-section of the methods used.

	Dependent Variable		
	Rate of Return		Trading Volume
	OLS	VAR	VAR
average daily temperature			2
daily sum of precipitation	1		2
insolation		1	
duration of rainfall			1
mean daily overall cloudiness		1	
average daily wind speed			3
average daily relative humidity		3	1
mean daily sea level pressure	1		

5. Discussion and Conclusions

Summarizing the conclusions drawn, it can be stated that:

- not all analytical methods are equally effective. This can be clearly seen in the example of the VAR analysis. The number of positive indications obtained in this case significantly exceeds the relevance of weather factors in the OLS modeling. The limitations of the OLS methodology are described earlier;
- co-integration analysis with the use of the Engle–Granger and Johansen tests is in a sense unjustified as it always indicates relationships between variables. It, therefore, disrupts the general view of causal relationships;
- determining the direction of the relationship weather factor $\rightarrow$ rate of return may be based only on the values of the tangent of the angle of inclination in the regression analysis;
- the most common determinants of the rate of return and the volume of trading include the weather factor in the form of average daily relative humidity and average wind speed. The first of these weather variables is the cause of the percentage changes in stock prices, the second is responsible for changes in the trading volume. The results in this case are different from those obtained in the studies of the team of Tarczyński et al. [54] with the use of ARCH class models. Then, the variable with a clear influence was atmospheric pressure. Therefore, it can be assumed that the considered causality depends on the methodology used;
- the trading volume seems to be much more susceptible to possible influences from independent variables;
- impulse response functions turn out to be extremely useful in VAR analyses, allowing additional inference in the case of dependence from meteorological time series;
- it would be advisable to continue research in the causality area and to verify the effectiveness of indications in the context of the distribution of the analyzed variables. Perhaps the nature of the distribution itself (convergence or divergence with the normal distribution) may affect the final indications;
- in the case of some weather variables, consideration should also be given to take into account the phenomenon of seasonality over a longer period of time;
- in general, it can be said that the indications depend on the type of method used and the nature of the test itself. Hence, weather variables play a role in modeling the mood of stock market investors investing in the energy sector, mainly in the context of Granger causality analysis.

Summarizing all of the research conducted, it is impossible to notice that there is no unique systematic and permanent correlation in the weather–stock market relationship. While in the case of one study we have confirmation of a certain state and assumptions, when studying subsequent studies we may have a different opinion, or previous beliefs may not be so important. Therefore, when summarizing the information presented in this paper, it should be noted that there are many potential reasons for the lack of this consensus.

Time is a factor that must certainly be considered. As already stated, Chang S.C. et al. [55] proved the influence of changes in weather factors during the market opening. On the other hand, Akhtari M. [56] discovered that there was not only the time of day that needed to be considered but also specific moments of time. Additionally, he noticed a certain cyclical nature of the whole process over the years. Trombley M.A. [57] and Lee Y.M and Wang K.M. [58] additionally stated that the analyzed relationships tended to fluctuate somewhat throughout the year.

Location may be another important element in this regard. For instance, Keef S.P. and Roush M.L [44] showed that the significance of various weather factors depends to a large extent on the analyzed (specific) location. In their opinion, it is one of the most critical factors in this type of analysis as climatic conditions vary from place to place. Moreover, there is a certain degree of heterogeneity with regard to individuals as well as their psychological characteristics in different cultures and regions; therefore, it is very likely that this could trigger different market reactions in response to the same weather conditions depending on the location of the stock markets. This fact was also confirmed by Simeonidis L., Daskalakis G. and Markellos R. [39]. The team of Loughran T. and Schultz P. [35] indicated that the volume of trading in shares differs depending on the place of residence of stock investors, which also confirms the previous thesis. Therefore, different conclusions from similar studies can be explained by the significant variation in variables and locations.

Moreover, attention should be paid to the definitions and hypotheses used. They often turn out to be crucial. It is not only the way of formulating the null hypothesis that is important but also the definitions [59]. This has a direct impact on the final conclusions.

In addition, the type of investors is important. Levy O. and Galili I. [60] and Shu H. [61] proved that the significance of the meteorological determinant largely depends on the type of investors. Moreover, the test procedures and statistics applied may give erroneous results [57,60].

Generally, whether a given weather factor is significant or not depends on many components: time, place, weather definition, formulation of hypotheses, type of investor, the procedure used and test statistics used in the research.

In addition, modeling the impact of weather factors should also take into account other aspects that may often result in complex, often contradictory, results of research on analyzed weather–investor behavior relations:

- many studies of this type do not take into account the phenomenon of the seasonality of twists. This may cause a distortion of the causal relationships between investor behavior and meteorological factors;
- some analyses did not investigate whether variables take into account other weather indicators. For instance, the effect of a certain temperature level on mood may depend on whether or not it is raining;
- as a rule, this type of analysis does not take into account the behavior of investors at different times of the year. There may be a dependence that higher winter/summer temperatures may positively/negatively affect decision-making processes;

There are many questions and possible doubts. Therefore, there remains the issue of seeking answers that will actually confirm the relations considered. Thus, any question marks appearing in the article constitute the background for further research on the analyzed problem.

Author Contributions: Conceptualization, U.M.; methodology, W.T. and U.M.; validation, U.M.; formal analysis, U.M.; investigation, U.M.; resources, U.M.; data curation, U.M.; writing—original draft preparation, U.M.; writing—review and editing, W.T., G.M. and U.S.; visualization, W.T., G.M. and U.S.; supervision, W.T., U.M., G.M. and U.S.; project administration, W.T. and U.M. All authors have read and agreed to the published version of the manuscript.

Funding: This research received no external funding.

Institutional Review Board Statement: Not applicable.

Informed Consent Statement: Not applicable.

Data Availability Statement: The data is part of the research carried out by Urszula Mentel in her doctoral dissertation entitled 'Weather conditions as a determinant of volatility in the stock market. Behavioral quantitative analysis'.

Acknowledgments: Not applicable.

Conflicts of Interest: The authors declare no conflict of interest.

References

1. Mehra, R.; Prescott, E.C. The equity premium: A puzzle. *J. Monet. Econ.* **1985**, *15*, 145–161. [CrossRef]
2. Kahneman, D.; Knetsch, J.L.; Thaler, R. Fairness as a constraint on profit seeking: Entitlements in the market. *Am. Econ. Rev.* **1986**, *76*, 728–741. [CrossRef]
3. Allen, M.A.; Fischer, G.J. Ambient Temperature Effects on Paired Associate Learning? *Ergonomics* **1978**, *21*, 95–101. [CrossRef]
4. Hu, T.-Y.; Xie, X.; Li, J. Negative or positive? The effect of emotion and mood on risky driving. *Transp. Res. Part F Traffic Psychol. Behav.* **2013**, *16*, 29–40. [CrossRef]
5. Keller, M.C.; Fredrickson, B.L.; Ybarra, O.; Cote, S.; Johnson, K.; Mikels, J.; Conway, A.; Wager, T. A Warm Heart and a Clear Head: The Contingent Effects of Weather on Mood and Cognition. *Psychol. Sci.* **2005**, *16*, 724–731. [CrossRef]
6. Howarth, E.; Hoffman, M.S. A multidimensional approach to the relationship between mood and weather. *Br. J. Psychol.* **1984**, *75*, 15–23. [CrossRef] [PubMed]
7. Loewenstein, G. Emotions in Economic Theory and Economic Behavior. *Am. Econ. Rev.* **2000**, *90*, 426–432. [CrossRef]
8. Saunders, E.M. Stock prices and the Wall Street weather. *Am. Econ. Rev.* **1993**, *83*, 1337–1345.
9. Hirshleifer, D.; Shumway, T. Good Day Sunshine: Stock Returns and the Weather. *J. Financ.* **2003**, *58*, 1009–1032. [CrossRef]
10. Floros, C. Stock market returns and the temperature effect: New evidence from Europe. *Appl. Financ. Econ. Lett.* **2008**, *4*, 461–467. [CrossRef]
11. Shim, H.; Kim, H.; Kim, J.; Ryu, D. Weather and stock market volatility: The case of a leading emerging market. *Appl. Econ. Lett.* **2014**, *22*, 987–992. [CrossRef]
12. Pardo, A.; Valor, E. Spanish stock returns: Where is the weather effect? *Eur. Financ. Manag.* **2003**, *9*, 117–126. [CrossRef]
13. Cao, M.; Wei, J. Stock market returns: A note on temperature anomaly. *J. Bank. Financ.* **2005**, *29*, 1559–1573. [CrossRef]
14. Chang, T.; Nieh, C.-C.; Yang, M.J.; Yang, T.-Y. Are stock market returns related to the weather effects? Empirical evidence from Taiwan. *Phys. A Stat. Mech. Appl.* **2005**, *364*, 343–354. [CrossRef]
15. Dowling, M.; Lucey, B.M. Robust global mood influences in equity pricing. *J. Multinatl. Financ. Manag.* **2008**, *18*, 145–164. [CrossRef]
16. Wright, W.F.; Bower, G.H. Mood effects on subjective probability assessment. *Organ. Behav. Hum. Decis. Process.* **1992**, *52*, 276–291. [CrossRef]
17. Bagozzi, R.P.; Gopinath, M.; Nyer, P.U. The Role of Emotions in Marketing. *J. Acad. Mark. Sci.* **1999**, *27*, 184–206. [CrossRef]
18. Isen, A.M.; Shalker, T.E.; Clark, M.; Karp, L. Affect, accessibility of material in memory, and behavior: A cognitive loop? *J. Pers. Soc. Psychol.* **1978**, *36*, 1–12. [CrossRef]
19. Forgas, J.P.; Bower, G.H. Mood effects on person-perception judgments. *J. Personal. Soc. Psychol.* **1987**, *53*, 53–60. [CrossRef]
20. Clore, G.L.; Schwarz, N.; Conway, M. Affective causes and consequences of social information processing. In *Handbook of Social Cognition: Basic Processes; Applications*; Wyer, W.R.S., Srull, T.K., Eds.; Lawrence Erlbaum Associates, Inc.: Hillsdale, NJ, USA, 1994; pp. 323–417.
21. Forgas, J.P. Mood and judgment: The affect infusion model (AIM). *Psychol. Bull.* **1995**, *117*, 39–66. [CrossRef]
22. Bless, H.; Schwarz, N.; Kemmelmeier, M. Mood and Stereotyping: Affective States and the Use of General Knowledge Structures. *Eur. Rev. Soc. Psychol.* **1996**, *7*, 63–93. [CrossRef]
23. Isen, A. Positive affect and decision making. In *Handbook of Emotions*; Lewis, W.M., Haviland-Jones, J., Eds.; The Guilford Press: New York, NY, USA, 2000; pp. 261–277.
24. Schwarz, N. Feelings as information: Informational and motivational functions of affective states. In *Handbook of Motivation and Cognition*; Sorrentino, W.R., Higgins, E.T., Eds.; Guildford Press: New York, NY, USA, 1990; Volume 2, pp. 527–561.
25. Petty, R.E.; Gleicher, F.; Baker, S.M. Multiple roles for affect in persuasion. In *International Series in Experimental Social Psychology. Emotion and Social Judgments*; Forgas, W.J.P., Ed.; Pergamon Press: Oxford, UK, 1991; pp. 181–200.
26. Sinclair, R.C.; Mark, M.M. The effects of mood state on judgemental accuracy: Processing strategy as a mechanism. *Cogn. Emot.* **1995**, *9*, 417–438. [CrossRef]
27. Bless, H.; Clore, G.L.; Schwarz, N.; Golisano, V.; Rabe, C.; Wolk, M. Mood and the use of scripts: Does being in a happy mood really lead to mindlessness? *J. Personal. Soc. Psychol.* **1996**, *71*, 665–679. [CrossRef]
28. Johnson, E.J.; Tversky, A. Affect, generalization, and the perception of risk. *J. Personal. Soc. Psychol.* **1983**, *45*, 20–31. [CrossRef]

29. Arkes, H.R.; Herren, L.T.; Isen, A.M. The role of potential loss in the influence of affect on risk-taking behavior. *Organ. Behav. Hum. Decis. Process.* **1988**, *42*, 181–193. [CrossRef]
30. Loewenstein, G.F.; Weber, E.U.; Hsee, C.K.; Welch, N. Risk as feelings. *Psychol. Bull.* **2001**, *127*, 267–286. [CrossRef]
31. Slovic, P.; Finucane, M.L.; Peters, E.; MacGregor, D.G. The affect heuristic. *Eur. J. Oper. Res.* **2007**, *177*, 1333–1352. [CrossRef]
32. Frijda, N.H. The laws of emotion. *Am. Psychol.* **1988**, *43*, 349–358. [CrossRef]
33. Wilson, T.D.; Schooler, J.W. Thinking too much: Introspection can reduce the quality of preferences and decisions. *J. Pers. Soc. Psychol.* **1991**, *60*, 181–192. [CrossRef]
34. Schwarz, N.; Clore, G.L. Mood, misattribution, and judgments of well-being: Informative and directive functions of affective states. *J. Pers. Soc. Psychol.* **1983**, *45*, 513–523. [CrossRef]
35. Loughran, T.; Schultz, P. Weather, Stock Returns, and the Impact of Localized Trading Behaviour. *J. Financ. Quant. Anal.* **2004**, *39*, 343–364. [CrossRef]
36. Kamstra, M.J.; Kramer, L.A.; Levi, M.D. Losing sleep at the market: The daylight saving anomaly. *Am. Econ. Rev.* **2000**, *90*, 1005–1011. [CrossRef]
37. Kang, S.H.; Jiang, Z.; Yoon, S.M. Weather Effects on the Returns and Volatility of Hong Kong and Shenzhen Stock Markets. 2010. Available online: http://citeseerx.ist.psu.edu/viewdoc/download?doi=10.1.1.466.5379&rep=rep1&type=pdf (accessed on 14 August 2021).
38. Dowling, M.; Brian, M.L. Weather, Biorhythms, Beliefs and Stock Returns Some—Preliminary Irish Evidence. *Int. Rev. Financ. Anal.* **2005**, *14*, 337–355. [CrossRef]
39. Symeonidis, L.; Daskalakis, G.; Markellos, R.N. Does the weather affect stock market volatility? *Financ. Res. Lett.* **2010**, *7*, 214–223. [CrossRef]
40. Goetzmann, W.N.; Zhu, N. Rain or Shine: Where is the Weather Effect? *Eur. Financ. Manag.* **2005**, *11*, 559–578. [CrossRef]
41. Hong, H.; Kubik, J.D.; Stein, J.C. Social Interaction and Stock-Market Participation. *J. Financ.* **2004**, *59*, 137–163. [CrossRef]
42. Vlady, S.; Tufan, E.; Hamarat, B. Causality of Weather Conditions in Australian Stock Equity Returns. *Rev. Tinerilor Econ.* **2011**, *1*, 187–194.
43. Worthington, A. An Empirical Note on Weather Effects in the Australian Stock Market. *Econ. Pap. A J. Appl. Econ. Policy* **2009**, *28*, 148–154. [CrossRef]
44. Keef, S.P.; Roush, M.L. Daily weather effects on the returns of Australian stock indices. *Appl. Financ. Econ.* **2007**, *17*, 173–184. [CrossRef]
45. Wang, Y.-H.; Lin, C.-T.; Lin, J.D. Does weather impact the stock market? Empirical evidence in Taiwan. *Qual. Quant.* **2011**, *46*, 695–703. [CrossRef]
46. Tuna, G. Analyzıng Weather Effect on Istanbul Stock Exchange: An Empırıcal Analysıs for 1987–2006 Period. *Econ. Financ. Rev.* **2014**, *3*, 17–25.
47. Silva, P.; Almeida, L. *Weather and Stock Markets: Empirical Evidence from Portugal*; MPRA Paper 54119; University Library of Munich: Munich, Germany, 2011.
48. Zadorozhna, O. Does Weather Affect Stock Returns Across Emergıng Markets? Master's Thesis, Kyiv School of Economics, Kyiv, Ukraine, 2009.
49. Engle, R.F.; Granger, C.W.J. Co-Integration and Error Correction: Representation, Estimation, and Testing. *Econometrica* **1987**, *55*, 251. [CrossRef]
50. Johansen, S. Statistical analysis of cointegration vectors. *J. Econ. Dyn. Control.* **1988**, *12*, 231–254. [CrossRef]
51. Maddala, G.S. *Ekonometria*; PWN: Warszawa, Poland, 2006.
52. Hamulczuk, M.; Grudkowska, S.; Gędek, S.; Klimkowski, C.; Stańko, S. *Essential Econometric Methods of Forecasting Agricultural Commodity Prices*; Institute of Agricultural and Food Economics—National Research Institute: Warszawa, Poland, 2013.
53. Kusideł, E. *Dane Panelowe i Modelowanie Wielowymiarowe w Badaniach Ekonomicznych*; Absolwent: Łódź, Poland, 2000.
54. Tarczyński, W.; Majewski, S.; Tarczyńska-Łuniewska, M.; Majewska, A.; Mentel, G. The Impact of Weather Factors on Quotations of Energy Sector Companies on Warsaw Stock Exchange. *Energies* **2021**, *14*, 1536. [CrossRef]
55. Chang, S.-C.; Chen, S.-S.; Chou, R.; Lin, Y.-H. Weather and intraday patterns in stock returns and trading activity. *J. Bank. Financ.* **2008**, *32*, 1754–1766. [CrossRef]
56. Akhtari, M. Reassessment of the Weather Effect: Stock Prices and Wall Street Weather. *Undergrad. Econ. Rev.* **2011**, *7*, 1–25.
57. Trombley, M.A. Stock Prices and Wall Street Weather: Additional Evidence. *Q. J. Bus. Econ.* **1997**, *36*, 11–21.
58. Lee, Y.-M.; Wang, K.-M. The effectiveness of the sunshine effect in Taiwan's stock market before and after the 1997 financial crisis. *Econ. Model.* **2011**, *28*, 710–727. [CrossRef]
59. Krämer, W.; Runde, R. Stocks and the Weather: An Exercise in Data Mining or Yet Another Capital Market Anomaly? *Empir. Econ.* **1997**, *22*, 637–641. [CrossRef]
60. Levy, O.; Galili, I. Stock purchase and the weather: Individual differences. *J. Econ. Behav. Organ.* **2008**, *67*, 755–767. [CrossRef]
61. Shu, H.-C. Investor mood and financial markets. *J. Econ. Behav. Organ.* **2010**, *76*, 267–282. [CrossRef]

energies

MDPI

Article

The Nexus of Energy, Green Economy, Blue Economy, and Carbon Neutrality Targets

Suleman Sarwar [1], Rida Waheed [2], Ghazala Aziz [3] and Simona Andreea Apostu [4,5,*]

1 Finance and Economics Department, University of Jeddah, Jeddah 21959, Saudi Arabia
2 Finance and Economics Department, College of Business, University of Jeddah, Jeddah 23445, Saudi Arabia
3 Department of Business Administration, College of Administrative and Financial Sciences, Saudi Electronic University, Jeddah 23442, Saudi Arabia
4 Department of Statistics and Econometrics, Bucharest University of Economic Studies, 010552 Bucharest, Romania
5 Institute of National Economy, 050771 Bucharest, Romania
* Correspondence: simona.apostu@csie.ase.ro

Abstract: The aim of current study is to investigate the significance of green and blue economic activities to mitigate the carbon emission in Saudi Arabia. We use the time series data which covers the period from 1990 to 2019. For empirical estimations, we use nonlinear ARDL approach which confirms that energy indicators and blue economic indicators are not mature enough to achieve carbon neutrality objectives. However, after Vision 2030 empirics, positive shocks in green indicators are turning down the carbon level. The findings of energy and blue indicators are useful for policy recommendations which help to achieve the sustainable environmental goals of Vision 2030.

Keywords: energy; green indicators; blue indicators; carbon neutrality; nonlinear ARDL

Citation: Sarwar, S.; Waheed, R.; Aziz, G.; Apostu, S.A. The Nexus of Energy, Green Economy, Blue Economy, and Carbon Neutrality Targets. *Energies* **2022**, *15*, 6767. https://doi.org/10.3390/en15186767

Academic Editors: Grzegorz Mentel and Sebastian Majewski

Received: 16 August 2022
Accepted: 8 September 2022
Published: 16 September 2022

1. Introduction

Carbon neutrality is the term used to explain that the overall carbon dioxide emission and its offset are equal. It is a fact that a huge amount of carbon is released into the atmosphere due to economic, as well as social activities. However, the same amount of carbon squeezed and makes net carbon emission nearly zero is carbon neutrality, which may guarantee environmental protection. While discussing about the sustainable economic development, carbon neutrality seems important, and majority of the countries around the world are giving high priority to carbon neutrality so that economic development can be achieved without deteriorating the environment.

Though carbon neutrality is important for all countries, the same is the case for Saudi Arabia. The Saudi economy is the largest economy of the middle east and also on the list of top 20 economies of the world, due to its high level of natural resources including oil and petroleum. Alongside, industrialization, as well as urbanization, is also increasing in the country. All these factors contributed to the overall economic growth of the country; however, this also lead to a high level of carbon emission. Figure 1 shows the severity of the issue as it can be seen that carbon emission in the country is increasing significantly since 1965. Although many efforts are made by the government to reduce this intensity and to create carbon sinks to squeeze excess carbon [1,2] to achieve carbon neutrality. However, up till now, the problem is still prevailing and needs immediate attention so that new ways could be found to help the country attain sustainable economic development along with carbon neutrality.

After realizing the importance of carbon neutrality for sustainable economic development, researchers tried to explore different factors which can help in the reduction of carbon emission and achieving carbon neutrality. In his regard, it is noted that renewable energy can be an important measure to achieve carbon neutrality [3–6]. According to these studies, carbon neutrality in selected countries is enhanced by reducing carbon emissions using renewable energy sources. Similarly, energy intensity is also noted to be a significant factor toward carbon neutrality [7–10]. Besides these solutions, innovations are also presented by previous studies as a potential solution for carbon-related problems of the environment and it is noted that through innovative and eco-friendly products, carbon neutrality can be enhanced [11–14]. In terms of the importance of oceans for carbon neutrality, researchers tried to explore the impact of the fishery industry on carbon neutrality and found that the development of this industry plays a vital role in the reduction of carbon emission as well as its absorption [15–17]. These research conclude that fishery is an important source toward the achievement of carbon neutrality.

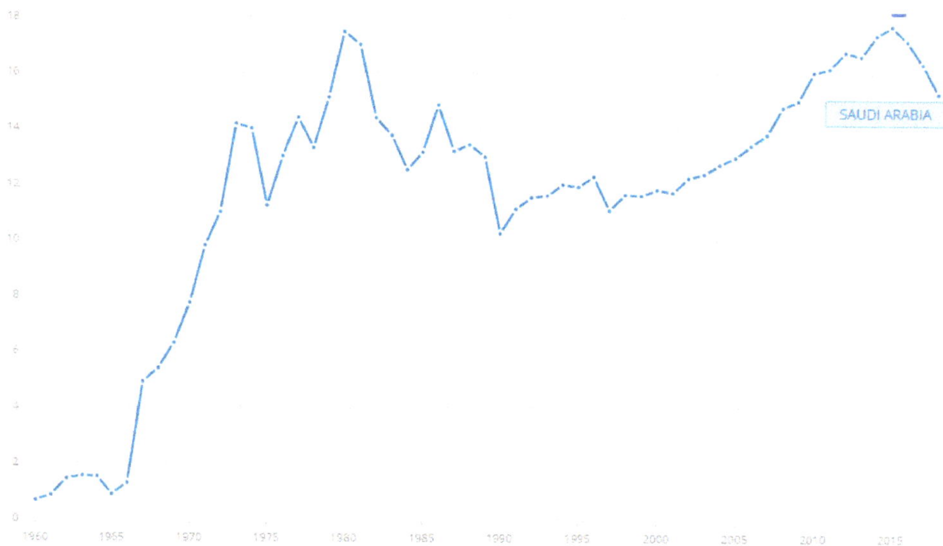

Figure 1. CO_2 emissions—Saudi Arabia (Source: World Development Bank).

Critical analysis of previous literature revealed that none of the studies explore the joint effect of energy components, green components, and blue components on carbon neutrality. Some studies explore few components including renewable energy, innovations and fishery; however, previous studies used only one component. Hence, in light of previous research, the current study is an attempt to fill the gap by addressing four contributions: first, the study explores the impact of the technical grant on carbon neutrality which is not yet explored in empirical research. Technical grants can play a significant role toward carbon neutrality as huge funds are required to develop technically sophisticated but eco-friendly products. This grant enables emerging economies such as Saudi Arabia to enhance research and development in the country to develop sustainable products. These grants help in human wellbeing by providing them clean environment. Currently, no research directly explores the impact of technical grants on carbon neutrality. This study is a pioneer in this regard and uses Saudi Arabia's data to explore this nexus.

The second contribution of the study is exploring the joint effect of blue factors including fishery, and marine trade on carbon neutrality. There is no research done to explore if marine trade can also play any role in the reduction of carbon emissions. The importance of marine trade cannot be ignored because the majority of global trade is done through water and a huge amount of energy is consumed in this sector. Although currently it is noted that the contribution of this sector in carbon emission is low [18] as compared to other forms of trade. This shows that incorporation of this factor in research is important to check the role of marine trade in Saudi Arabia, in terms of carbon neutrality, so that policy implications can be formulated ahead of time.

The third contribution is the investigation of energy factors, green factors, and blue factors simultaneously, to know the combined impact of these factors on carbon neutrality. As far as green economic factors are concerned, Du et al. [19] argue that technological innovations can help in the reduction of carbon emission. This argument motivates us to check if these green economic factors can help in achieving carbon neutrality or not. In light of these arguments, it seems vital to examine the impact of energy, green and blue economic factors on carbon neutrality; it would be beneficial as it will allow us to put forward important policy implications in terms of ways to achieve carbon neutrality.

The last contribution of this research is to check the impact of Vision 2030 on the carbon neutrality targets of Saudi Arabia. The main reason is the fact that fossil fuel consumption is the main energy source of the country; however, under Vision 2030, it is committed that usage of sustainable energy sources, such as renewable energy, will be implemented to meet the carbon neutrality targets. Hence, it is important to check how many sustainable goals of Saudi Vision 2030 are going to be met. The data set is divided according to the availability since Vision 2030 implementation so that direction of this initiative toward carbon neutrality can be noticed. The findings of the study would be helpful for policymakers to examine the directions and success of Vision 2030, in terms of offsetting the carbon emission and achieving the carbon neutrality goals. For data analysis purposes, the nonlinear ARDL technique is used to check the impact of energy factors, green factors, and blue factors to achieve carbon neutrality in Saudi Arabia.

To account the above-mentioned contributions, the study draws five objectives to address the research questions. The first objective is to examine the impact of energy factors on carbon neutrality. The second objective is to investigate the relationship between green economic factors and carbon neutrality. The third objective is to address the consequences of blue factors on carbon neutrality. The fourth objective is to compare the effect of energy factors, green factors, and blue factors on carbon neutrality to know which factors are most beneficial in terms of achieving carbon neutrality in Saudi Arabia. The fifth objective is to check the importance of Vision 2030 for the carbon neutrality targets of Saudi Arabia.

2. Literature Review

The literature is divided into three sections; energy factors and carbon neutrality, green indicators and carbon neutrality, and blue economic factors and carbon neutrality.

2.1. Impact of Energy Factors on Carbon Neutrality

When it comes to carbon neutrality, energy factors are the first to be considered because it is noted that energy consumption is the main source of carbon-related environmental issues. Currently, researchers are trying to explore how energy intensity is related to carbon neutrality and in this regard Gil and Bernardo [20] explored the measures which can be taken in Portugal to achieve carbon neutrality. They assess the road map toward carbon neutrality and asserts that reduction in energy intensity is the pathway toward achieving carbon neutrality. Similarly, Wang et al. [21] used Chinese data from the year 2007 to 2019 to investigate how much carbon neutrality targets age achieved in the country through effective use of energy by reducing energy intensity in this period. They found that energy intensity is high in China and it is recommended that strict measures should be taken against industries that are energy-intensive so that carbon neutrality goals can be achieved. Likewise, another study is done by Andersson and Karpestam [22] to check how carbon neutrality can be achieved by examining the determinants of energy intensity which is the main source of carbon emission. They used data from two emerging and two developed countries and asserts that although capital is the main source of carbon emission, however, carbon neutrality can be achieved through a reduction in energy intensity by making each unit of capital more energy efficient.

Besides energy intensity, renewable energy can also be related to carbon neutrality as Li et al. [5] used Chinese data from year 1989 to 2019 to check the relationship between renewable energy and carbon neutrality. They found that China's carbon neutrality targets can be achieved through using renewable energy sources as it reduces carbon emissions significantly. Additionally, Li et al. [4] conducted a study in top exporting countries of the world to examine if renewable energy can be a source of carbon neutrality in these countries. They found a positive association between renewable energy and carbon neutrality and recommend that to achieve carbon neutrality, countries should promote the use of renewable energy sources. Similarly, Juan Lin et al. [23] explored the nexus between renewable energy sources and carbon neutrality in BRICS countries from the year 1980–2018. Their results also indicate that renewable energy consumption is a viable option to achieving carbon neutrality in these countries.

2.2. Impact of Green Factors on Carbon Neutrality

Carbon neutrality can also be achieved through green economic factors including innovations and technical grants. As far as innovation is concerned, it is the growth of energy-efficient, green, and clean products which leads to carbon neutrality. In this regard, Sun et al. [13] researched Turkey to investigate how ecological innovations can help in achieving carbon neutrality targets in emerging countries. By applying the QARDL approach to Turkish data from 1995 to 2018, they found that these innovations have a significant and positive impact on carbon neutrality targets. Another study by Iqbal et al. [24] investigated the nexus between innovation and carbon neutrality in OECD economies from the year 1970 to 2019. Through the augmented mean group approach, they found that innovations play a significant role in helping these economies to achieve their carbon neutrality targets. Similarly, Shao et al. [25] employed (CS-ARDL) test on data from N-11 countries from 1980 to 2018 to check if the positive relationship between innovation and carbon neutrality is valid. Their result confirmed that green technological innovations play a significant and positive role in the carbon neutrality of these countries.

Although it is a top priority of countries around the world to reduce carbon footprint and achieve carbon neutrality, however, it is also a fact that technologies required for this goal are expensive and countries need support for this. The technical grant helps in the development of sophisticated and updated products and systems for environmental protection as well as the introduction of programs for human development. In this regard, the green climate fund is established by United Nations to help developing countries to achieve carbon neutrality by reducing carbon emissions. According to United Nations Environment Program [26] these funds help in the reduction of carbon emission of 4.4 million tons. Moreover, this grant creates jobs that affect the lives of people directly and many other areas will also be impacted in terms of the 2030 Agenda for Sustainable development [27]. Likewise, another technical grant is also approved by the US government for carbon reduction so that until the mid of century, carbon neutralization can be achieved. It shows that technical grants can play a positive role in carbon neutrality.

2.3. Impact of Blue Factors on Carbon Neutrality

In terms of the blue economy, fishery, as well as marine trade, are important factors. As far as the fishery is concerned, the economic impact of this industry is very prominent and it is believed that the fishing industry has the least carbon footprint as compared to other types of food [28]. The main reason is the fact that fishery does not need livestock care as well as farmland Mcdermott [29]; hence, previously it is believed that for carbon neutrality, the fishery can play a role. Logically this seems valid and research literature also validates this point. According to Wan et al. [30], although it is believed that energy consumption in the fishery is a carbon contributing element, however, through marine ranching, Zhou et al. [17], a significant amount of carbon can be absorbed from the atmosphere, Wan et al. [30], which can balance the overall carbon in the air. Another channel through which fishery helps in carbon neutralization is through the cultivation of seaweed as according to Feng et al. [16].

This reduced a significant amount of carbon in China proving that there is a huge potential in ocean-based solutions for climate change.

As far as the marine trade is concerned, it is noted that almost 80% of the trade is done through this channel. This shows the importance of marine trade for the economy; however, as with all types of economic activities, environmental considerations for marine trade are also important. According to Ozer et al. [31] assessment of different types of gas emissions by marine trade is crucial in terms of the implementation of regulations. In this regard, Taghvaee et al. [32] conducted a study to investigate the difference in carbon emission by water transport and other forms of transport. Their analysis revealed that marine trade is the most energy-effective transport and that is why the overall carbon emission by marine trade is significantly lower. The same type of argument is also made by Fratila et al. [33] who mentioned that in terms of carbon emission, marine trade proved to be the most efficient type of transportation. However, Ben Jebli and Belloumi [34] found that there exists bidirectional causality between marine trade and carbon emission. They used data from Tunisia and concluded that marine trade contributes to carbon emission.

The above-mentioned literature highlights that there is not even a single study that checks the impact of marine trade on carbon neutrality. However, it is important to check this nexus because if carbon emission can be affected by marine trade then it is important to check how and through which channel it can help in achieving carbon neutrality. Moreover, empirical research on nexus between technical grant and carbon neutrality is missing. It is essential to fill this research gap by using empirical data to investigate the role of technical grants to attain carbon neutrality. Alongside, different elements of energy factors, blue economic factors, and green economic factors show some impact on carbon neutrality. However, the combined impact of these factors on carbon neutrality is missing in the literature. Hence, it is important to fill this research gap so that major policy implications can be advised.

3. Data and Methodology

3.1. Data

For empirics, the data of carbon, energy, green and blue indicators are covering the period of 1990–2019. According to Union of Concerned Statistics (https://www.ucsusa.org/resources/each-countrys-share-co2-emissions, accessed on 12 July 2022), Saudi Arabia is the higher per capital carbon emitter in the world; however, it is urgent to study the significant measures that may help to minimize the environmental externalities. To address this environmental issue, the study addresses the role of energy factors, green factors, and blue factors to attain carbon neutrality. For energy factors, energy intensity and renewable energy are used as a proxies for empirical estimations. Green factors included the innovation and technical grant. Fishery and ocean trade are used as a proxy of blue indicators. The data of carbon emission, energy intensity, innovation and technical grants are obtained from World Development Indicators (WDI), renewable energy is from International Renewable Energy Agency (IRENA), fishery and ocean trade is collected from Food and Agriculture Organization of the United States (FAO). The definitions and notations are reported in Table 1.

The studied models for energy, green and blue indicators are as below:

Model 1: $CE = f\ (EI,\ RE)$
Model 2: $CE = f\ (INNOV,\ GRANT)$
Model 3: $CE = f\ (FISHERY,\ OTRADE)$

The models are formed by using the studies of [35–37] which have ignored the economic growth and population in the studied models. Iqbal et al. [24] has not included the population in econometric model, whereas, Yue et al. [14] has not used the economic growth. However, we have formed the model specifications to confirm the reliability of models, such as used the omitted variable bias test (we use STATA command "ovtest" for omitted variable bias. The F-statistics for Model 1 is 0.07, Model 2 is 0.24, and Model 3 is 0.94, with p-values 0.977, 0.865, and 0.438, respectively. However, it is concluded that the models have no omitted variable bias). Moreover, to investigate the implications of Saudi Vision 2030, we split the data into two sub sets; before-Vision and after-Vision. This assists to evaluate the impacts of plans that are formulated under the umbrella of Vision 2030, which helps to determine the future plans and directions that lead toward the carbon neutrality.

Table 1. Data and source.

Description	Notation	Proxy	Source	
Carbon Emission	CE	CO_2 emissions (kt)	WDI	World Development Indicators
Energy Intensity	EI	Energy intensity level of primary energy (MJ/2011 USD PPP)	WDI	World Development Indicators
Renewable Energy	RE	Renewable energy consumption (TJ)	IRENA	International Renewable Energy Agency
Innovation	INNOV	Total Patent (Patent applications, nonresidents + Patent applications, residents)	WDI	World Development Indicators
Grants	GRANT	Technical cooperation grants (BoP, current US$)	WDI	World Development Indicators
Fishery	FISHERY	Total fisheries production (metric tons)	FAO	Food and Agriculture Organization of the United Nations
Ocean Trade	OTRADE	Commodity Trade and Production: Value (USD, 000)	FAO	Food and Agriculture Organization of the United Nations

3.2. *Methodology*

Autoregressive Distributed Lag (ARDL)

$$\overbrace{CE}^{Carbon\ Emission} = \overbrace{EI \quad RE}^{Energy\ Factors}, \overbrace{INNOV \quad GRANT}^{Green\ Factors}, \overbrace{FISHERY \quad OTRADE}^{Blue\ Factors} \tag{1}$$

$$lnCE_t = \alpha_0 + \beta_1\, lnEI_t + \beta_2\, lnRE_t + \varepsilon_t \tag{2}$$

$$lnCE_t = \alpha_0 + \beta_1\, lnINNOV_t + \beta_2\, lnGRANT_t + \varepsilon_t \tag{3}$$

$$lnCE_t = \alpha_0 + \beta_1\, lnFISHERY_t + \beta_2\, lnOTRADE_t + \varepsilon_t \tag{4}$$

CE is carbon neutrality in the above equations, EI is energy efficiency, RE is renewable energy, $INNOV$ is innovation, $FISHERY$ is the fishery, and $OTRADE$ is ocean trade. The error term is shown by ε_t, whereas β_1, β_2 are elasticity coefficients for the long term. For the cointegration test, short-run elasticities but long-run cointegration is checked in these equations. This is done because the only long-run impact of explanatory variables on carbon neutrality is checked in these equations.

ARDL form of Equations (2)–(4) takes the following form:

$$\Delta lnCE_t = \alpha_0 + \sum_{i=1}^{n} \mu_1 \Delta lnCE_{t-i} + \sum_{i=0}^{n} \mu_2 \Delta lnEI_{t-i} + \sum_{i=0}^{n} \mu_2 \Delta lnRE_{t-i} + \gamma_0 lnCE_{t-1} + \gamma_1 lnEI_{t-1} + \gamma_2 lnRE_{t-1} + \omega_t \tag{5}$$

$$\begin{aligned} \Delta lnCE_t = {} & \alpha_0 + \sum_{i=1}^{n} \mu_1 \Delta lnCE_{t-i} + \sum_{i=0}^{n} \mu_2 \Delta lnINNOV_{t-i} + \sum_{i=0}^{n} \mu_2 \Delta lnGRANT_{t-i} + \gamma_0 lnCE_{t-1} \\ & + \gamma_1 lnINNOV_{t-1} + \gamma_2 lnGRANT_{t-1} + \omega_t \end{aligned} \tag{6}$$

$$\Delta lnCE_t = \alpha_0 + \sum_{i=1}^{n} \mu_1 \Delta lnCE_{t-i} + \sum_{i=0}^{n} \mu_2 \Delta lnFISHERY_{t-i} + \sum_{i=0}^{n} \mu_2 \Delta lnOTRADE_{t-i} + \gamma_0 lnCE_{t-1} + \gamma_1 lnFISHERY_{t-1} + \gamma_2 lnOTRADE_{t-1} + \omega_t \quad (7)$$

By converting the above equations in matrix form by making all variables as dependent variables, Equations (5)–(7) can be formed. It is essential to measure the long as well as short-run cointegration and do this. The hypothesis is formulated where, in the long run, the null hypothesis of no cointegration is $[H0 : \gamma 11\ to\ \gamma 33 = 0]$, and the alternative hypothesis is $[H0 : \gamma 11\ to\ \gamma 33 \neq 0]$. Similarly, the short-run null hypothesis is $[H0 : \mu 11\ to\ \mu 33 = 0]$ and the alternative hypothesis is $[H0 : \mu 11\ to\ \mu 33 \neq 0]$.

$$(1-B)\begin{bmatrix} lnCE \\ lnEI \\ lnRE \end{bmatrix} = \begin{bmatrix} \alpha_{01} \\ \alpha_{02} \\ \alpha_{03} \end{bmatrix} + \sum_{i=1}^{k} 1 - B \begin{bmatrix} lnCE \\ lnEI \\ lnRE \end{bmatrix}_{t-i} \times \begin{bmatrix} \mu_{11} & \mu_{12} & \mu_{13} \\ \mu_{21} & \mu_{22} & \mu_{23} \\ \mu_{31} & \mu_{32} & \mu_{33} \end{bmatrix} + \begin{bmatrix} lnCE \\ lnEI \\ lnRE \end{bmatrix}_{t-1} \times \begin{bmatrix} \gamma_{11} & \gamma_{12} & \gamma_{13} \\ \gamma_{21} & \gamma_{22} & \gamma_{23} \\ \gamma_{31} & \gamma_{32} & \gamma_{33} \end{bmatrix} + \begin{bmatrix} \omega \\ \omega \\ \omega \end{bmatrix}_t \quad (8)$$

$$(1-B)\begin{bmatrix} lnCE \\ lnINNOV \\ lnRRANT \end{bmatrix} = \begin{bmatrix} \alpha_{01} \\ \alpha_{02} \\ \alpha_{03} \end{bmatrix} + \sum_{i=1}^{k} 1 - B \begin{bmatrix} lnCE \\ lnINNOV \\ lnRRANT \end{bmatrix}_{t-i} \times \begin{bmatrix} \mu_{11} & \mu_{12} & \mu_{13} \\ \mu_{21} & \mu_{22} & \mu_{23} \\ \mu_{31} & \mu_{32} & \mu_{33} \end{bmatrix} + \begin{bmatrix} lnCE \\ lnINNOV \\ lnGRANT \end{bmatrix}_{t-1} \times \begin{bmatrix} \gamma_{11} & \gamma_{12} & \gamma_{13} \\ \gamma_{21} & \gamma_{22} & \gamma_{23} \\ \gamma_{31} & \gamma_{32} & \gamma_{33} \end{bmatrix} + \begin{bmatrix} \omega \\ \omega \\ \omega \end{bmatrix}_t \quad (9)$$

$$(1-B)\begin{bmatrix} lnCE \\ lnFISHERY \\ lnOTRADE \end{bmatrix} = \begin{bmatrix} \alpha_{01} \\ \alpha_{02} \\ \alpha_{03} \end{bmatrix} + \sum_{i=1}^{k} 1 - B \begin{bmatrix} lnCE \\ lnFISHERY \\ lnOTRADE \end{bmatrix}_{t-i} \times \begin{bmatrix} \mu_{11} & \mu_{12} & \mu_{13} \\ \mu_{21} & \mu_{22} & \mu_{23} \\ \mu_{31} & \mu_{32} & \mu_{33} \end{bmatrix} + \begin{bmatrix} lnCE \\ lnFISHERY \\ lnOTRADE \end{bmatrix}_{t-1} \times \begin{bmatrix} \gamma_{11} & \gamma_{12} & \gamma_{13} \\ \gamma_{21} & \gamma_{22} & \gamma_{23} \\ \gamma_{31} & \gamma_{32} & \gamma_{33} \end{bmatrix} + \begin{bmatrix} \omega \\ \omega \\ \omega \end{bmatrix}_t \quad (10)$$

where, Δ = 1st difference operator, μ_1 to μ_3 = short run elasticity operators, γ_1 to γ_3 = long run elasticity operators, α_0 = constant, ω_t = noise.

F-statistics and critical values are used to decide whether to accept or reject the hypothesis. The critical values presented by Narayan [38] and Pesaran et al. [39] are used for conclusive statements regarding cointegration.

Nonlinear ARDL

Cointegration tests are based on the linear association between independent and dependent variables. However, this association can be negative and positive, and to cater this asymmetric association, a nonlinear ARDL approach is used [38].

According to Bahmani-Oskooee and Mohammadian [40] and Delatte and lopez-Villavicencio [41], independent variable should be decomposed into two more set of series by considering the positive and negative changes. Following this, Equations (11)–(13) take the following form:

$$\begin{cases} POS(EF)_t = \sum_{L=1}^{t} lnEF_L^+ = \sum_{L=1}^{T} MAX(\Delta lnEF_L, 0) \\ NEG(EF)_t = \sum_{L=1}^{t} lnEF_k^- = \sum_{L=1}^{T} MAX(\Delta lnEF_L, 0) \end{cases} \quad (11)$$

$$\begin{cases} POS(GF)_t = \sum_{L=1}^{t} lnGF_L^+ = \sum_{L=1}^{T} MAX(\Delta lnGF_L, 0) \\ NEG(GF)_t = \sum_{L=1}^{t} lnGF_k^- = \sum_{L=1}^{T} MAX(\Delta lnGF_L, 0) \end{cases} \quad (12)$$

$$\begin{cases} POS(BF)_t = \sum_{L=1}^{t} lnBF_L^+ = \sum_{L=1}^{T} MAX(\Delta lnBF_L, 0) \\ NEG(BF)_t = \sum_{L=1}^{t} lnBF_k^- = \sum_{L=1}^{T} MAX(\Delta lnBF_L, 0) \end{cases} \quad (13)$$

After incorporating the positive and negative changes, Equations (8)–(10) takes the following form:

$$\Delta lnCE_t = \alpha_0 + \sum_{i=1}^{n} \mu_1 \Delta lnCE_{t-i} + \sum_{i=0}^{n} \mu_2^{+} \Delta lnPOS(EF)_{t-i} + \sum_{i=0}^{n} \mu_2^{-} \Delta lnNEG(EF)_{t-i} + \gamma_0 lnCE_{t-1} + \gamma_1^{+} lnPOS(EF)_{t-1} + \gamma_1^{-} lnNEG(EF)_{t-1} + \omega_t \quad (14)$$

$$\Delta lnCE_t = \alpha_0 + \sum_{i=1}^{n} \mu_1 \Delta lnCE_{t-i} + \sum_{i=0}^{n} \mu_2^{+} \Delta lnPOS(GF)_{t-i} + \sum_{i=0}^{n} \mu_2^{-} \Delta lnNEG(GF)_{t-i} + \gamma_0 lnCE_{t-1} + \gamma_1^{+} lnPOS(GF)_{t-1} + \gamma_1^{-} lnNEG(GF)_{t-1} + \omega_t \quad (15)$$

$$\Delta lnCE_t = \alpha_0 + \sum_{i=1}^{n} \mu_1 \Delta lnCE_{t-i} + \sum_{i=0}^{n} \mu_2^{+} \Delta lnPOS(BF)_{t-i} + \sum_{i=0}^{n} \mu_2^{-} \Delta lnNEG(BF)_{t-i} + \gamma_0 lnCE_{t-1} + \gamma_1^{+} lnPOS(BF)_{t-1} + \gamma_1^{-} lnNEG(BF)_{t-1} + \omega_t \quad (16)$$

where, μ_1 and μ_2 = coefficients of short run elasticity, γ_0 and γ_1 = coefficients of long run elasticity, CE_t = carbon neutrality, EF = energy factors, GF = green factors, BF = blue factors.

The Wald test is applied to measure the short run and long run asymmetries, whereas, for optimal lag determination, the Akaike information criterion is used. It is also worth mentioning that bound test is also necessary [42] to check the long-run cointegration by comparing f-statistics and critical value. In this scenario, the null hypothesis can be presented through $\gamma 0 = \gamma 1+ = \gamma 1- = 0$.

4. Results

4.1. Descriptive Statistics

Descriptive statistics for main variables are presented in Table 2, and it can be noted that grant has the highest mean value and energy intensity has the lowest mean value. In the case of standard deviation, the highest standard deviation is for innovation showing that innovation is highly volatile, whereas energy intensity is least volatile.

Table 2. Summary of statistics.

Variable	Obs	Mean	Std. Dev.	Min	Max
CE	30	12.677	0.390	12.022	13.238
EI	30	1.691	0.084	1.435	1.832
RE	30	5.585	0.305	5.088	6.285
INNOV	30	6.893	0.610	6.120	8.203
GRANT	30	16.537	0.185	16.182	16.912
FISHERY	30	11.184	0.341	10.656	11.911
OTRADE	30	12.355	0.973	10.876	13.806

Notes: CE shows the carbon neutrality, EI and RE are energy intensity and renewable energy respectively, which are the energy factors. INNOV and GRANT represents the innovation and technical grants respectively, which reflects the green indicators. FISHERY and OTRADE mentions the fishery and ocean trade that reflects the blue factors.

4.2. Structural Break Test

It is necessary to check if there are any structural breaks in the data and for this purpose Chow structural break test is used. The null hypothesis represents no structural break, whereas the alternative hypothesis represents that there is a structural break. The result of the Chow structural break test is presented in Table 3. It can be seen that the value of F-statistics is insignificant. Hence, it can be said that there is no structural break, and coefficients are constant across the sample. In addition to Chow structural break test, we used the Quandt-Andrew structural break test to confirm the findings of Chow test. The null hypothesis of Quandt-Andrew breakpoint test represents that there exists no break point [43,44]. The findings are reported in second section of Table 3, which fail to reject the null hypothesis. However, both breakpoint tests mention the non-existence of structural breaks in data.

Table 3. Structural break test.

Chow structural break			
F-Statistics	2.491	Prob. F(7,16)	0.062
Quandt-Andrews structural break test	Maximum LR	Expected LR	Average LR
F-Statistics	129.285	120.552	100.749

Note: The null hypothesis represents that coefficients are constant across sample (no structural break). Chow F-statistics is calculated by using this formula "[{RSSr-(RSS1 + RSS2)}/k]/[(RSS1 + RSS2)/(N-2k)]". There is no evidence of structural break. Null hypothesis for Quandt–Andrews test is "no breakpoint". As the insignificant results of maximum LR, expected LR and average LR are insignificant, however, null hypothesis is accepted.

4.3. Unit Root Test

In order to check the stationarity of the data, two unit root tests are used, including DF-GLS and KSUR. Table 4 presents the results of these tests. It is evident that unit root is present at a level in the majority of the variables, which is because test statistics are lower than the critical value. This means data are non-stationary at the level. Hence, the first difference is required. However, all variables are stationary at the first difference at a 1% significance level. It is evident that unit root is present at a level for CE, RE, INNOV, FISHERY, as test statistics are lower than the critical value. Grant and ocean trade are stationary at a level for GF-GLS unit root test. In case of KSUR, energy intensity and grant are stationary at a level. Hence, the first difference is required. However, all variables are stationary at the first difference at a 1% significance level. In the presence of mix results of unit root, we must use bond cointegration test, which is used to confirm the long run existence of the studied models.

Table 4. Unit root test.

	DF-GLS				KSUR			
DF-GLS	**Level**		**Diff**		**Level**		**Diff**	
Variable	**Stat**		**Stat**		***p*-Value**		***p*-Value**	
CE	−1.228	IS	−4.609 ***	Sig	0.911	IS	0.003 ***	Sig
EI	−2.564	IS	−8.015 ***	Sig	0.005 ***	Sig	0.006 ***	Sig
RE	−2.601	IS	−5.479 ***	Sig	0.348	IS	0.000 ***	Sig
INNOV	−1.522	IS	−4.225 ***	Sig	0.968	IS	0.006 ***	Sig
GRANT	−4.496 ***	Sig	−6.776 ***	Sig	0.001 ***	Sig	0.001 ***	Sig
FISHERY	−1.859	IS	−4.882 ***	Sig	0.995	IS	0.004 ***	Sig
OTRADE	−3.098 *	Sig	−5.958 ***	Sig	0.94	IS	0.005 ***	Sig

Note: GF-GLS does not assume stationary nonlinear, whereas KSUR unit root test presented by Kapetanios & Shin [45] considers stationary nonlinear. The null hypothesis assumes the presence of unit root. ***, * represent the level of significance of 1%, and 10% respectively.

4.4. Cointegration Test

ARDL bounds cointegration test is used to check cointegration between variables, and results are presented in Table 5. It can be noticed that F-statistics in all models is higher than the upper bound critical value at 1%, 5%, and 10% critical values. This shows that cointegration exists between variables, and the null hypothesis of no cointegration is rejected. In this case, with stationarity of variables at level and first differences the ARDL test is appropriate to check the short-run and the long-run association between variables.

Table 5. ARDL bounds cointegration test.

ARDL Bounds Cointegration Test	F-Stat	Result
CE = f (EI, RE)	7.786	Cointegration
CE = f (INNOV, GRANT)	10.381	Cointegration
CE = f (FISHERY, OTRADE)	4.637	Cointegration
Lower-bound critical value at 1%		4.13
Upper-bound critical value at 1%		5.00
Lower-bound critical value at 5%		3.1
Upper-bound critical value at 5%		3.87
Lower-bound critical value at 10%		2.63
Upper-bound critical value at 10%		3.35

4.5. BDS Test

After examining the cointegration test, we have to use the linearity and nonlinearity test. The purpose is to analyze the existence of nonlinearity in the data series. We have used the BDS test, proposed by Broock et al. [46] with null hypothesis of "series are linearly dependent". Table 6 confirms the significance of series at each dimension, mentioning that the variables are nonlinearly dependent. However, we have to apply nonlinear ARDL test, instead of simple ARDL test.

Table 6. BDS test.

Dimension	2	3	4	5	6
CE	0.151 ***	0.236 ***	0.441 ***	0.536 ***	0.457 ***
EI	0.210 ***	0.566 ***	0.223 ***	0.542 ***	0.506 ***
RE	0.181 ***	0.231 ***	0.437 ***	0.514 ***	0.596 ***
INNOV	0.158 ***	0.578 ***	0.624 ***	0.127 ***	0.541 ***
GRANT	0.355 ***	0.874 ***	0.685 ***	0.403 ***	0.480 ***
FISHERY	0.262 ***	0.413 ***	0.025 ***	0.718 ***	0.629 ***
OTRADE	0.135 ***	0.136 ***	0.108 ***	0.380 ***	0.558 ***

Notes: *** represents the level of significance at 1%.

4.6. Estimation

4.6.1. Non-Linear ARDL Estimates

In order to check the impact of positive and negative shocks in independent variables on dependent variables, the non-linear ARDL method is used, the results are presented in Tables 7 and 8. Afterward, the data are divided in two parts; before implementation of Vision 2030 and after Vision 2030 to investigate the effect of Vision 2030 policies to minimize the environmental externalities. This will help the policymakers to check if there is a need to alter the environmental policies in Vision 2030.

Table 7. Nonlinear ARDL estimation.

Long-Run	Model-1	Model-2	Model-3
CE_{t-1}	0.105	−0.195	−0.709 ***
EI^{+}_{t-1}	−0.232		
EI^{-}_{t-1}	0.288		
RE^{+}_{t-1}	0.217		
RE^{-}_{t-1}	0.104		
$INNOV^{+}_{t-1}$		−0.042	
$INNOV^{-}_{t-1}$		−0.057	
$GRANT^{+}_{t-1}$		0.029 ***	
$GRANT^{-}_{t-1}$		−0.047	

Table 7. *Cont.*

Long-Run	Model-1	Model-2	Model-3
$FISHERY^{+}{}_{t-1}$			0.254 *
$FISHERY^{-}{}_{t-1}$			0.538 *
$OTRADE^{+}{}_{t-1}$			0.378 ***
$OTRADE^{-}{}_{t-1}$			−2.122 **
Short-run			
Δ CE_{t-1}	−0.108	−0.296	0.142
Δ $EI^{+}{}_{t-1}$	0.059		
Δ $EI^{-}{}_{t-1}$	0.416 *		
Δ $RE^{+}{}_{t-1}$	0.091		
Δ $RE^{-}{}_{t-1}$	0.082		
Δ $INNOV^{+}{}_{t-1}$		0.025	
Δ $INNOV^{-}{}_{t-1}$		−0.107	
Δ $GRANT^{+}{}_{t-1}$		−0.025	
Δ $GRANT^{-}{}_{t-1}$		0.097	
Δ $FISHERY^{+}{}_{t-1}$			0.338
Δ $FISHERY^{-}{}_{t-1}$			0.568 *
Δ $OTRADE^{+}{}_{t-1}$			0.269 **
Δ $OTRADE^{-}{}_{t-1}$			−3.166 **
Constant	−1.169	2.451	7.638 ***

Notes: CE shows the carbon neutrality, EI and RE are energy intensity and renewable energy respectively, which are the energy factors. INNOV and GRANT represent the innovation and technical grants respectively, which reflect the green indicators. FISHERY and OTRADE mention the fishery and ocean trade that reflects the blue factors. Δ is the difference.***, **, * represent the level of significance at 1%, 5%, and 10% respectively.

Table 8. Asymmetric and model diagnostics.

	Long Run (+)	Long Run (-)	Long Run Asymmetry (*p*-Value) W_{LR}	Short Run Asymmetry (*p*-Value) W_{LR}
CE = f (EI, RE)				
EI	2.204	2.738	0.344	0.163
RE	−2.062	0.99	0.571	0.145
Cointegration			F-Stat	1.763
Portmanteau			*p*-value	0.681
Heteroskedasticity			*p*-value	0.525
Ramsey test			*p*-value	0.167
J-B test			*p*-value	0.671
CE = f (INNOV, GRANT)				
INNOV	−0.217	0.293	0.809	0.233
GRANT	0.15	0.241	0.196	0.154
Cointegration			F-Stat	2.377
Portmanteau			*p*-value	0.7839
Heteroskedasticity			*p*-value	0.1515
Ramsey test			*p*-value	0.2176
J-B test			*p*-value	0.0012
CE = f (FISHERY, OTRADE)				
FISHERY	0.358	−0.759 *	0.054 *	0.221
OTRADE	0.532 ***	2.991 ***	0.001 ***	0.013 **
Cointegration			F-Stat	2.729
Portmanteau			*p*-value	0.7887
Heteroskedasticity			*p*-value	0.0010
Ramsey test			*p*-value	0.1075
J-B test			*p*-value	0.0759

Notes: CE shows the carbon neutrality, EI and RE are energy intensity and renewable energy respectively, which are the energy factors. INNOV and GRANT represent the innovation and technical grants respectively, which reflect the green indicators. FISHERY and OTRADE mention the fishery and ocean trade that reflects the blue factors. Δ is the difference. ***, **, * represent the level of significance at 1%, 5%, and 10% respectively.

4.6.2. Full Sample Analysis

Long Run Estimates

Results from full sample analysis are reported in Table 7, where energy efficiency, renewable energy, and innovation have insignificant coefficients. In achieving carbon neutrality in Saudi Arabia, these factors do not play any role. These results contradict the previous studies, including [10,25,27]. The explanation for this insignificant effect could be that Saudi Arabia is the major oil producer, and the country has the highest oil reserves, which makes energy sources irrelevant due to lower prices and lower production costs. Likewise, the negative and positive shocks in innovation coefficients are negative but insignificant. This suggests that although innovation adversely affects carbon neutrality, its impact is insignificant. One possible explanation could be that the country has not achieved a level at which innovation can help to reduce the carbon emission. Due to its dependence on oil, Saudi Arabia had focused less toward industrial innovation and promoting entrepreneurs [47,48].

As far as the grant is concerned, the positive shocks in the grant have significant and positive coefficient, which suggest that 0.029% carbon can be enhanced by a 1% increase in the grant. However, negative shocks in grants do not affect carbon neutrality. The results cannot be compared with the results of other researchers. However, the insignificant relationship could be because, in Saudi Arabia, a major portion of the grant is not used for solving related environmental issues. Moreover, Saudi Arabia recently established different funds to invest in carbon neutralization and targets to achieve carbon neutrality in 2060. Hence, these grants and funds are not showing any impact.

In the case of blue factors, the positive and negative shocks in the fishery have a significant and positive impact on carbon neutrality; however, the strength of the effect is not strong. This confirms that increase (decrease) in fishery increases (decreases) carbon emission. This result is aligned with the results of Wang et al. [31]. Moreover, positive shocks in ocean trade have a significant and positive coefficient, suggesting that any positive shock increases carbon emission. The negative shocks in ocean trade have a significant and negative coefficient, suggesting that negative shocks in ocean trade reduce carbon emission. The results regarding the blue factors confirm that Saudi Arabia understands the importance of marine sector. The country's strategic location links three continents, and it is surrounded by many coasts; hence, a significant portion of global trade is done through it. Moreover, Saudi Arabia is struggling to attain sustainable marine development which leads to carbon neutrality.

Short Run Estimates

The results regarding the short-run analysis confirm that negative shocks in energy efficiency enhance carbon externalities, as the coefficient is significant and positive. This result is aligned with the findings of Wang et al. [10], and the main reason is that being a leader in oil production, the country's energy needs are met by fossil fuels. However, recent reforms ensure that strict measures are taken against energy-intensive industries. Moreover, economic diversification is in progress to reduce the dependence on the oil sector. This results in a reduction of energy intensity helping the country to meet carbon neutrality targets. In terms of green factors, it can be seen that these factors do not impact carbon neutrality. This suggests that green reforms are not mature enough to accomplish carbon neutrality targets. The results for blue factors show that negative shocks in the fishery enhance carbon emission. However, positive shocks have a positive coefficient for ocean trade, and negative shocks have a negative coefficient. This suggests that any increase in ocean trade increases carbon, in contrast, negative shocks in ocean trade reduce carbon emission. This shows that the efforts of the government to reduce the carbon footprint of the ocean trade sector are showing benefits.

4.6.3. Before Vision 2030 Analysis

The results regarding the before Vision 2030, for non-linear ARDL, are presented in Tables 9 and 10. For energy efficiency, long-run and short-run estimates show that the positive shocks in energy intensity have a positive coefficient. In contrast, negative shocks in energy intensity only have a significant positive coefficient in the short run. This means an increase in energy intensity improves carbon emission, in the long run. These results are opposite to previous studies [10,20,22]. It is observed that the economic development of Saudi Arabia is dependent on energy-intensive industries, and the lifestyle is also energy-intensive because energy prices are low. Due to this, although energy intensity is high before Vision 2030, carbon emission is increased due to other measures. For renewable energy, coefficients are significant only in the short run, showing that positive shocks in renewable energy consumption significantly enhance carbon emission. In comparison, negative shocks in renewable energy consumption reduce carbon emission. This result is aligned with the findings of Li et al. [4].

Table 9. Nonlinear ARDL estimation (before Vision 2030).

Long-Run	Model-1	Model-2	Model-3
CE_{t-1}	−0.004 ***	−0.002 ***	−0.003 ***
EI^{+}_{t-1}	0.007 ***		
EI^{-}_{t-1}	−0.001		
RE^{+}_{t-1}	−0.012		
RE^{-}_{t-1}	0.024		
$INNOV^{+}_{t-1}$		0.023	
$INNOV^{-}_{t-1}$		−0.032	
$GRANT^{+}_{t-1}$		−0.006	
$GRANT^{-}_{t-1}$		0.005 ***	
$FISHERY^{+}_{t-1}$			0.012 ***
$FISHERY^{-}_{t-1}$			0.015
$OTRADE^{+}_{t-1}$			0.095 **
$OTRADE^{-}_{t-1}$			0.010
Short-run			
ΔCE_{t-1}	0.975 ***	0.965 ***	0.973 ***
ΔEI^{+}_{t-1}	0.497 ***		
ΔEI^{-}_{t-1}	0.313 ***		
ΔRE^{+}_{t-1}	0.025 ***		
ΔRE^{-}_{t-1}	−0.024 ***		
$\Delta INNOV^{+}_{t-1}$		−0.022 ***	
$\Delta INNOV^{-}_{t-1}$		−0.092	
$\Delta GRANT^{+}_{t-1}$		0.024 **	
$\Delta GRANT^{-}_{t-1}$		0.007	
$\Delta FISHERY^{+}_{t-1}$			0.110 **
$\Delta FISHERY^{-}_{t-1}$			0.215 ***
$\Delta OTRADE^{+}_{t-1}$			0.053 ***
$\Delta OTRADE^{-}_{t-1}$			−0.009
Constant	0.048 ***	0.024 ***	0.038 ***

Notes: CE shows the carbon neutrality, EI and RE are energy intensity and renewable energy respectively, which are the energy factors. INNOV and GRANT represent the innovation and technical grants respectively, which reflects the green indicators. FISHERY and OTRADE mentions the fishery and ocean trade that reflects the blue factors. Δ is the difference. ***, ** represent the level of significance at 1%, and 5% respectively.

The same is true for innovations where coefficients are significant only in the short run. In the short run, positive shocks in innovation increase carbon neutrality and negative shocks in innovation are insignificant. The finding contradicts with Iqbal et al. [24], which reported that innovative products are vital in reaching carbon neutrality targets. The main reason for contradiction is that previously Saudi Arabia has focused on oil sector and was less motivated toward high tech, innovation, and environment. As far as the grant is concerned, long-run estimates show that negative shocks enhance carbon neutrality.

However, in the short-run positive shocks in grant surge the carbon level. This suggests that before Vision 2030, the full benefits of technical grants were not achieved because there was no policy to use technical grants to improve the country's carbon neutrality targets. The blue factors suggest that positive shocks in the fishery enhance carbon, in both the short and long run. However, negative shocks improve carbon neutrality in the short run. Before Vision 2030, the country's economy was dependent on the oil sector, and other sectors, including ocean resources, were neglected [36]. Hence, part of these sectors is limited in the country's major sustainable issues. For ocean trade, only positive shocks impact carbon positively in both the short-run and long run.

Table 10. Asymmetric and model diagnostics (before Vision 2030).

	Long Run (+)	**Long Run (-)**	**Long Run Asymmetry (*p*-Value) W_{LR}**	**Short Run Asymmetry (*p*-Value) W_{LR}**
CE = f (EI, RE)				
EI	1.818 ***	0.163	0.000 ***	0.231
RE	−0.031	−0.059 *	0.090 *	0.641
Cointegration			F-Stat	8.621
Portmanteau			*p*-value	0.743
Heteroskedasticity			*p*-value	0.196
Ramsey test			*p*-value	1.330
J-B test			*p*-value	0.181
CE = f (INNOV, GRANT)				
INNOV	0.119	0.160	0.036 **	0.424
GRANT	−0.030	0.254	0.044 ***	0.215
Cointegration			F-Stat	3.273
Portmanteau			*p*-value	0.964
Heteroskedasticity			*p*-value	0.583
Ramsey test			*p*-value	0.465
J-B test			*p*-value	0.869
CE = f (FISHERY, OTRADE)				
FISHERY	0.372 *	−0.465	0.829	0.784
OTRADE	0.298 ***	−0.032	0.305	0.112
Cointegration			F-Stat	5.359
Portmanteau			*p*-value	0.688
Heteroskedasticity			*p*-value	0.365
Ramsey test			*p*-value	0.868
J-B test			*p*-value	0.746

Notes: CE shows the carbon neutrality, EI and RE are energy intensity and renewable energy respectively, which are the energy factors. INNOV and GRANT represent the innovation and technical grants respectively, which reflects the green indicators. FISHERY and OTRADE mention the fishery and ocean trade that reflect the blue factors. Δ is the difference. ***, **, * represent the level of significance at 1%, 5%, and 10% respectively.

4.6.4. After Vision 2030 Analysis

In Tables 11 and 12, results for non-linear ARDL estimates after Vision 2030 are presented. It can be noticed that the coefficient of energy intensity is negative and positive shocks are significant, which shows that energy intensity plays a significant role in carbon emission. However, the coefficient of negative shocks in energy intensity gets significant and positive after Vision 2030, showing the positive effect of energy-related policies on carbon neutrality after the introduction of Vision 2030. This change can be attributed to increased energy products and diversification of the economy to the non-oil sector. Moreover, the government tried to introduce technically enhanced and low energy-intensive products, which reduced the energy intensity and enhanced the overall carbon neutrality targets. In the case of renewable energy consumption, in both short-run and long-run estimates, positive shocks enhance carbon, aligned with previous studies [22,24].

Table 11. Nonlinear ARDL estimation (after Vision 2030).

Long-Run	Model-1	Model-2	Model-3
CE_{t-1}	−0.011 ***	−0.016 ***	−0.007 ***
EI^{+}_{t-1}	0.005 ***		
EI^{-}_{t-1}	0.006 ***		
RE^{+}_{t-1}	0.019 ***		
RE^{-}_{t-1}	−0.241		
$INNOV^{+}_{t-1}$		0.001	
$INNOV^{-}_{t-1}$		−0.011	
$GRANT^{+}_{t-1}$		−0.001 ***	
$GRANT^{-}_{t-1}$		−0.194	
$FISHERY^{+}_{t-1}$			0.039 ***
$FISHERY^{-}_{t-1}$			1.786 ***
$OTRADE^{+}_{t-1}$			0.002 ***
$OTRADE^{-}_{t-1}$			1.091 ***
Short-run			
ΔCE_{t-1}	1.151 ***	1.070 ***	1.090 ***
ΔEI^{+}_{t-1}	−0.023		
ΔEI^{-}_{t-1}	0.210 ***		
ΔRE^{+}_{t-1}	0.064 ***		
ΔRE^{-}_{t-1}	16.195		
$\Delta INNOV^{+}_{t-1}$		−0.879 ***	
$\Delta INNOV^{-}_{t-1}$		−1.529 ***	
$\Delta GRANT^{+}_{t-1}$		−5.351 ***	
$\Delta GRANT^{-}_{t-1}$		6.123 ***	
$\Delta FISHERY^{+}_{t-1}$			1.064 ***
$\Delta FISHERY^{-}_{t-1}$			0.053 ***
$\Delta OTRADE^{+}_{t-1}$			0.050 ***
$\Delta OTRADE^{-}_{t-1}$			0.064 **
Constant	−0.325	−0.208 **	0.305 ***

Notes: CE shows the carbon neutrality, EI and RE are energy intensity and renewable energy respectively, which are the energy factors. INNOV and GRANT represent the innovation and technical grants respectively, which reflects the green indicators. FISHERY and OTRADE mentions the fishery and ocean trade that reflects the blue factors. Δ is the difference. ***, ** represent the level of significance at 1%, and 5% respectively.

The impact of innovation is insignificant in the long run, but both negative and positive shocks have a negative coefficient in the short run. This means increases (decrease) in innovation increase (decrease) the carbon neutrality. This result is in line with Shao et al. [25], which presents the efforts of Saudi government to introduce the significant policies regarding innovation which on the one hand trigger the economic growth and, on other hand, lead toward carbon neutrality in Saudi Arabia. However, in the long run, the magnitude is low. Hence, it is needed that efforts should be diverted to protect innovations through patents in a way that low carbon emission goals should also be made sure by these patents.

The coefficients for positive and negative shocks are negative and positive respectively, for technical grants, showing that any shock in technical grants reduces carbon emission. Wang et al. [49] explains that development aids are sometimes used to achieve the goals of donor countries; hence misallocation of funds takes place. Due to these reasons, the original intention of the funds, such aa reduction in carbon emission, is deviated, leading to a reduction of the effect of these grants on climate.

As far as blue factors are concerned, the coefficients of positive and negative shocks in the fishery are positive, showing that any increase (decrease) in fishery increase (decrease) the carbon level. Fishery sector requires nonrenewable energy consumption, which increases the carbon emission. However, up till now, the Saudi government has not achieved to get renewable energy sources to support the fishery sector, such as renewable source of energy for boating [50]. In terms of ocean trade, positive shocks reduce the carbon level, in the long and short run. Negative shocks in ocean trade reduce carbon level, in the long run and short run. Before Vision 2030, the role of ocean trade toward carbon neutrality is not significant. However, after Vision 2030, the alignment of ocean trade policies with sustainability is not achieved yet. However, to fulfil the carbon neutrality targets of Saudi Arabia, a prompt action is required.

Table 12. Asymmetric and model diagnostics (after Vision 2030).

	Long Run (+)	Long Run (-)	Long Run Asymmetry (p-Value) W_{LR}	Short Run Asymmetry (p-Value) W_{LR}
CE = f (EI, RE)				
EI	0.507 ***	−0.539 ***	0.885	0.000 ***
RE	0.010 ***	22.408	0.533	0.610
Cointegration			F-Stat	261.986
Portmanteau			p-value	0.547
Heteroskedasticity			p-value	0.253
Ramsey test			p-value	0.128
J-B test			p-value	0.528
CE = f (INNOV, GRANT)				
INNOV	0.073	0.730	0.199	0.000 ***
GRANT	−0.072	12.323 ***	0.000 ***	0.000 ***
Cointegration			F-Stat	2017.971
Portmanteau			p-value	0.6139
Heteroskedasticity			p-value	0.1171
Ramsey test			p-value	0.225
J-B test			p-value	0.7704
CE = f (FISHERY, OTRADE)				
FISHERY	0.561 ***	259.907 ***	0.000 ***	0.036 **
OTRADE	0.307 ***	158.795 ***	0.000 ***	0.000 ***
Cointegration			F-Stat	364.141
Portmanteau			p-value	0.657
Heteroskedasticity			p-value	1.573
Ramsey test			p-value	1.649
J-B test			p-value	0.844

Notes: CE shows the carbon neutrality, EI and RE are energy intensity and renewable energy respectively, which are the energy factors. INNOV and GRANT represent the innovation and technical grants respectively, which reflect the green indicators. FISHERY and OTRADE mention the fishery and ocean trade that reflects the blue factors. Δ is the difference. ***, ** represent the level of significance at 1%, and 5% respectively.

5. Conclusions

The first objective of this study is related to the nexus between energy factors and carbon neutrality. It is found that in full sample estimation, energy factors are irrelevant to carbon neutrality. However, before vison 2030, energy factors including energy intensity and renewable energy consumption enhance carbon neutrality, which gets significant after vison 2030. Saudi Arabia is a leading producer and exporter of petroleum products; hence energy prices are low. The main focus is on other measures to enhance carbon neutrality, which is the reason for the positive impact of energy intensity on carbon neutrality. However, in Vision 2030, attention is diverted toward the diversification of the economy to low energy intensity sectors and the introduction of energy-efficient products.

The second objective is regarding the green factors and carbon neutrality, and it is noticed that innovation is irrelevant to carbon neutrality in the full sample, whereas, after vison 2030, both positive and negative shocks in innovation enhance carbon neutrality, in the short run. For technical grants, the effect on carbon neutrality is enhanced after the introduction of Vision 2030 because any shock results in increased carbon neutrality. In comparison, the third objective related to the impact of blue factors on carbon neutrality is addressed. In the full sample, positive and negative shocks in fishery negatively affected carbon neutrality. The results regarding ocean trade show that positive shocks and negative shocks enhance and reduce carbon emission respectively, for full sample and after Vision 2030.

While addressing the fourth objective, which is about the combined impact of energy factors, green factors, and blue factors, in the full sample as well as before and after Vision 2030 estimations, green factors proved to be the most significant in terms of the effect on carbon neutrality. The importance of blue sector is not attributed yet, however, they are unable to achieve carbon neutrality. Similarly, energy factors are either irrelevant or are in the opposite direction.

In light of these findings, some important policy implications are diverting the attention toward low-intensity energy products to achieve green environmental objectives. It is recommended that innovations in terms of low-intensity products should be promoted and alternative types of energy products should be introduced in the country. This will reduce the problem of energy intensity and reduce the country's dependence on fossil fuel energy. The marine sector needs to introduce modern mechanisms which push up the fishery and marine trade, such as solar power boating. Solar boating will be useful in Saudi Arabia, as the sea shores have sunshine throughout the year.

As for the limitations of the study, we were unable to collect the pre 1990 data to create a longer time series data. In case of Saudi Arabia, monthly and quarterly data are not available. For future directions, researchers can focus on GCC countries and carry out a comparative study. Similarly, to increase the scope of study, the future research can be carried out on North African and Middle Eastern countries. There are fewer studies on Blue economy; however, I suggest that upcoming works should include more variables of blue economy.

Author Contributions: Conceptualization, S.S., R.W., G.A. and S.A.A.; methodology, S.S., R.W., G.A. and S.A.A.; software, S.S., R.W., G.A. and S.A.A.; validation, S.S., R.W., G.A. and S.A.A.; formal analysis, S.S., R.W., G.A. and S.A.A.; investigation, S.S., R.W., G.A. and S.A.A.; resources, S.S., R.W., G.A. and S.A.A.; data curation, S.S., R.W., G.A. and S.A.A.; writing—original draft preparation, S.S., R.W., G.A. and S.A.A.; writing—review and editing, S.S., R.W., G.A. and S.A.A.; visualization, S.S., R.W., G.A. and S.A.A.; supervision, S.S., R.W., G.A. and S.A.A.; project administration, S.S., R.W., G.A. and S.A.A.; funding acquisition, S.S., R.W., G.A. and S.A.A. All authors have read and agreed to the published version of the manuscript.

Funding: This work was funded by the University of Jeddah, Jeddah, Saudi Arabia, under grant No. (UJ-22-DR-43). The authors, therefore, acknowledge with thanks the University of Jeddah for its technical and financial support.

Institutional Review Board Statement: Not applicable.

Informed Consent Statement: Not applicable.

Acknowledgments: The authors, therefore, acknowledge with thanks the University of Jeddah for its technical and financial support.

Conflicts of Interest: The authors declare no conflict of interest.

References

1. Shaltout, K.H.; Ahmed, M.T.; Alrumman, S.A.; Ahmed, D.A.; Eid, E.M. Evaluation of the carbon sequestration capacity of arid mangroves along nutrient availability and salinity gradients along the Red Sea coastline of Saudi Arabia. *Oceanologia* **2020**, *62*, 56–69. [CrossRef]
2. The Independent. From Waste to Resource: Can Recycling CO_2 Help Saudi Arabia Go Green? Available online: https://www.independent.co.uk/climate-change/sgi/carbon-dioxide-recycling-saudi-arabia-b1890563.html (accessed on 12 July 2022).
3. Lahiani, A.; Mefteh-Wali, S.; Shahbaz, M.; Vo, X.V. Does financial development influence renewable energy consumption to achieve carbon neutrality in the USA? *Energy Policy* **2021**, *158*, 112524. [CrossRef]
4. Li, H.-S.; Geng, Y.-C.; Shinwari, R.; Yangjie, W.; Rjoub, H. Does renewable energy electricity and economic complexity index help to achieve carbon neutrality target of top exporting countries? *J. Environ. Manag.* **2021**, *299*, 113386. [CrossRef] [PubMed]
5. Li, M.; Ahmad, M.; Fareed, Z.; Hassan, T.; Kirikkaleli, D. Role of trade openness, export diversification, and renewable electricity output in realizing carbon neutrality dream of China. *J. Environ. Manag.* **2021**, *297*, 113419. [CrossRef]

6. Zheng, F.; Zhou, X.; Rahat, B.; Rubbaniy, G. Carbon neutrality target for leading exporting countries: On the role of economic complexity index and renewable energy electricity. *J. Environ. Manag.* **2021**, *299*, 113558. [CrossRef]
7. Akash, A.R.; Rao, A.B.; Chandel, M.K. Relevance of Carbon Capture & Sequestration in India's Energy Mix to Achieve the Reduction in Emission Intensity by 2030 as per INDCs. *Energy Procedia* **2017**, *114*, 7492–7503. [CrossRef]
8. Nam, E.; Jin, T. Mitigating carbon emissions by energy transition, energy efficiency, and electrification: Difference between regulation indicators and empirical data. *J. Clean. Prod.* **2021**, *300*, 126962. [CrossRef]
9. Wang, H.; Yang, Y.; Keller, A.A.; Li, X.; Feng, S.; Dong Y, n.a.n.; Li, F. Comparative analysis of energy intensity and carbon emissions in wastewater treatment in USA, Germany, China and South Africa. *Appl. Energy* **2016**, *184*, 873–881. [CrossRef]
10. Wang, R.; Wang, Q.; Yao, S. Evaluation and difference analysis of regional energy efficiency in China under the carbon neutrality targets: Insights from DEA and Theil models. *J. Environ. Manag.* **2021**, *293*, 112958. [CrossRef]
11. Chien, F.; Ananzeh, M.; Mirza, F.; Bakar, A.; Vu, H.M.; Ngo, T.Q. The effects of green growth, environmental-related tax, and eco-innovation towards carbon neutrality target in the US economy. *J. Environ. Manag.* **2021**, *299*, 113633. [CrossRef]
12. Shan, S.; Genç, S.Y.; Kamran, H.W.; Dinca, G. Role of green technology innovation and renewable energy in carbon neutrality: A sustainable investigation from Turkey. *J. Environ. Manag.* **2021**, *294*, 113004. [CrossRef] [PubMed]
13. Sun, Y.; Duru, O.A.; Razzaq, A.; Dinca, M.S. The asymmetric effect eco-innovation and tourism towards carbon neutrality target in Turkey. *J. Environ. Manag.* **2021**, *299*, 113653. [CrossRef] [PubMed]
14. Yue, X.G.; Liao, Y.; Zheng, S.; Shao, X.; Gao, J. The role of green innovation and tourism towards carbon neutrality in Thailand: Evidence from bootstrap ADRL approach. *J. Environ. Manag.* **2021**, *292*, 112778. [CrossRef] [PubMed]
15. Ebarvia, M.C.M. Economic Assessment of Oceans for Sustainable Blue Economy Development. *J. Ocean Coast. Econ.* **2016**, *2*, 7. [CrossRef]
16. Feng, C.; Ye, G.; Jiang, Q.; Zheng, Y.; Chen, G.; Wu, J.; Feng, X.; Si, Y.; Zeng, J.; Li, P.; et al. The contribution of ocean-based solutions to carbon reduction in China. *Sci. Total Environ.* **2021**, *797*, 149168. [CrossRef]
17. Zhou, X.; Zhao, X.; Zhang, S.; Lin, J. Marine ranching construction and management in East China Sea: Programs for sustainable fishery and aquaculture. *Water* **2019**, *11*, 1237. [CrossRef]
18. Whieldon, E. *Your Climate Change Goals May Have a Maritime Shipping Problem*; S & P Global; Available online: https://www.spglobal.com/esg/insights/your-climate-change-goals-may-have-a-maritime-shipping-problem (accessed on 15 July 2022).
19. Du, K.; Li, P.; Yan, Z. Do green technology innovations contribute to carbon dioxide emission reduction? Empirical evidence from patent data. *Technol. Forecast. Soc. Change* **2019**, *146*, 297–303. [CrossRef]
20. Gil, L.; Bernardo, J. An approach to energy and climate issues aiming at carbon neutrality. *Renew. Energy Focus* **2020**, *33*, 37–42. [CrossRef]
21. Wang, Y.; Liao, M.; Xu, L.; Malik, A. The impact of foreign direct investment on China's carbon emissions through energy intensity and emissions trading system. *Energy Econ.* **2021**, *97*, 105212. [CrossRef]
22. Andersson, F.N.G.; Karpestam, P. CO_2 emissions and economic activity: Short- and long-run economic determinants of scale, energy intensity and carbon intensity. *Energy Policy* **2013**, *61*, 1285–1294. [CrossRef]
23. Juan lin Shen, Y.; Li, X.; Hasnaoui, A. BRICS carbon neutrality target: Measuring the impact of electricity production from renewable energy sources and globalization. *J. Environ. Manag.* **2021**, *298*, 113460. [CrossRef] [PubMed]
24. Iqbal, N.; Abbasi, K.R.; Shinwari, R.; Guangcai, W.; Ahmad, M.; Tang, K. Does exports diversification and environmental innovation achieve carbon neutrality target of OECD economies? *J. Environ. Manag.* **2021**, *291*, 112648. [CrossRef] [PubMed]
25. Shao, X.; Zhong, Y.; Liu, W.; Li, R.Y.M. Modeling the effect of green technology innovation and renewable energy on carbon neutrality in N-11 countries? Evidence from advance panel estimations. *J. Environ. Manag.* **2021**, *296*, 113189. [CrossRef] [PubMed]
26. United Nations Environment Program. Technology Support Helps Deliver Millions in Climate Funding, Emissions Reductions. Available online: https://www.unep.org/news-and-stories/story/technology-support-helps-deliver-millions-climate-funding-emissions (accessed on 10 July 2022).
27. Institute for Marine and Antarctic Studies. *Beef, Lamb, Lobster or Fish? Fisheries Study Shows Impact of Food Choice on Carbon Emissions*; University of Tasmania: Hobart, Australia, 2019.
28. Wan, X.; Li, Q.; Qiu, L.; Du, Y. How do carbon trading platform participation and government subsidy motivate blue carbon trading of marine ranching? A study based on evolutionary equilibrium strategy method. *Mar. Policy* **2021**, *130*, 104567. [CrossRef]
29. Mcdermott, A. Eating Seafood Can Reduce Your Carbon Footprint, but Some Fish Are Better than Others. Available online: https://oceana.org/blog/eating-seafood-can-reduce-your-carbon-footprint-some-fish-are-better-others/#:~{}:text=Worldwide%2Cmorethan25percentemissionscomefromfoodproduction.&text=Somefoodscostmorethan,farmlandorcareoflivestock (accessed on 21 July 2022).
30. Wan, X.; Xiao, S.; Li, Q.; Du, Y. Evolutionary policy of trading of blue carbon produced by marine ranching with media participation and government supervision. *Mar. Policy* **2021**, *124*, 104302. [CrossRef]
31. Ozer, M.; Canbay, S.; Kirca, M. The impact of container transport on economic growth in Turkey: An ARDL bounds testing approach. *Res. Transp. Econ.* **2020**, *88*, 101002. [CrossRef]
32. Taghvaee, S.M.; Omaraee, B.; Taghvaee, V.M. Maritime transportation, environmental pollution, and economic growth in Iran: Using dynamic log linear model and granger causality approach. *Iran. Econ. Rev.* **2017**, *21*, 185–210. [CrossRef]

33. Fratila, A.; Gavril, I.A.; Nita, S.C.; Hrebenciuc, A. The importance of maritime transport for economic growth in the european union: A panel data analysis. *Sustainability* **2021**, *13*, 7961. [CrossRef]
34. Ben Jebli, M.; Belloumi, M. Investigation of the causal relationships between combustible renewables and waste consumption and CO_2 emissions in the case of Tunisian maritime and rail transport. *Renew. Sustain. Energy Rev.* **2017**, *71*, 820–829. [CrossRef]
35. Chaabouni, S.; Zghidi, N.; Mbarek, M.B. On the causal dynamics between CO_2 emissions, health expenditures and economic growth. *Sustain. Cities Soc.* **2016**, *22*, 184–191.
36. Sarwar, S.; Alsaggaf, M.I. The role of governance indicators to minimize the carbon emission: A study of Saudi Arabia. *Manag. Environ. Qual. Int. J.* **2021**, *32*, 970–988. [CrossRef]
37. Waheed, R. The Significance of Energy Factors, Green Economic Indicators, Blue Economic Aspects towards Carbon Intensity: A Study of Saudi Vision 2030. *Sustainability* **2022**, *14*, 6893. [CrossRef]
38. Narayan, P.K. *Reformulating Critical Values for the Bounds F-Statistics Approach to Cointegration: An Application to the Tourism Demand Model for Fiji*; Monash University: Clayton, VIC, Australia, 2004; pp. 1441–5429.
39. Pesaran, M.H.; Shin, Y.; Smith, R.J. Bounds testing approches to analysis of long run relationships. *J. Appl. Econom.* **2001**, *16*, 289–326. [CrossRef]
40. Bahmani-Oskooee, M.; Mohammadian, A. Asymmetry Effects of Exchange Rate Changes on Domestic Production: Evidence from Nonlinear ARDL Approach. *Aust. Econ. Pap.* **2016**, *55*, 181–191. [CrossRef]
41. Delatte, A.L.; López-Villavicencio, A. Asymmetric Exchange rate pass-through: Evidence from major countries. *J. Macroecon.* **2012**, *34*, 833–844. [CrossRef]
42. Qamruzzaman, M.; Jianguo, W. Nexus between financial innovation and economic growth in South Asia: Evidence from ARDL and nonlinear ARDL approaches. *Financ. Innov.* **2018**, *4*, 20. [CrossRef]
43. Richmond, S.; Andrews, M. Orthodontic treatment standards in Norway. *Eur. J. Orthod.* **1993**, *15*, 7–15. [CrossRef]
44. Andrews, D.W.; Ploberger, W. Optimal tests when a nuisance parameter is present only under the alternative. *Econometrica* **1994**, *62*, 1383–1414. [CrossRef]
45. Kapetanios, G.; Shin, Y. GLS detrending-based unit root tests in nonlinear STAR and SETAR models. *Econ. Lett.* **2008**, *100*, 377–380. [CrossRef]
46. Broock, W.A.; Scheinkman, J.A.; Dechert, W.D.; LeBaron, B. A test for independence based on the correlation dimension. *Econom. Rev.* **1996**, *15*, 197–235. [CrossRef]
47. Shin, Y.; Yu, B.; Greenwood-Nimmo, M. Modelling Asymmetric Cointegration and Dynamic Multipliers in a Nonlinear ARDL Framework. *SSRN Electron. J.* **2012**. [CrossRef]
48. Kahia, M.; Omri, A.; Jarraya, B. Green Energy, economic growth and environmental quality nexus in Saudi Arabia. *Sustainability* **2021**, *13*, 1264. [CrossRef]
49. Wang, Q.; Guo, J.; Li, R. Official development assistance and carbon emissions of recipient countries: A dynamic panel threshold analysis for low- and lower-middle-income countries. *Sustain. Prod. Consum.* **2022**, *29*, 158–170. [CrossRef]
50. Sunaryo Syahrihaddin, A.; Imfianto, P.S. Solar Energy for a Traditional Coastal Fishing Platform. *J. Mar. Sci. Appl.* **2019**, *18*, 366–371. [CrossRef]

energies

MDPI

Article

Perceived Barriers to Nearly Zero-Energy Housing: Empirical Evidence from Kilkenny, Ireland

Cynthia Souaid *, Harry van der Heijden and Marja Elsinga

Faculty of Architecture and the Built Environment, Delft University of Technology, Julianalaan 134, 2628 BL Delft, The Netherlands
* Correspondence: c.souaid@tudelft.nl

Abstract: In 2010, the Energy Performance of Buildings Directive announced that all new buildings are to be nearly zero-energy as of January 2021. Having reached year 2022, it can be said that the transition has proven to be slower than anticipated. Transition research has long acknowledged the potential impact of the human factor in the process of change. While there is a relative wealth of literature on end-users and their perceptions as recipients of change within the demand end of the market, research on professionals and their perceptions as actors in the process of change is limited. Thus, this study looks at the human factor in the supply end of the market by bringing professionals' perceptions to the forefront in its investigation of barriers to the implementation and uptake of nearly zero-energy housing in practice. As part of the project entitled Housing 4.0 Energy: Affordable and Sustainable Housing through Digitization, data were collected through a focus group and semi-structured interviews with housing professionals in Kilkenny, Ireland. Descriptive coding, inferential coding, and fact tracing revealed several identified barriers to be perceptions and not actual barriers to nearly zero-energy housing. Additionally, information dissemination and assimilation between policy and industry was identified as an overarching barrier. Therefore, the paper ends with recommendations to reduce delay factors at the supply end of the market, thus contributing to closing the gap between the development of policies and their implementation.

Keywords: nearly zero-energy housing; NZEB; barriers; perceptions; housing professionals; sustainability transition

Citation: Souaid, C.; van der Heijden, H.; Elsinga, M. Perceived Barriers to Nearly Zero-Energy Housing: Empirical Evidence from Kilkenny, Ireland. *Energies* **2022**, *15*, 6421. https://doi.org/10.3390/en15176421

Academic Editors: Grzegorz Mentel and Sebastian Majewski

Received: 10 August 2022
Accepted: 29 August 2022
Published: 2 September 2022

1. Introduction

In 2010, the European Parliament announced through Article 9(1) of the Energy Performance of Buildings Directive (EPBD) 2010/31/EU that all new buildings are to be nearly zero-energy as of January 2021 [1]. Back then, it was assumed that a decade is enough time for policy, industry, and society to assimilate this change [2] and take necessary action to make the transition toward a (nearly) zero-energy built environment. To facilitate this transition, European Member States (MS) were required to submit National Action Plans on nearly zero-energy buildings (NZEBs) at an early stage and to include intermediate targets for 2015. The review of submitted action plans in 2013 already called attention to an initial potential delay in the transition process toward NZEBs [3]. Consequently, in a preventative effort, the Directive required of European MS "a minimum percentage of new buildings" to be NZEBs by 2015 in its publication of recommendations and guidelines on the promotion of NZEBs. The publication even clearly refers to the implementation of NZEBs as an "obligation" stating that "[. . .] citizens buying newly constructed buildings or apartments in 2021 would expect the market to have evolved in line with these targets and buildings to be NZEBs" (p.L208/51) [4]. Yet, by 2018, notwithstanding the added emphasis on the mandatory compliance and urgency of accelerated action, 24% of European MS still did not have a detailed definition of NZEBs stated in legal documents [5]. Thus, it may well be argued that the transition toward NZEBs has been slower than anticipated even after taking into account the latest required submission of updated National Action Plans in 2019 [6,7]. More importantly, this brings into question why the transition toward NZEBs has proven to be slower than anticipated despite the given decade for preparation and adjustment and the corresponding facilitating measures implemented throughout.

It has been argued and now recognized that energy or sustainability transitions entail societal and cultural changes just as much as technical changes [8–10]. This is reflected in transition research across disciplines where it has long been acknowledged that, to develop a proper understanding of the process of change, research needs to go beyond the particular subject of study and take into account the potential impact of people, otherwise known as the human factor, in their investigations [11,12]. This recognition of the human factor and the potential impact of characteristics such as perceptions, habits, and practices has particularly been growing in energy and sustainability transition research. Studies accounting for and investigating the interrelations between technological and social change are increasing. Within the context of NZEBs, after mono-disciplinary studies plateaued in the technical advancements around the performance of sustainability measures, research was directed namely to the investigation of end-users as the human factor obstructing change. End-users were approached as recipients of change, and studies centered around end-user behavior [9]. This underlines two main research gaps. First, while the assimilation of the role and importance of the human factor has become more common in NZEB research, the focus has been mainly on people on the receiving end, involved in the use of energy measures. Research has focused less on people on the delivering end, involved in the provision of energy measures within the overarching institutional context [9] resulting in fewer studies on the perceptions of professionals involved in the provision of NZEBs. Yet, the societal aspects of the institutional context where a sustainability measure is to be implemented are not restricted to market demand but also include market supply. That is to say, perceptions, habits, and practices are as impactful throughout the provision and implementation processes of sustainability measures as they are throughout their use [9–13]. In addition, it is important to establish a simultaneous understanding of the practices of both professionals and end-users in the study of change [9]. Second, interdisciplinary research argued that changing approaches and considering individuals as *actors* within a system, that is, their surroundings, would provide a better understanding of their practices within the mechanism toward change [13]. The distinction of individuals as *actors for change* from individuals *as recipients of change* maintains the importance of taking into account characteristics underpinning practices, such as perceptions, but it also allows the investigation of the potential impact one has on the other. Most importantly, this reversed approach purposely emphasizes the importance and potentially significant impact of people's actions, underpinned by their perceptions, in the process of change. This is equally applicable to professionals as it is to end-users considering they too could play a pivotal role within that process.

One of the primary and most common approaches to the evaluation of new policies and their implementation is the study of challenges or barriers [14]. In fact, one way to define a barrier is as an explanation for the reluctance to adopt change [12]. This makes the investigation of barriers particularly relevant to studies around energy or sustainability transitions. That said, with an overall aim to unravel the potential impact of the human factor within the provision of NZEBs, this study seeks to address the following main research question: *To what extent do the perceptions of housing professionals affect the identification of barriers to the implementation of NZEBs?*

Section 2 of this paper starts by setting the background around sustainability transitions by presenting the literature reviewed on barriers to the implementation and uptake of sustainability measures including NZEBs. It also highlights the predominant overlooking of professionals' perceptions in previous investigations of barriers. Section 3 traces the different ways perceptions were included in the few studies that did take them into account. Section 4 describes the iterative research process adopted in this study alternating between desk research, data collection, and data analysis. Section 5 then presents the research methods behind the qualitative data collection. Section 6 describes the different approaches within the data analysis while simultaneously presenting the study outcomes. Section 7 discusses these outcomes in relation to previous studies. Section 8 covers policy implications, introduces corresponding recommendations, and concludes the paper by highlighting its contribution, identifying its limitations, and providing suggestions for future research.

2. Background

2.1. General Barriers to Sustainability Measures including NZEBs

To trace the development of the challenges faced in the implementation and uptake of sustainability measures in general including NZEBs in particular, the literature reviewed deliberately comprised research conducted at different points in time, spanning across different geographical contexts, covering different scopes, and adopting different perspectives (Table 1). With the exception of study number 5, all of these studies investigating barriers to the implementation and uptake of sustainability measures do so in consultation with a wide range of professionals. These include varying combinations of experts in regulation, social housing, local authorities and government agencies, architects, engineers, designers, consultants, developers, (sub)contractors, researchers, teachers, and policy makers. In other words, it can be said that the investigation of barriers to the implementation and uptake of sustainability measures including NZEBs has been extensively covered from all perspectives involved in their provision. What becomes noticeable then is that experts with different professional backgrounds identified a considerable number of similar barriers. Consequently, instead of tracing the development of challenges across time and across policy changes, what became evident through this combination of previous studies is actually the recurrence and persistence of a specific group of barriers despite the different professional perspectives adopted in their investigation. Table 2 lists the 10 most common barriers identified in previous literature. In this matrix, the most common barriers are entered as rows and the previous studies as columns (numbered 1 to 25, as they are listed in Table 1). An occurrence is marked by an "X" and the total number of occurrences is the addition of these marks. The barrier that has the highest number of occurrences is ranked 1, and the barrier that has the lowest number of occurrences is ranked, in this case, 5. When two barriers have the same number of occurrences, they are given the same rank.

Table 1. Summary list of studies included in the literature review.

Study Number	Publication Year	Study Location	Research Keywords	Research Perspective	Research Methods	Reference
1	2013	Europe	Sustainability, European energy policy, Energy efficiency in buildings	Regulation experts working within academic institutions, private companies, and public authorities such as ministries and energy agencies	Questionnaire	[15]
2	2007	UK	Legislation, Building specifications	Experts within the Royal Institute of British Architects (RIBA) involved in architectural practices in the UK	Questionnaire	[16]
3	2015	Spain	Sustainable urban transformation, Low-carbon transitions	Stakeholders from different levels of decision making with authority or interest in energy matters	Q methodology, interviews, review of relevant literature	[17]

Table 1. *Cont.*

Study Number	Publication Year	Study Location	Research Keywords	Research Perspective	Research Methods	Reference
4	2019	Australia	Sustainability transition, Low carbon, Green buildings	Sustainability consultants and advocates, energy and sustainability assessors, architects, and experts involved in teaching and research	Focus groups	[18]
5	2013	Germany	Energy efficiency, Low and zero carbon technologies	Private homeowners of single and semi-detached homes who carried out refurbishment measures	Questionnaire	[19]
6	2019	Chile	Energy policy, Nearly zero energy building	Local experts within the construction industry and the Chilean state including building professionals and researchers [1]	Literature review and focus groups	[20]
7	2018	International	Net zero energy buildings	Book—N/A	N/A	[21]
8	2014	Europe	Zero energy buildings	ZEBRA 2020 EU-funded project—N/A	N/A	[22]
9	2017	Southern Europe	Nearly zero energy building, Net zero energy building	Experts in national nearly zero-energy building regulations	Literature review and questionnaire	[23]
10	2021	Europe	Nearly zero energy buildings, European energy policy	Overview on the progress of the NZEB development in Europe—N/A	Desk study and literature review	[24]
11	2017	International	Sustainability, Housing	Experts in the prefab industry including consultants, architects/engineers, builders/subcontractors, developers, and manufacturers/distributors [1]	Literature review and questionnaire	[25]
12	2019	Brighton, UK	Low-energy, Housing	Local and national policy makers, housing associations, researchers, and not-for-profit practitioners	Literature review and expert interviews	[26]
13	2015	Sweden	Low-energy buildings, Passive houses	Experts within construction companies that build low-energy buildings	Interviews	[10]
14	2012	UNECE Region	Low-carbon transitions, Residential buildings	Policy framework	N/A	[27]
15	2016	England and Wales, UK	Sustainability, Zero carbon, Homes,	Practitioners within the Home Builders Federation (HBF) particularly involved in the construction of houses	Literature review and questionnaire	[28]

Table 1. *Cont.*

Study Number	Publication Year	Study Location	Research Keywords	Research Perspective	Research Methods	Reference
16	2017	International	Barriers to energy-efficient technologies, Building energy	Systematic literature review—N/A	Systematic literature review	[29]
17	2009	England, UK	Barriers, Zero carbon homes	Experts working within house building companies	Questionnaire and semi-structured interviews	[30]
18	2011	England, UK	Challenges, Low carbon, Housing refurbishment	Architects with housing refurbishment experience	Desk study, questionnaire, and semi-structured interviews	[31]
19	2015	UK	Barriers, Zero carbon homes	Developers, contractors, architecture and design consultants, experts within local authority and government agency with experience in low carbon homes	Semi-structured interviews	[32]
20	2007	England, UK	Barriers, Sustainability, Building	Experts in land use and planning regulations and in development and construction [2]	Literature review and interviews	[33]
21	2020	International	Critical barriers, Sustainable housing	Experts in affordable and sustainable housing studies	Literature review and questionnaire	[34]
22	2002	Netherlands	Institutional barriers, Sustainable construction	Institutions in the building and real estate sector	N/A	[35]
23	2018	Ghana	Barriers, Green building technologies	Engineers, architects, quantity surveyors, and project/contract managers with green building experience	Questionnaire	[36]
24	2017	Singapore	Barriers, Sustainable development	Project managers, consultants, quantity surveyors, design and facilities managers involved in green building projects (including residential projects)	Literature review, questionnaire, and follow-up interviews	[37]
25	2018	Chongquing, China	Barriers, Prefabrication	Experts with experience in off-site construction including professors, contractors, engineer project managers, and design directors	Questionnaire	[7]

[1] This study also included three European experts representing Germany, Spain, and Belgium (out of a total of 60 participants). [2] These studies also include end-users; however, the majority of the participants consulted remain experts involved in the field of study.

Table 2. Most common barriers identified in literature (adapted from [38]).

List of Barriers	Occurrence of Barrier in Previous Studies (Study Number)																									Total	Rank
	1	2	3	4	5	6	7	8	9	10	11	12	13	14	15	16	17	18	19	20	21	22	23	24	25		
Higher costs	X		X	X	X	X		X	X	X		X		X	X	X	X		X	X	X	X		X	X	19	1
Lack of awareness		X	X		X	X	X		X	X		X	X		X	X	X		X	X	X	X	X	X	X	19	1
Lenient building regulations			X		X	X			X			X	X	X	X	X	X	X	X	X	X				X	15	2
Shortage of skills	X	X		X		X	X			X		X	X			X	X		X	X		X		X	X	15	2
Cultural preferences	X		X		X	X			X		X		X	X	X		X		X		X	X		X		14	3
Lack of knowledge	X		X		X	X			X	X		X	X		X		X			X	X		X	X		14	3
Lack of adequate financial incentives		X	X	X		X		X	X	X					X			X	X	X	X	X		X		14	3
Business-as-usual mindset		X	X		X				X			X			X	X	X		X	X	X		X		X	13	4
Uncertainty and risks of innovation	X	X	X			X		X	X		X	X		X	X		X			X	X					13	4
Payback period and return on investment			X			X		X	X				X	X		X	X			X	X			X	X	12	5

2.2. The Factor of Perception

In previous studies on barriers to the implementation and uptake of sustainability measures including NZEBs, the terms perspective and perception are often used interchangeably. Lexically, a perspective is commonly defined as a way of thinking, an angle, or a viewpoint [39] while a perception is defined as a belief that is formulated based on impressions, appearances, and/or how things are seen [39–41]. Generally, perspective is more likely to influence perception. In other words, it can be assumed that individuals with different perspectives are more likely to have different perceptions of things. However, considering that perceptions are based on how things appear to be, the possibility for individuals with different perspectives to have similar perceptions cannot be dismissed. In the context of NZEBs, adopting the definition of perspective as a viewpoint can be translated into professionals constituting one perspective in comparison to end-users. Perspectives can also be more specific and the group of professionals itself can include different perspectives such as experts involved in housing policy, housing design, housing construction, or housing research among others. Distinctively, adopting the definition of perception as a belief that is based on how things appear, the identification of higher costs can constitute a perception in the context of NZEBs when it is based on an impression rather than a proper comparative investigation [26]. Accordingly, while current studies cover various perspectives through professionals with different expertise, the majority do not mention perceptions, and only a few focus on actually capturing the perceptions of professionals in their investigation. In other words, a possible explanation for the reaching similar outcomes despite adopting different perspectives could be the non-distinction between perceived identified barriers and actual identified barriers.

2.3. Study Contribution

The fact that most of the studies on the barriers to sustainability measures including NZEBs consult professionals in their investigation makes professionals' input significantly deterministic of the recommendations and action plans these studies reach for better implementation and uptake. This only reinforces the importance of investigating and articulating professionals' perceptions in addition to adopting different perspectives. Recalling the importance of the human factor and characteristics such as perceptions in a transition process, a clear distinction must be drawn between the terms perspective and perception in the investigation of barriers to better gauge the latter and reach overall distinct outcomes. With that in mind, this study mainly questions why previous research predominantly undermined the potential impact of professionals' perceptions and has not dedicated a certain amount of attention to developing a proper understanding of them, especially within studies around the investigation of barriers. Considering the slower than anticipated transition toward a (nearly) zero-energy built environment, this paper aims to investigate and identify current barriers to the implementation and uptake of nearly zero-energy housing from the perspective of housing professionals. However, taking into account the role of professionals as actors and the potential impact of their perceptions in the process of change, this paper also aims to bring professionals' perceptions to the forefront throughout its process. It is not restricted to adopting different perspectives of professionals in its investigation but contributes to the discussion around barriers to NZEBs by going further and dedicating special attention to perceptions in the supply end of the market.

3. Professionals' Perceptions in Previous Studies

Acknowledging perceptions in the investigation of barriers can have different forms. Within the few past studies that did acknowledge perceptions in their investigations, some distinguished perceptions from perspectives when reporting their outcomes. Others recognized the importance of professionals' perceptions at an early stage, prior to gathering their data, and incorporated it into their methodology. Thus, this paper proceeds by identifying the different ways professionals' perceptions were included in previous studies on the barriers to sustainability measures including NZEBs. Two main categorizations were established, and studies were grouped accordingly.

3.1. Inclusion of Perceptions on an Empirical Level

On an empirical level, the most common way perceptions were included in the investigation of barriers to sustainability measures is through an explicit concurrence. This is the case when studies pre-identify barriers at an initial stage of the research based on the existing literature. Then, professionals participate at a later stage where they are asked to rate and/or discuss the pre-identified barriers that are given to them. In these cases, the perceptions captured are mostly around the significance, criticality, and importance of existing barriers [17,18,27,28,35,42]. While it is important to identify the barriers that are perceived to be most obstructive to professionals, this approach can have a limiting effect as it potentially influences professionals' input by providing them with pre-identified barriers from the outset. In other words, issues that have already been identified and addressed by previous studies are being repetitively referred to when there is a need for research to investigate more closely the reasons why previously identified barriers persist and why their corresponding remedial measures have also persistently failed to redress the situation.

Another way of including perceptions on an empirical level is to consider all barriers identified by professionals as perceived. Here, very few studies follow up their data collection phase with a fact-tracing phase. The most common barriers that were linked to professionals' perceptions were higher costs and the risk and uncertainty that are linked to the implementation of novel designs and technologies. In other words, when reporting higher costs as a barrier, it was recognized that professionals identified this barrier based on their impression and not on a thorough investigation of actual costs [26,30,35,37]. More particularly, this was based on the belief that anything outside of business as usual would result in more expenses [26]. In fact, professionals' perception that the business-as-usual approach is adequate enough was identified as a barrier itself in previous research on the implementation and uptake of energy-efficient technologies [37].

3.2. Inclusion of Perceptions on a Theoretical Level

On a more theoretical level of analysis, a study on the barriers and drivers to energy performance building labels recognized the potential impact of perceptions prior to their data collection and incorporated it into their methodology. Based on the diffusion of innovation theory, perceptions of housing professionals were linked to the rate of diffusion of the labels arguing that a successful diffusion depends on how advantageous it is perceived rather than on the actual objective advantages. The perceptions of professionals were then gauged through a questionnaire formulated based on this theoretically developed model [43,44]. A study on a city's low-carbon transition focused on professionals' perceptions of themselves in their investigation into the complexity of sustainability transitions and the role and interactions of professionals throughout. The study identified four different conceptual profiles of actors involved in the process of change: the follower, the visionary, the pragmatist, and the skeptic actors. It explained that while the follower believes change is more likely to be achieved following a top-down approach, the visionary believes that formal institutions are failing to address the urgent need for change and that a bottom-up approach supported by energy transition regulatory frameworks is more effective. The pragmatist recognizes the potential impact individuals have in the process of change; however, they accord a higher level of trust to public institutions and governance processes. Finally, the skeptic does not believe climate change is caused by human-related influences and is only driven by economic motives to achieve change. With these distinctions, the study highlighted the extent to which professionals who fall into the follower and skeptic discourse could obstruct others who fall into the visionary and pragmatist discourse and who are key to the initiation of change. Overall, through these four profiles, the study described how the perceptions professionals have of themselves could act as an incentive or as a deterrent to change [36]. Last but not least, an interdisciplinary categorization of theoretical barriers to energy efficiency that reflects the nested hierarchy of the model of socio-technical change repeatedly highlighted the potential impact of professionals' perceptions in the process of change. This impact was most prominent in the barriers that fall under the socio-technical regime category where outcomes are most influenced by the human actors and where the occurrence of change is the slowest. Particularly, the barrier of bounded rationality describes professionals as decision makers who overlook energy efficiency measures based on their embedded knowledge and previously established rules of thumb. Similarly, the barrier of inertia describes how professionals could actively oppose change by falling back on their habits and previously established routines in the workplace in an effort to avoid uncertainty and potential issues which could in turn result in the overlooking of adequate energy efficiency measures [12].

4. Research Process

Whether empirical or theoretical, having reviewed the different ways professionals' perceptions were included in previous research, this study engages in the discussion through several means. First, it prevents influencing professionals' contribution by purposely not adopting the explicit concurrence approach. It aims at initially seeking out the raw perceptions and knowledge of professionals around current barriers thus contributing to the need for research to investigate barriers more closely and gaining insight into the reasons behind their recurrence. Second, this study establishes a balance between empiricism and theory by recognizing perceptions throughout its process, from inception through to implementation and analysis of outcomes. Third, it adopts an iterative approach that alternates between desk research, data collection, and data analysis. The research process follows the initial explorative literature review and focus group with fact tracing and semi-structured interviews for the validation and finalization of outcomes. This is what enables the distinction of professionals' perceptions in its outcomes. This is of particular importance seeing as it is these implicit characteristics, namely perceptions, established habits, and embedded knowledge of professionals, that are the most difficult to identify and articulate and yet that could significantly disrupt the process of change [45].

Figure 1 depicts this iterative approach by illustrating how the study alternates between desk research, data collection, and data analysis through its different research stages along with a brief description of each stage. The following Section 5 describes in more detail the methods implemented throughout.

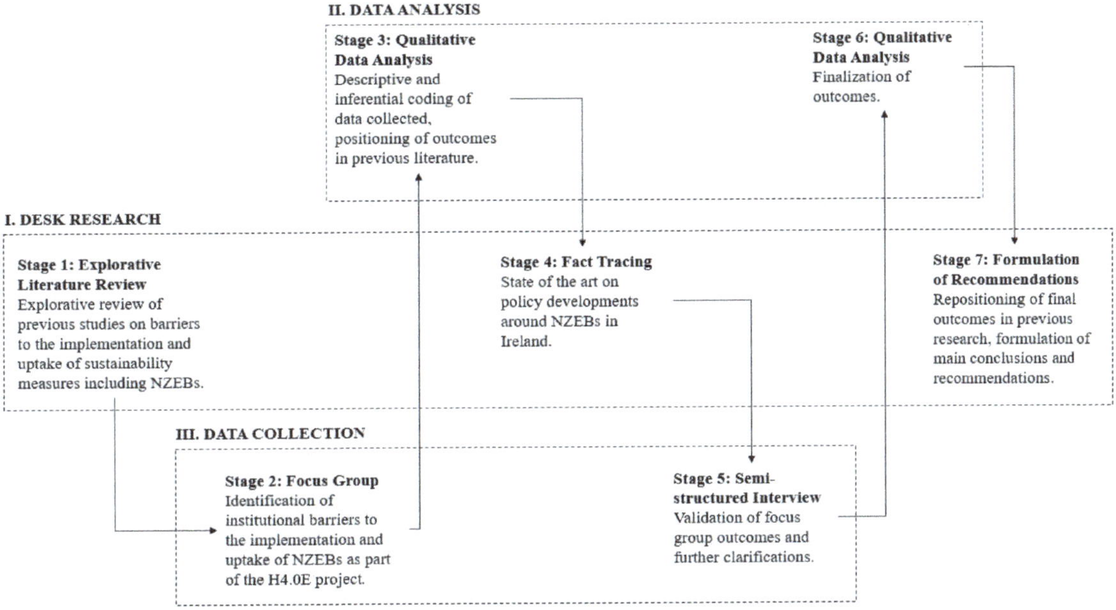

Figure 1. Iterative research process.

5. Research Methods

5.1. Desk Research

Overall, a wide range of documents were consulted in this study. In an initial stage, the desk research consisted of an explorative review of the literature to establish an understanding of the development of barriers to the implementation and uptake of NZEBs. For that, three main research concepts were used: institutional barriers, the built environment, and energy efficiency. The main keywords derived from these concepts and used in the search queries are as follows: challenges, obstacles, hindrances, together with building and/or housing and low-energy, low carbon, (near) zero-energy, zero-carbon. The main search engines consulted are Scopus, Google Scholar, and the Delft University of Technology search engine. The main sorting principle that determined whether or not an article was included in this study was the explicit address of barriers in its text. In other words, studies that did not explicitly address barriers in their text were discarded. This selection process resulted in 25 references ranging from academic journal articles and conference proceedings to textbooks. The outcomes of this initial explorative review are presented in Section 2.1, Table 1, where previous studies are listed according to their year of publication, study location, main keywords, research perspectives, and methods. Figure 2 depicts how the collection of keywords used in these 25 references falls within the research concepts of this study. At a later stage, the desk research revolved around establishing the state of the art on policy development around NZEB implementation in Ireland. To that end, different types of documents were consulted such as government publications, reports, and European projects' websites. In total, 7 main documents were referred to. These include the Irish Climate Action Plan, the Irish National Energy and Climate Action Plan, its following quarterly progress reports, the European Commission Assessment Report, and a report published by Ireland's Expert Group on Future Skills Needs.

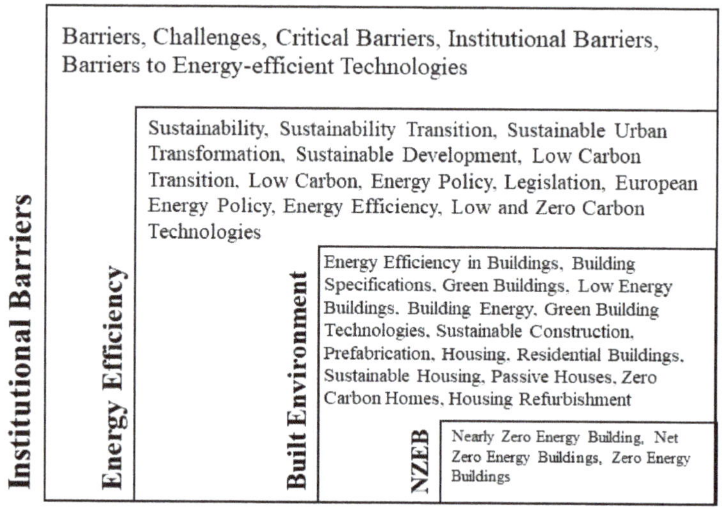

Figure 2. Main research concepts and derived study keywords.

5.2. Data Collection

The qualitative data of this study were collected through the conduction of focus groups and semi-structured interviews as part of a larger ongoing research project entitled Housing 4.0 Energy: Affordable and Sustainable Housing through Digitization (H4.0E) funded by Interreg North-West Europe [46]. Data collection was conducted in Kilkenny, Ireland, and it was carried out between the months of April and December 2019.

5.2.1. Focus Group

Focus groups are recognized to enable the collection of data that are dense in content and rich in details, even more so when the topic addressed is complex and requires a nuanced and granulated understanding [47]. This is particularly valuable to this study where the aim is to capture professionals' perceptions, an implicit characteristic that was found difficult to pin down by previous research. Focus groups are also known to allow participants to openly discuss and share different views on the research topic [48], another aspect that is of value to this study where the aim is to make a clear distinction of perceptions amongst various perspectives.

Focus group participants were recruited by nomination [49] which allowed the selection to include experts representing housing associations, social housing, local and regional authorities, the governmental housing department, financial institutions, and researchers, engineers, and architects in the field. In other words, the focus group gathered decision makers involved in housing regulation, design, implementation, and local and regional provision thus ensuring an overall balanced and representative composition. In the end, a total of 9 housing professionals were present falling within the recommended average range of 8 to 12 participants and not exceeding the maximum of 15 [50]. Table 3 provides the different profiles of the focus group participants by listing them according to their expertise, years of experience, and the professional sector they represent. Due to cancellations, developers representing the private housing sector were missing which is recognized as a potential limitation to this study.

Table 3. Focus group participant profiles.

Participant Code	Expertise	Years of Experience	Professional Sector
FGP01	Retail management, Mortgage advisory	15 years	Financial Institution
FGP02	Engineering	12 years	Housing and Planning, Local Government
FGP03	Business management, EU projects officer	23 years	Regional Authority
FGP04	Engineering	Undisclosed	Local Authority
FGP05	Research and organizational development	13 years	Social Housing
FGP06	Architecture	30 years	Construction
FGP07	Property and project management, Building surveying	21 years	Social Housing
FGP08	Building information modeling training and certification	Undisclosed	Design Standards
FGP09	Engineering, energy, and sustainability management	10 years	Non-profit Energy Agency

As previously mentioned, this study did not provide participants with the previously established list of the most common barriers identified throughout the literature review. Both to avoid bias and to allow the generation of new insights, the focus group content consisted of open-ended, explorative, and engaging questions around the following key themes: housing policy, planning and land use policy, financial schemes, energy policy, building regulations and standards, and cultural habits and preferences (Table 4). Additionally, the focus group discussion was divided into two rounds. The first round explicitly addressed the current implementation of nearly zero-energy dwellings in Kilkenny. The second round addressed the general upscaling of nearly zero-energy housing within Ireland which entailed a change of location, ownership, tenure type, target group, and income range. With the participants' consent, the focus group discussion was recorded and transcribed, and a summary of preliminary outcomes was created.

Table 4. Focus group guiding questions.

Category	Theme	Focus Group Open-Ended Questions
Institutional Barriers	Housing policy	What are the potential housing laws, regulations or policies that would prohibit/inhibit the realization of near zero-energy dwellings?
	Planning and land use policy	What are the planning or land use policies that would hinder/ facilitate the realization of near zero-energy dwellings?
Financial Barriers	Financial schemes	Which economic policies or financial schemes could prohibit/ inhibit the realization of near zero-energy dwellings?
Technical Barriers	Energy policy	What energy policies or standards are positively or negatively affecting the implementation of such projects?
	Building regulations and standards	What are the current general and technical building regulations prohibiting/inhibiting the realization of near zero-energy dwellings?
Cultural Barriers	Cultural habits and preferences	What are the cultural norms, habits or preferences that would prohibit/ inhibit a successful implementation of near zero-energy dwellings?
Miscellaneous	N/A	What are the additional barriers or inhibitors faced in the upscaling of near zero-energy dwellings?

5.2.2. Semi-Structured Interviews

After data were generated from the interactions of the different housing professionals, two follow-up semi-structured interviews were conducted with two H4.0E pilot representatives involved in the implementation of the H4.0E dwellings in Ireland (Table 5). Consulting pilot representatives after gathering initial data from external housing professionals explicitly opposed general input gained from industry to input gained based on an existing, ongoing project (H4.0E). This facilitated the distinction between actual barriers and perceived barriers. The interview proceedings enabled H4.0E pilot representatives to clarify and/or validate focus group data, provide more details on the design and construction of the H4.0E nearly zero-energy dwellings in Ireland, and elaborate more on the barriers that are being encountered in the process. The summary of preliminary outcomes was focal to the content of the interviews as the aim was, first, to prevent any misinterpretations and, second, to build upon the data that were collected during the focus group. Accordingly, interviewees were free to build the conversation and the list of interview questions was formulated thereafter, based on the validation or additional clarification of preliminary outcomes. Together with the summary of preliminary outcomes, it was shared two weeks prior giving interviewees enough time to prepare their feedback. The semi-structured interviews were organized in the form of online meetings followed by email exchanges, and with the interviewees' consent, exchanges were transcribed and documented for analysis.

Table 5. Interview participant profiles.

Participant Code	Expertise	Years of Experience	Professional Sector
SIP01	Energy Engineering	17 years	Non-profit energy agency
SIP02	Architectural Engineering	13 years	Non-profit energy agency

6. Data Analysis and Results

This section describes the different stages of the data analysis and gives detailed examples of the reasoning leading to the final study outcomes. It starts with descriptive and inferential coding which focuses on the analysis of the focus group discussion. It then moves to fact tracing where, through another desk study, focus group outcomes were cross-checked with the simultaneous policy developments. Lastly, it presents the validation and clarification of outcomes through the analysis of the follow-up semi-structured interview discussions.

6.1. Descriptive and Inferential Coding

The qualitative data analysis process mirrors this research's iterative approach alternating between data analysis, data collection, and desk research. At the outset, an initial screening of focus group outcomes allowed the recognition of the most common barriers that were pre-identified in the literature review and that recurred in this study. In that way, the pre-identified most common barriers listed in Table 2 served as the main thematic groupings throughout what is known to be the *descriptive coding* phase [51]. Descriptive coding was followed by *inferential coding* where second and third data screenings were conducted [51]. The implications of the inferential coding phase were twofold. First, it allowed the identification of barriers implicitly inferred in participants' statement. In some instances, implicit indications of barriers were dominant which is a direct manifestation of the density and high level of detail known to be characteristic of qualitative data [47]. Second, it also highlighted the extent to which barriers are intra- and interrelated to each other. Statement 1 demonstrates how both explicit and implicit barriers can be extracted out of one focus group participant statement.

> "The other thing is, we are making houses more airtight, we are bringing mechanical forms of ventilation (but) it is still out there whether that is actually good for the person living in the property. [. . .] I know you mentioned air quality and I don't know the question is out there for me." (FGP06, FG Statement 1)

This statement explicitly manifests an uncertainty and reluctance in the adoption of new technologies. Implicitly, this statement suggests an underlying preference for the business-as-usual approach. Overall, it does imply a potential lack of awareness with regard to the urgency of action when it comes to the implementation of measures to facilitate the transition toward a nearly zero-energy built environment. Following both descriptive and inferential coding, this initial phase of data analysis revealed that all the most common barriers listed in Table 2 recurred one way or another in the focus group outcomes. The codebook presented in Table 6 demonstrates how these pre-identified barriers extracted from past literature recurred in the focus group. It lists the barriers' codes, descriptions, and corresponding participants' statements. With regard to the number of occurrences, while some would argue that the most significant barriers are the ones that are mentioned the most [10], others state that importance does not go hand in hand with frequency. There are barriers that, although not as frequently mentioned, would lead to a significant obstruction to the implementation of a sustainability measure when they occur [26]. As such, significance is not attached nor restricted to frequency in this study. Nevertheless, the number of comments per barrier is included in Table 6. Overall, this presentation of results sheds light on the fact that previously known factors or challenges to the implementation of NZEBs were still perceived as challenging in 2019. More importantly, keeping in mind that the pre-identified list of barriers were not disclosed to participants, this supports the assumption that a possible explanation to the reaching similar outcomes could be the non-distinction between perspectives and perceptions of housing professionals.

6.2. Fact Tracing

At this stage of the study, it was important to establish an updated understanding of the state of the art with regard to the most recent policy developments around measures addressing the transition toward NZEBs. Accordingly, the descriptive and inferential coding phase was followed by a fact-tracing phase [21]. The particular focus of this second desk research was government proceedings, reports, and websites that are most relevant to the development of NZEBs within the Irish context [3,52–54]. Statement 2 demonstrates how focus group participants stated that current building regulations are not established as per a nearly zero-energy performance. This was identified as a potential barrier since aiming for zero energy is not mandatory.

> "You are expected to meet building regulations, you can exceed them but this becomes like any other project it is assessed based on an individual basis." (FGP03, FG Statement 2)

However, referring to governmental proceedings, the Irish National Energy and Climate Action Plan (NECP) states that, starting the first of November 2019, all new dwellings will be built to NZEB standards. The implementation of more stringent building regulations is mentioned again under existing measures [52]. Additionally, Action 56 of the Irish Climate Action Plan concerning the publication of "a methodology for compliance to NZEB in all new buildings" was reported as complete in the first progress report covering all actions within quarters 2 and 3 of 2019 [53]. Thus, it could be argued that this barrier is perceived rather than actual considering it contradicts the policy developments that were occurring simultaneously. In turn, this perception itself becomes the barrier to the implementation and uptake of NZEBs.

By adopting the same approach, the opposite can be said about the shortage of skills barrier as it can indeed be categorized as an actual barrier according to most recent policy documents (FG Statement 3).

> "After the last downturn, we lost a lot of skills." (FGP06, FG Statement 3)

Even though the shortage of skills has been addressed in the Irish Climate Action Plan and the Irish NECP [52,55], it was still recognized as constraining in the 2020 assessment report of the European Commission [56]. This was also confirmed by Ireland's Expert Group on Future Skills Needs in 2020 which indicates that this barrier persists [57]. In that manner, fact tracing weighed in on the distinction between barriers that have already been addressed in policy documents and existing barriers that remain to be addressed. Accordingly, Table 7 lists barriers that were addressed in Irish policy documents by providing the corresponding references and listing the policy action numbers where applicable. It also provides the justification such as an example of the corresponding policy measure to address the barrier in question. It states its latest policy status, where applicable, all leading to its final classification as a perceived or actual barrier. Considering the intra- and interrelations between all barriers, in some cases, there are several actions or measures that address a single barrier. In other words, the classification of barriers as actual or perceived is not a straightforward process as it entails a combination of measures acting together. However, this process still allows the formation of a preliminary understanding on the balance between housing professionals' perceptions and current policy developments.

Table 6. Codebook for the analysis of focus group transcript (adapted from [38]).

Barrier	Description	Example Statement	Number of Comments
Higher costs	Additional costs of implementing sustainability measures compared to standard construction and measures imposed by current policy and regulations (includes hidden, maintenance, and conservation costs).	"[. . .] you do not get funding for exceeding building regulations [. . .]." (FGP03)	31
Lack of awareness	The event when people, be it end-users or professionals, do not realize the magnitude of climate change consequences and the urgency of action. It can be manifested as a lack of demand for sustainability measures.	"I think that the need for housing at the moment is pushing everything on at a particular speed and the urgency to get houses built and to get people into houses." (FGP02)	10
Lenient building regulations	Less stringent current regulations that do not require the sustainability measure in question.	"You are expected to meet building regulations, you can exceed them but this becomes like any other project [. . .] based on an individual basis [. . .]." (FGP03)	5
Shortage of skills	Concerns the implementation of sustainability measures within the construction sector. Includes the lack of training.	"After the last downturn, we lost a lot of skills." (FGP06)	17
Cultural preferences	Unwillingness to stray away from traditional designs, technologies, or materials and accept or adopt new ones.	"[. . .] there is a mind-set about timber frame in this country." (FGP04)	17
Lack of knowledge	The non-consideration of sustainability measures that go beyond existing policies and regulations generally associated with a lack of interest in sustainability.	"We are building to building regulations as far as we're warranted [. . .]" (FGP07)	8
Business-as-usual approach	Applicable when the decision making is based on established rules of thumb due to the reluctance to go beyond what is already known or required by current policy and regulations.	"The department of housing in the government is more focused on traditional construction." (FGP02)	11
Uncertainty and risks of innovation	Reluctance to adopt new methods and designs and use new materials and technologies due to insufficient testing and lack of experience in implementation, maintenance, and management.	"New innovative technologies and techniques means unforeseen issues." (FGP05)	13
Lack of adequate financial incentives	Reluctance to loan partly reinforced by insufficient testing and lack of supporting evidence resulting in the absence of adequate and supporting schemes.	"We cannot give money upfront unless the architect or engineer signed off and works have been completed." (FGP01)	13
Payback period and return on investment	Specifically applicable to developers or investors including financial institutions.	"If the first thing they learn is that the value of their security will be 0 in 15 years that will have a big bearing on their willingness to lend against the property." (FGP03)	18

Table 7. Perceived versus actual barriers addressed in Irish policy and other official documents.

Barrier	References *	Action	Justification	Status **	Outcome
Higher costs	1–5	N/A	The European Commission requires the determination of NZEB regulations based on the cost optimization method. This requirement has been addressed in several EU MS action plans.	N/A	Perceived
Lack of awareness	6	68	Promote awareness and understanding of EPC [1] and provide Project Assistance Grants, training, and other support to public and private sector organizations to implement EPC projects.	Ongoing	Actual
Lenient building regulations	6	56	Measure: publish methodology for compliance with NZEB in all new buildings.	Complete	Perceived
Shortage of skills	6	50	Support relevant professional bodies in the development of training specifications/ courses for the design of NZEB and deep retrofit buildings.	Ongoing	Actual
Uncertainty and risks of innovations	7 and 8	N/A	The technology behind NZEBs is available and proven. Technology is going even further, and the main focus now is shifting toward energy-plus buildings that contribute to energy generation rather than break even.	N/A	Perceived
Lack of adequate financial incentives	6	44, 54	Establish a Steering Committee and Working Group to design a new financing scheme to provide easier-to-access tailored finance for SMEs [2] and residential energy efficiency investment utilizing the European Commission's Smart Finance for Smart Buildings loan scheme.	Complete	Perceived
Payback period and return on investment	6	45	Develop a tool to deliver a roadmap to individual homes to achieve BER [3] B2, cost-optimal, and NZEB.	Complete	Perceived

* 1: [3], 2: [4], 3: [58], 4: [52], 5: [59], 6: [60], 7: [6], 8: [61]. ** The focus group was conducted in April 2019. Accordingly, the statuses of actions mentioned in this table were based on the progress reports published in 2019. 1 EPC: Energy Performance Contracting. 2 SME: Small and Medium Enterprise. 3 BER: Building Energy Rating.

6.3. Validation and Clarification of Outcomes

As previously mentioned, interviewing H4.0E pilot representatives enabled input that is based on actual current experiences happening during the H4.0E project. Consequently, the data collected at this stage of the research process allowed a straightforward identification and/or confirmation of *actual* barriers. For example, interview statement 1 is an indication of the general lack of knowledge barrier amongst housing providers manifested through the non-consideration of sustainability measures that go beyond existing policies and regulations at the time. This renders the lack of knowledge an *actual* barrier to NZEBs. Implicitly, this statement also indicates a general lack of awareness on the urgent need to shift toward a zero-energy built environment that is manifested through that same lack of effort in exceeding the mandatory requirements. Thus, this reconfirms a lack of awareness as another *actual* barrier to NZEBs.

> "In this Technical Guideline (TGD) is outlined a minimum standard that all buildings must comply with. Unfortunately, the LAs (local authorities) took and take this minimum requirement as a benchmark." (SIP01, IW Statement 1)

Other examples can be found in interview statement 2. On the one hand, this statement is an explicit example of the extent to which the reluctance to adopt innovative measures of design or construction obstructs and delays the project implementation. It is a direct manifestation of the perception of uncertainty and risks linked to innovation rendering this barrier a *perceived* barrier to NZEBs. On the other hand, it also exposes the business-as-usual approach and its potentially obstructive effect amongst individuals in the sector rendering it an *actual* barrier to the implementation and uptake of NZEBs.

> " [. . .] individuals do not want to be held responsible if a new type of design fails, so they are very cautious [. . .]. Even it would be in their favour [. . .]"
> (SIP02, IW Statement 2)

Overall, the iterative research process followed in this study and the combination of methods implemented succeeded in distinguishing the perceptions of housing professionals. It differentiated between barriers that are based on perceptions and actual barriers. Table 8 demonstrates how both perceptions and actual barriers were validated by pilot representatives in the semi-structured interviews by listing barrier codes, descriptions, and participant statements. Table 9 provides a summary of this study's outcomes where it can be seen that more than half of the most common barriers that recurred in focus group outcomes were based on perceptions and were not actual barriers.

Table 8. Codebook for the analysis of semi-structured interview transcript.

Barrier	Description (Listed in Table 6)	Example Statement	Outcome
Lack of awareness	The event when people, be it end-users or professionals, do not realize the magnitude of climate change consequences and the urgency of action. It can be manifested as a lack of demand for sustainability measures.	"Even it would be in their favour it takes a lot of time and effort to [. . .] convince the LAs for adapting highly efficient, low energy and low carbon options [. . .]" (SIP02)	Actual
Cultural preferences	Unwillingness to stray away from traditional designs, technologies, or materials and accept or adopt new ones.	"Even the fact that the quality of recent build timber construction is up to a high-quality standard the old picture of a failed timber frame house is shaping the behavior and opinion." (SIP02)	Perception
Lack of knowledge	The non-consideration of sustainability measures that go beyond existing policies and regulations generally associated with a lack of interest in sustainability.	"In this Technical Guideline (TGD) is outlined a minimum standard that all buildings must comply with. Unfortunately, the LAs (local authorities) took and take this minimum requirement as a benchmark." (SIP01)	Actual
Business-as-usual mindset	Applicable when the decision making is based on established rules of thumb due to the reluctance to go beyond what is already known or required by current policy and regulations.	"[. . .] we need to take on extra time and effort to convince the responsible auteurs to take on better values and to invest in future proved buildings" (SIP01)	Actual
Uncertainty and risks of innovation	Reluctance to adopt new methods and designs and use new materials and technologies due to insufficient testing and lack of experience in implementation, maintenance, and management.	" [. . .] individuals do not want to be held responsible if a new type of design fails, so they are very cautious [. . .]. Even it would be in their favour [. . .]" (SIP02)	Perception

Table 9. Summary table of outcomes.

Barrier	Method		Outcome
	Fact Tracing	Follow-Up Interviews	
Higher costs	⊗		Perception
Lack of awareness	⊗	⊗	Actual
Lenient building regulations	⊗		Perception
Shortage of skills	⊗		Actual
Cultural preferences		⊗	Perception
Lack of knowledge		⊗	Actual
Business-as-usual mindset		⊗	Actual
Uncertainty and risks of innovation	⊗	⊗	Perception
Lack of adequate financial incentives	⊗		Perception
Payback period and return on investment	⊗		Perception

7. Discussion

7.1. A Shift in the Model Composition: Housing Professionals' Perceptions as the Obstacle

In an investigation of barriers, one can distinguish three main features composing the overall barrier model: the obstacle, the subject, and the action. The obstacle is defined as the obstructive entity, the subject consists of the entity that is affected by the obstruction, and the action comprises the phenomenon that is being prevented [12]. In this study, implementing and upscaling nearly zero-energy housing would qualify as *the action*. This action would have an impact on the environment altogether which includes virtually everyone rendering all people the *subject* of obstruction. The consultation of housing professionals in the process of identifying barriers, or *obstacles*, insinuates they are an objective and external entity to the model composition, unaffected by or unaffecting the overall investigation. While this research approach does generate valuable insight on the transition process, shifting the model composition and looking at housing professionals as a subjective element with subjective perceptions having the potential to become obstacles themselves reveals an entirely different list of impediments. This study allowed the distinction of these perceptions and demonstrated several times over how a shift in approach could potentially lead to a change in outcome.

In this study, the barrier of higher costs that describes concerns around the extra costs specific to nearly zero-energy housing due to all the added energy efficiency measures and that underlines a trade-off between energy performance and affordability is a manifestation of participants' perceptions because it was formulated with reference to the costs of traditional dwelling designs as a benchmark. Instead, if the costs of new-build housing designs complying with the soon-to-be mandatory building regulations were considered as the benchmark, higher costs may not have been identified as a barrier. Additionally, the affordability of new-build nearly zero-energy housing is currently being addressed in policy documents and the development of NZEB regulations [4]. This echoes findings from previous studies recognizing this same barrier as based on an impression rather than an investigation of actual costs [26,30,35,37]. The barrier of uncertainty and risks of innovation that describes in this particular study participants' concerns around airtightness and mechanical ventilation systems was revealed to be a manifestation of perceptions. Current research has surpassed uncertainties about technologies within nearly zero-energy housing, and the literature is now focusing on energy-plus housing [61]. The barrier of lenient building regulations that portrays nearly zero-energy housing as exceeding current mandatory requirements was also revealed to be a perception seeing as policy documents state that NZEB regulations are to be enacted starting the second half of 2019 [52]. Additionally, focus group statements describing lenient building regulations or governmental entities giving precedence to housing provision rather than a zero-energy performance can be said to portray a dependence of housing professionals on higher authorities. Recalling the follower-type depiction of professionals, this becomes a manifestation of professionals' perception of themselves believing that change is more likely to be achieved following a top-down approach. This was identified as a cognitive barrier itself in previous research [36]. Overall, given that these barriers, or perceptions, persist despite research and policy documents stating otherwise is an indication of the strength of the overarching preference for the business-as-usual approach, another finding that echoes previous study outcomes [26,37]. In fact, this recalls the theoretical barriers of bounded rationality and inertia that describe professionals falling back on previously established knowledge, resisting change to avoid uncertainty, potentially resulting in the overlooking of adequate energy efficiency measures [12].

7.2. The Overarching Barrier of Information Dissemination and Assimilation

This study's data collection was conducted throughout the year 2019. On a general level of analysis, it can be said that housing professionals were consulted about the implementation and uptake of nearly zero-energy housing in the same time frame as corresponding policy and regulations were being developed [60]. Relevant dates around the implementation of NZEB regulations and construction were already released. Even when final documents were still in progress, drafts and draft assessments were being published. In other words, NZEB information was available regardless of whether or not it was still under review, and it was only a matter of months before the NZEB regulations were enacted. This parallelism underlines a potential gap between (inter)national policy makers and local practice. It suggests a lack of awareness and knowledge of the soon-to-be mandatory, more stringent building regulations. The fast development of technology potentially leading to the unawareness of professionals has already been flagged by previous research as impeding the "future success of delivering a more sustainable built environment" ([26], p. 144). Indeed, an earlier study on the feasibility of zero-carbon homes marked a 6-year gap between industry's expectations and actual policy goals when asking professionals about their perceptions on a realistic timeline for the transition [17]. Another study attributed the increasing gap between industry, technology, and policy to professionals' perceptions of their own overestimated level of knowledge on current designs and technologies [36]. In hindsight, this begs the question: Is the gap between policy developments and local practice caused by a lack of awareness of housing professionals and a persistence of the business-as-usual approach? Or does the overarching barrier behind this gap lie within information dissemination? Or perhaps a combination of both? What is certain is that a successful transition toward a nearly zero-energy built environment requires policy and industry to coincide. While a top-down approach has been recognized as most effective for the implementation of new regulations, the current gap suggests that it might not be enough and highlights a potential flaw in how information is being transferred.

7.3. The Role of Information Dissemination in a Transition Process

The importance of information dissemination and the critical role it plays in a transition process has been raised in many previous studies. Corresponding measures and recommendations have already been identified and previously formulated [12,22,23,62,63]. However, the majority of these recommendations were initially directed at end-users, and very few in comparison had housing professionals as their target audience. Meanwhile, the transfer of information, new policies, and regulations to relevant housing professionals can be as challenging as the transfer of information to end-users [7]. Intensive knowledge transfer between housing professionals is known to be essential to achieve actual rather than incremental change [64], even more so when recalling the fragmented decision-making process present in complex sustainability transitions such as the shift toward a zero-energy built environment [25]. Thus, a lot can be learned if these same findings were directed toward housing professionals. For instance, when it comes to learning new information, it is argued that people are selective about which information to accept and assimilate. They are passive rather than active information seekers [12]. Keeping in mind the fast-developing technologies/policies and the overwhelming amount of information available, looking at this study's outcomes through this lens could explain why focus group participants were potentially not up to date with the latest policy developments around NZEBs. Another example concerns the rational-actor assumption that accounts people as actors who respond rationally to the information that is made available to them. Previous research on end-user behavior revealed that reasoning is ineffective [62,63]. Within the context of this study, this could justify why the lack of awareness of housing professionals is still a barrier even though the NZEB concept was introduced more than a decade ago and the urgency to transition toward a zero-energy built environment is continuously increasing. Last but not least, research on end-users' decision-making process suggested that a timely and measured integration of information provision throughout the process is most effective for the actual implementation of desired outcomes [22]. Within the context of this study, the absence of key actors to effectively transfer the most recent policy developments could explain the desynchronization between policy developments and the knowledge and awareness within local practice.

8. Conclusions and Policy Implications

8.1. Policy Implications and Recommendations

All in all, there is a need for innovation in information dissemination within the provision end of the market be it on a general level between policy and local practice or on a more detailed level between housing professionals themselves. Maintaining the shift in model composition and referring back to the insights gained from previous research directed at end-users leads to the formulation of several suggestions specific to housing professionals and the provision of NZEBs. First, the provision of NZEB information should be more consistently and systematically linked to concrete situations and/or opportunities in a particular context. Just like information provision should be integrated into end-users' decision-making process [22], policy and regulatory information provision should be integrated into the process of new housing provision through the inclusion and training of key intermediaries. These trained experts should be incorporated at key decision-making moments that local authorities, social housing associations, private developers, or other housing professionals encounter throughout the process of housing provision.

Second, recalling the formulation of information that is vivid, clear, concise, and customized to the specific context in question [12,22,23,62,63], the distributed NZEB information should be personalized and tailored to the situational context of its targeted audience for a more impactful dissemination. Within the communication channels amongst housing professionals, this would entail varying necessary NZEB information to fit the professional field it is addressing. Just like the successful diffusion of labels for highly efficient housing required a formulation that is contextually compatible with the professionals implementing them [43], policy regulations and expert knowledge need to be actively translated to tailor the expertise and interests of the targeted audience of housing professionals: architects, engineers, contractors, developers, and local authorities, among others. Thus, the training of intermediaries would not only cover NZEB-related information and regulation but also communication skills to develop the ability to address different housing professionals according to their different interests and goals. Additionally, developing an understanding of housing professionals' different expertise and interests is of particular importance in the attempt to overcome the challenging, complex, and fragmented decision-making process that occurs in practice when implementing NZEBs.

Third, referring back to the introduction of sustainability champions that would increase the likelihood of creating an environmentally aware community [22,62,65], the number of NZEB practices should significantly increase through industry advocates or pilot projects within local authorities. If the rational-actor reasoning is applied to housing professionals, it can be expected that the availability of information on NZEB design, benefits, regulations, and the overall urgency of action would provoke concern and result in the smooth adoption of the relevant changes. However, focus group outcomes revealed the prevalence of the business-as-usual approach despite very soon to be mandatory regulations, an occurrence confirmed by previous research stating that raising awareness is not enough to change long-established perceptions and habits [62]. Thus, implementing the reverse hypothesis that starts with the implementation of environmental practices underlines the need for a bottom-up approach to work in tandem with the top-down regulations. In other words, imposing new regulations alone on housing professionals is not enough, and there is a need to simultaneously shift the business-as-usual approach through industry advocates and pilot projects to achieve a successful transition of the industry as a whole. This reversed approach would particularly help increase the likelihood of raising openness within housing professionals to more effectively integrate NZEB information.

8.2. Concluding Remarks

The main aim of this paper was to demonstrate the importance and potential impact of the perceptions of professionals involved in the provision of NZEBs when identifying barriers to their implementation and uptake. In doing so, this study's engagement in the discussion of energy or sustainability transition is twofold. On a general level, not only did this study recognize the importance of the human factor in the process of change, but it also incorporated it in its investigation. Through its shift in model composition, individuals were involved as actors and not just recipients within the process of change. On a more specific level, this study contributed to narrowing the research gap around experts' behavior within the context of NZEBs by setting the perceptions of professionals as the focal point of its investigation of barriers to the implementation and uptake of nearly zero-energy housing.

Falling back on this paper's main outcomes, more than half of the identified barriers were revealed to be perceived and not actual barriers. That is to say, the explicit distinction of the factor of perception throughout the study's iterative research process did indeed succeed in articulating housing professionals' perceptions. First, purposely choosing not to adopt the explicit concurrence approach in the identification of barriers allowed the prevention of bias when gauging professionals' current knowledge and perceptions around existing barriers to the implementation and uptake of nearly zero-energy housing. Second, following up the qualitative data analysis with fact tracing allowed the establishment of an updated understanding of the state of the art with regard to the most recent policy developments addressing the transition toward NZEBs. This initiated the distinction between perceived and actual barriers. Namely, the barriers of higher costs, lenient building regulations, cultural preferences, uncertainty and risks of innovation, lack of adequate financial incentives, and the payback period and return on investment barriers were identified as perceptions and not actual barriers. Third, seeking out input from professionals involved in an ongoing project led to the validation of outcomes such as the negative perception of innovative sustainability measures or designs translated into the uncertainty and risks of innovation barrier. It also allowed the validation of overarching barriers such as the lack of awareness, the lack of knowledge, and the strength of the business-as-usual approach. Last but not least, distinguishing the factor of perception within the identification of barriers shed light on a potential significant gap between policy developments and local practice indicating an overarching potential barrier to information dissemination and assimilation. Thus, this paper called for innovation in information dissemination be it between policy and industry or between housing professionals themselves which in turn was the focus of the suggestions and recommendations formulated.

Finally, though insightful, this paper's outcomes are specific to the study context in question. Considering the scarcity of research on the human factor in the supply end of the NZEB market, precedence was given to identifying professionals' perceptions and to demonstrating their potential impact on the identification of barriers to nearly zero-energy housing. Rather than increasing sample size for more generalizable outcomes, the paper takes a closer look into the detailed qualitative data collected from a small sample. This is what allowed the distinction of perception, an implicit characteristic that is initially difficult to identify and articulate. Thus, having established this initial demonstration, future research can build upon this study to investigate professionals' perceptions across larger samples and within different contexts.

Author Contributions: Conceptualization, C.S.; methodology, C.S.; validation, H.v.d.H. and M.E.; formal analysis, C.S.; investigation, C.S. and H.v.d.H.; writing—original draft preparation, C.S.; writing—review and editing, H.v.d.H. and M.E.; visualization, C.S.; supervision, H.v.d.H. and M.E.; project administration, M.E.; funding acquisition, M.E. All authors have read and agreed to the published version of the manuscript.

Funding: This research was conducted as part of a larger project entitled Housing 4.0 Energy: Affordable and Sustainable Housing through Digitization (H4.0E) funded by Interreg North-West Europe [46] [NWE 705].

Institutional Review Board Statement: Ethical review and approval were waived for this study due to the following reasons:

- The study did not include participants who were unable to provide informed consent.
- No videos, pictures or other identifiable data are stored.
- The study did not include participants that are in dependent positions to the investigator.
- It was not necessary for participants to participate in the study without their knowledge or consent at the time.
- The study does not actively deceived participants.
- The study does not collect personal sensitive data such as financial data, location data, data relating to children or vulnerable groups.
- No substances are used in the study, no blood or tissue samples are taken, no pain is inflicted as a result of the study and the study does not risk causing psychological stress or anxiety.

Informed Consent Statement: Informed consent was obtained from all subjects involved in the study.

Data Availability Statement: The data presented in this study are available upon request from the corresponding author.

Conflicts of Interest: The authors declare no conflict of interest. The funders had no role in the design of the study; in the collection, analyses, or interpretation of data; in the writing of the manuscript; or in the decision to publish the results.

References

1. The European Parliament. *Directive 2010/31/EU Of The European Parliament and of the Council of 19 May 2010 on the Energy Performance of Buildings*; Official Journal of the European Union; The European Union: Maastricht, The Netherlands, 2010; pp. 13–35.
2. Build Up. BUILD UP The European Portal for Energy Efficiency in Buildings. Available online: https://www.buildup.eu/en/learn/ask-the-experts/there-difference-between-shall-and-should-epbd-0 (accessed on 16 May 2022).
3. D'Agostino, D.; Zangheri, P.; Cuniberti, B.; Paci, D.; Bertoldi, P. *Synthesis Report on the National Plans for Nearly Zero Energy Buildings (NZEBs): Progress of Member States towards NZEBs*; Joint Research Center (JRC) Publications: Ispra, Italy, 2016; ISSN 1018-5593/1831-9424. Available online: https://op.europa.eu/en/publication-detail/-/publication/082cbbb3-0205-11e6-b713-01aa75ed71a1/language-en (accessed on 4 August 2021).
4. European Commission. Commission Recommendation (EU) 2016/1318 of 29 July 2016 on Guidelines for the Promotion of Nearly Zero-Energy Buildings and Best Practices to Ensure that, by 2020, All New Buildings Are Nearly Zero-Energy Buildings. 2016; pp. 46–57. Available online: http://data.europa.eu/eli/reco/2016/1318/oj (accessed on 16 May 2022).
5. Erhorn, H.; Erhorn-Kluttig, H. *New Buildings & NZEBs—2018 Status in February 2018*; Fraunhofer Institute for Building Physics: Stuttgart, Germany, 2018; Available online: https://www.epbd-ca.eu/wp-content/uploads/2018/04/CA-EPBD-CT1-New-buildings-NZEBs.pdf (accessed on 2 November 2021).
6. European Commission. National Energy and Climate Plans EU Countries' 10-Year National Energy and Climate Plans for 2021–2030. Available online: https://ec.europa.eu/info/energy-climate-change-environment/implementation-eu-countries/energy-and-climate-governance-and-reporting/national-energy-and-climate-plans_en (accessed on 19 May 2022).
7. Martiskainen, M.; Kivimaa, P. Role of knowledge and policies as drivers for low-energy housing: Case studies from the United Kingdom. *J. Clean. Prod.* **2019**, *215*, 1402–1414. [CrossRef]
8. Stephenson, J.; Barton, B.; Carrington, G.; Doering, A.; Ford, R.; Hopkins, D.; Lawson, R.; McCarthy, A.; Rees, D.; Scott, M.; et al. The energy cultures framework: Exploring the role of norms, practices and material culture in shaping energy behaviour in New Zealand. *Energy Res. Soc. Sci.* **2015**, *7*, 117–123. [CrossRef]
9. Pellegrino, M.; Musy, M. Seven questions around interdisciplinarity in energy research. *Energy Res. Soc. Sci.* **2017**, *32*, 1–12. [CrossRef]
10. Martek, I.; Hosseini, M.R.; Shrestha, A.; Edwards, D.J.; Durdyev, S. Barriers inhibiting the transition to sustainability within the Australian construction industry: An investigation of technical and social interactions. *J. Clean. Prod.* **2019**, *211*, 281–292. [CrossRef]

11. Williamson, O.E. The New Institutional Economics: Taking Stock, Looking Ahead. *J. Econ. Lit.* **2000**, *38*, 595–613. [CrossRef]
12. Thollander, P.; Palm, J.; Rohdin, P. Categorizing barriers to energy efficiency: An interdisciplinary perspective. *Energy Effic.* **2010**, *11*, 49–63. [CrossRef]
13. Stephenson, J. Sustainability cultures and energy research: An actor-centred interpretation of cultural theory. *Energy Res. Soc. Sci.* **2018**, *44*, 242–249. [CrossRef]
14. Stern, N. Stern Review: The Economics of Climate Change. 2006. Available online: https://www.osti.gov/etdeweb/biblio/20838308 (accessed on 14 May 2022).
15. Attia, S. *Net Zero Energy Buildings (NZEB): Concepts, Frameworks and Roadmap for Project Analysis and Implementation*; Elsevier Science & Technology: San Diego, CA, USA, 2018. [CrossRef]
16. Dave, M.; Watson, B.; Prasad, D. Performance and perception in prefab housing: An exploratory industry survey on sustainability and affordability. *Procedia Eng.* **2017**, *180*, 676–686. [CrossRef]
17. Osmani, M.; O'Reilly, A. Feasibility of zero carbon homes in England by 2016: A house builder's perspective. *Build Environ.* **2009**, *44*, 1917–1924. [CrossRef]
18. Adabre, M.A.; Chan, A.P.C.; Darko, A.; Osei-Kyei, R.; Abidoye, R.; Adjei-Kumi, T. Critical barriers to sustainability attainment in affordable housing: International construction professionals' perspective. *J. Clean. Prod.* **2020**, *253*, 119995. [CrossRef]
19. Gan, X.; Chang, R.; Wen, T. Overcoming barriers to off-site construction through engaging stakeholders: A two-mode social network analysis. *J. Clean. Prod.* **2018**, *201*, 735–747. [CrossRef]
20. Mellegard, S.; Lund Godbolt, A.; Lappegard Hauge, A.; Klinski, M. ZEBRA 2020—Nearly Zero-Energy Building Strategy 2020 Deliverable D5.2: Market Actors' NZEB uptake—Drivers and Barriers in European Countries. 2016. Available online: https://www.sintef.no/en/publications/publication/?pubid=CRIStin+1453166 (accessed on 20 October 2021).
21. Attia, S.; Eleftheriou, P.; Xeni, F.; Morlot, R.; Menezo, C.; Kostopoulos, V.; Betsi, M.; Kalaitzoglou, I.; Pagliano, L.; Cellura, M.; et al. Overview and future challenges of nearly zero energy buildings (nZEB) design in Southern Europe. *Energy Build.* **2017**, *155*, 439–458. [CrossRef]
22. Stieß, I.; Dunkelberg, E. Objectives, barriers and occasions for energy efficient refurbishment by private homeowners. *J. Clean. Prod.* **2013**, *48*, 250–259. [CrossRef]
23. Persson, J.; Grönkvist, S. Drivers for and barriers to low-energy buildings in Sweden. *J. Clean. Prod.* **2015**, *109*, 296–304. [CrossRef]
24. D'Agostino, D.; Tzeiranaki, S.T.; Zangheri, P.; Bertoldi, P. Assessing Nearly Zero Energy Buildings (NZEBs) development in Europe. *Energy Strategy Rev.* **2021**, *36*, 100680. [CrossRef]
25. Van Bueren, E.M.; Priemus, H. Institutional barriers to sustainable construction. *Environ. Plan. B Plan. Des.* **2002**, *29*, 75–86. [CrossRef]
26. Williams, K.; Dair, C. What is stopping sustainable building in England? Barriers experienced by stakeholders in delivering sustainable developments. *Sustain. Dev.* **2007**, *15*, 135–147. [CrossRef]
27. Henderson, C.; Ganah, A.; Jon, G.A. Achieving sustainable homes by 2016 in the UK: The current status. *Environ. Dev. Sustain.* **2015**, *18*, 547–560. [CrossRef]
28. Darko, A.; Chan, A.P.C.; Yang, Y.; Shan, M.; He, B.-J.; Gou, Z. Influences of barriers, drivers, and promotion strategies on green building technologies adoption in developing countries: The Ghanaian case. *J. Clean. Prod.* **2018**, *200*, 687–703. [CrossRef]
29. Davies, P.; Osmani, M. Low carbon housing refurbishment challenges and incentives: Architects' perspectives. *Build Environ.* **2011**, *46*, 1691–1698. [CrossRef]
30. Heffernan, E.; Pan, W.; Liang, X.; de Wilde, P. Zero carbon homes: Perceptions from the UK construction industry. *Energy Policy* **2015**, *79*, 23–36. [CrossRef]
31. Annunziata, E.; Frey, M.; Rizzi, F. Towards nearly zero-energy buildings: The state-of-art of national regulations in Europe. *Energy* **2013**, *57*, 125–133. [CrossRef]
32. Piderit, M.B.; Vivanco, F.; van Moeseke, G.; Attia, S. Net Zero Buildings-A Framework for an Integrated Policy in Chile. *Sustainability* **2019**, *11*, 1494. [CrossRef]
33. Adeyeye, K.; Osmani, M.; Brown, C. Energy conservation and building design: The environmental legislation push and pull factors. *Energy Conserv.* **2007**, *25*, 375–390. [CrossRef]
34. Golubchikov, O.; Deda, P. Governance, technology, and equity: An integrated policy framework for energy efficient housing. *Energy Policy* **2012**, *41*, 733–741. [CrossRef]
35. Hwang, B.-G.; Zhu, L.; Tan, J.S.H. Green business park project management: Barriers and solutions for sustainable development. *J. Clean. Prod.* **2017**, *153*, 209–219. [CrossRef]
36. Olazabal, M.; Pascual, U. Urban low-carbon transitions: Cognitive barriers and opportunities. *J. Clean. Prod.* **2015**, *109*, 336–346. [CrossRef]
37. Yeatts, D.E.; Auden, D.; Cooksey, C.; Chen, C.-F. A systematic review of strategies for overcoming the barriers to energy-efficient technologies in buildings. *Energy Res. Soc. Sci.* **2017**, *32*, 76–85. [CrossRef]
38. Souaid, C.; van der Heijden, H.; Elsinga, M. Institutional Barriers to Near Zero-Energy Housing: A Context Specific Approach. *Sustainability* **2021**, *13*, 7135. [CrossRef]
39. Oxford Learner's Dictionaries. Oxford Learner's Dictionaries. Available online: https://www.oxfordlearnersdictionaries.com/definition/english/perception?q=perception (accessed on 18 May 2022).
40. Collins. Available online: https://www.collinsdictionary.com/dictionary/english/perception (accessed on 17 May 2022).

41. Cambridge Dictionary. Cambridge Dictionary. Available online: https://dictionary.cambridge.org/ (accessed on 15 October 2021).
42. Chan, A.P.C.; Darko, A.; Olanipekun, A.O.; Ameyaw, E.E. Critical barriers to green building technologies adoption in developing countries: The case of Ghana. *J. Clean. Prod.* **2018**, *172*, 1067–1079. [CrossRef]
43. Mlecnik, E.; Visscher, H.; van Hal, A. Barriers and opportunities for labels for highly energy-efficient houses. *Energy Policy* **2010**, *38*, 4592–4603. [CrossRef]
44. Mlecnik, E. Defining nearly zero-energy housing in Belgium and the Netherlands. *Energy Effic.* **2012**, *5*, 411–431. [CrossRef]
45. Hirsh, R.F.; Jones, C.F. History's contributio ns to energy research and policy. *Energy Res. Soc. Sci.* **2014**, *1*, 106–111. [CrossRef]
46. NWEurope. H4.0E—Housing 4.0 Energy. Available online: https://www.nweurope.eu/projects/project-search/h40e-housing-40-energy/ (accessed on 20 May 2022).
47. Kamberelis, G.; Dimitriadis, G. *Focus Groups: From Structured Interviews to Collective Conversations*; Routledge: London, UK, 2013. [CrossRef]
48. Morgan, D.L. Focus groups and social interaction. In *The Sage Handbook of Interview Research: The Complexity of the Craft*; Sage Publications: Thousand Oaks, CA, USA, 2012; Volume 2.
49. Stewart, D.W.; Shamdasani, P.N. *Focus Groups: Theory and Practice*; Sage publications: Thousand Oaks, CA, USA, 2014; Volume 20.
50. Powell, R.A.; Single, H.M. Focus groups. *Int. J. Qual. Health Care* **1996**, *8*, 499–504. [CrossRef]
51. Miles, M.B.; Huberman, A.M. *Qualitative Data Analysis: An Expanded Sourcebook*; Sage publications: Thousand Oaks, CA, USA, 1994.
52. Government of Ireland. Ireland Climate Action Plan 2019. Available online: https://www.gov.ie/en/publication/ccb2e0-the-climate-action-plan-2019/ (accessed on 14 October 2021).
53. Government of Ireland. Ireland Climate Action Plan 2019 Fifth Progress Report Q3 2020. Available online: https://www.google.com/url?sa=t&rct=j&q=&esrc=s&source=web&cd=&ved=2ahUKEwjju_LVp_H5AhVGX_EDHW81A08QFnoECAgQAQ&url=https%3A%2F%2Fassets.gov.ie%2F99563%2F79ef025c-5b83-489f-bec1-1900b19a4052.pdf&usg=AOvVaw13F4oRUSSw8jN0oMnMduDr (accessed on 14 October 2021).
54. European Commission. EU Countries' Nearly Zero-Energy Buildings National Plans. Available online: https://energy.ec.europa.eu/topics/energy-efficiency/energy-efficient-buildings/nearly-zero-energy-buildings_en (accessed on 22 May 2022).
55. Government of Ireland. *Ireland's National Energy and Climate Plan 2021–2030*; Government of Ireland: Department of the Environment, Climate and Communications: Dublin, Ireland. Available online: https://www.gov.ie/en/publication/0015c-irelands-national-energy-climate-plan-2021-2030/ (accessed on 25 September 2021).
56. European Commission. EUR-Lex Access to European Law. Available online: https://eur-lex.europa.eu/legal-content/EN/TXT/?qid=1565713062913&uri=CELEX:52019DC0285 (accessed on 18 May 2022).
57. EGFSN. *The Demand for Skills in Ireland's Built Environment Sector to 2030*; Department of Enterprise, Trade and Employement: Dubli, Ireland, 2020. Available online: https://www.enterprise.gov.ie/en/publications/building-future-skills.html (accessed on 15 August 2021).
58. Government of Belgium. *Belgian Integrated National Energy and Climate Plan 2021–2030 Section A: National Plan*; European Commission: Brussels, Belgium, 2019. Available online: https://www.google.com/url?sa=t&rct=j&q=&esrc=s&source=web&cd=&cad=rja&uact=8&ved=2ahUKEwjzy6eWw_P5AhUDO-wKHQCNDyIQFnoECBUQAQ&url=https%3A%2F%2Fenergy.ec.europa.eu%2Fsystem%2Ffiles%2F2020-09%2Fbe_final_necp_parta_en_0.pdf&usg=AOvVaw27hE-jhTvtIp0FK8CXCKrg (accessed on 16 August 2021).
59. Government of the Netherlands. *Climate Agreement*; Government of the Netherlands: The Hague, The Netherlands, 2019. Available online: https://www.google.com/url?sa=t&rct=j&q=&esrc=s&source=web&cd=&cad=rja&uact=8&ved=2ahUKEwi_uqXTxPP5AhUwsKQKHRz1ASoQFnoECA4QAQ&url=https%3A%2F%2Fwww.government.nl%2Ftopics%2Fclimate-change%2Fclimate-policy&usg=AOvVaw1W1IGpOa7iBcMRjFRds7DJ (accessed on 29 August 2021).
60. Government of Ireland. Climate Action Important Publications. Available online: https://www.gov.ie/en/publication/55fde-climate-action-important-publications/ (accessed on 20 May 2022).
61. Bointner, R.K.L.; Toleikyte, A. *Strategies for nZEB Market Transition on a National Level*; Buildings Performance Institute Europe: Brussels, Belgium, 2016; Available online: https://www.google.com/url?sa=t&rct=j&q=&esrc=s&source=web&cd=&cad=rja&uact=8&ved=2ahUKEwikrKjEw_P5AhUT7aQKHZbbAjEQFnoECAMQAQ&url=https%3A%2F%2Fzebra2020.eu%2Fpublications%2Fstrategies-for-nzeb-market-transition-on-national-level%2F&usg=AOvVaw13oc4_T2GcYfheD-9bq5yG (accessed on 20 August 2021).
62. Bartiaux, F. Does environmental information overcome practice compartmentalisation and change consumers' behaviours? *J. Clean. Prod.* **2008**, *16*, 1170–1180. [CrossRef]
63. Desmedt, J.; Vekemans, G.; Maes, D. Ensuring effectiveness of information to influence household behaviour. *J. Clean. Prod.* **2009**, *17*, 455–462. [CrossRef]
64. Mlecnik, E. Opportunities for supplier-led systemic innovation in highly energy-efficient housing. *J. Clean. Prod.* **2013**, *56*, 103–111. [CrossRef]
65. De Wilde, M. The sustainable housing question: On the role of interpersonal, impersonal and professional trust in low-carbon retrofit decisions by homeowners. *Energy Res. Soc. Sci.* **2019**, *51*, 138–147. [CrossRef]

energies

MDPI

Article

The Quality of Goodwill Disclosures and Impairment in the Financial Statements of Energy, Mining, and Fuel Sector Groups during the Pandemic Period—Evidence from Poland

Maciej Gierusz [1,*], Stanisław Hońko [2,*], Marzena Strojek-Filus [3] and Katarzyna Świetla [4]

1 Department of Accounting, The Faculty of Management, University of Gdańsk, 101 Armii Krajowej St., 80-309 Gdańsk, Poland
2 Department of Accounting, Institute of Economy and Finance, University of Szczecin, 64 Mickiewicza St., 71-101 Szczecin, Poland
3 Department of Accounting, Faculty of Finance, University of Economics in Katowice, 1 Maja 50, 40-287 Katowice, Poland
4 Department of Financial Accounting, Institute of Finance, Cracow University of Economics, 27 Rakowicka St., 31-510 Kraków, Poland
* Correspondence: mac_gie@wp.pl (M.G.); stanislaw.honko@usz.edu.pl (S.H.)

Abstract: The COVID-19 pandemic has strongly affected the economic situation of many countries, which is worth considering not only globally but also in the context of specific industries. An asset that is particularly sensitive to negative economic changes is goodwill. The aim of this study is to assess the impact of the pandemic on the quality of financial disclosures concerning goodwill in consolidated financial statements of groups of chosen strategic sectors in Poland. We investigated the implications of the pandemic on the frequency and scale of goodwill impairment in relation to 23 companies listed on the Warsaw Stock Exchange from the Energy, Mining, and Fuel Index. We identified the research gap in this area. For the purposes of this study, two research hypotheses were formulated: (H1) during the COVID-19 pandemic, there has been a slight improvement in the quality of goodwill disclosures in the consolidated financial statements of groups in the energy, fuel, and mining sectors; (H2) The COVID-19 pandemic caused a decrease in goodwill due to impairment losses in the consolidated financial statements of groups in the energy, fuel, and mining sectors. The hypotheses were verified on the basis of the above research sample. In order to verify the first hypothesis, we tested 81 consolidated financial statements for the years 2018–2021 based on a self-developed index of the quality of disclosed information. To verify the second hypothesis, we analyzed the frequency and scale of the estimated loss of goodwill during the COVID-19 pandemic and its impact on the deterioration of the financial condition of the same research sample. The conducted research shows that the quality of disclosures regarding the goodwill in the examined sample has changed slightly. Contrary to our expectations, the pandemic did not materially reduce the value of goodwill. This means that the first hypothesis was verified positively, while the second hypothesis had to be rejected.

Keywords: goodwill; impairment of goodwill; quality of financial statements; energy sector; mining; fuel; disclosures of financial information

Citation: Gierusz, M.; Hońko, S.; Strojek-Filus, M.; Świetla, K. The Quality of Goodwill Disclosures and Impairment in the Financial Statements of Energy, Mining, and Fuel Sector Groups during the Pandemic Period—Evidence from Poland. *Energies* **2022**, *15*, 5763. https://doi.org/10.3390/en15165763

Academic Editors: Grzegorz Mentel and Sebastian Majewski

Received: 10 June 2022
Accepted: 4 August 2022
Published: 9 August 2022

1. Introduction

The COVID-19 pandemic has had serious consequences in two dimensions: health and economics. COVID-19 has had a strong impact on human health and mortality worldwide. It has also caused problems and difficulties in the functioning of business entities in many sectors of the economy. This is the reason for the deterioration of financial results and, in some industries, also the bankruptcies of many companies. The impact of the pandemic on the financial situation of enterprises is also a consequence of the lack of predictability of future results and the instability of operations. One of the sectors with a high vulnerability to many social and economic factors is the energy sector, which has been experiencing strong turbulence for years due to, among other factors, climate policy and geopolitics. The energy sector in Poland is strongly connected with the hard coal and lignite mining industry and also strongly dependent on the policy of the authorities. Studies carried out in different countries also indicate the mutual coupling of this sector with the economic situation of the country and the region [1]. On the other hand, the economic situation is affected by special phenomena, including the pandemic, e.g., [2–7]. In this context, the question arises if the impact of the pandemic on the quality of information on goodwill and its write-offs in companies from the energy, mining, and fuel sectors in Poland can be identified.

The research also took into account two sectors: mining and fuel, which are proprietary to the energy sector in Poland. The mining and fuel sectors are heavily dependent on the geopolitical situation and factors such as the pandemic.

An asset that should serve as a litmus test for the crisis is goodwill. Simply, it can be assumed that it is a fixed estimated value of the market perception of an entity related to, for example, exceptional management qualifications, market share, or a unique business model. Goodwill is a special asset of groups reported as a result of the parent company acquiring control of other entities (subsidiaries) and is presented in the consolidated statements of its financial position [8–10]. Goodwill represents the parent's future expected economic benefits that are not attributable to the other assets reported in the consolidated accounts. It is the result of the occurrence and impact of other "invisible" assets of an intangible nature in a subsidiary [11]. The deterioration of the financial condition of the group is particularly emphasized by an impairment loss of goodwill. Companies in the energy sector belong to the largest business entities, often with extensive holding structures. Research conducted on entities in this sector in China indicates a strong negative effect of the pandemic on the situation of those entities that showed write-downs of goodwill in financial statements [3]. However, there is a deficit of research on this problem in other parts of the world, affected by the effects of the pandemic to varying degrees. The research gap identified in this article is the impact of the COVID-19 pandemic on potential impairment losses of goodwill in groups of the energy, mining, and fuel sectors in Poland.

The aim of the article is to assess the impact of the COVID-19 pandemic on the quality of consolidated financial statements, particularly in terms of goodwill and impairment losses in capital groups of the Warsaw Stock Exchange Index (WIG)-Energy, WIG-Fuel, and WIG-Mining.

The research results partially fill the gap regarding the impact of the pandemic on the goodwill presented in the consolidated financial statements of the energy sector groups and sectors strongly associated with it in the countries of Central and Eastern Europe. The research particularly focused on goodwill impairment as a result of the pandemic. Finally, our research complements the picture of the quality of disclosures of goodwill in the consolidated financial statements of groups in strategic sectors in Poland.

The structure of this article was adapted to the research purpose. The first part presents an overview of the literature on goodwill, with an emphasis on areas specified in the research hypotheses. The empirical part is preceded by a description of the research methods used. This section includes two main parts, aiming to verify research hypotheses. The last part consists of final conclusions and the indication of limitations and recommendations.

2. Literature Review

The review of the literature concerning the research problem was carried out in two parts corresponding to the stages of the empirical research. Firstly, research on the quality of disclosure of goodwill information was taken into account. Secondly, the remaining studies on goodwill and its write-offs were classified. Each part of the literature review ends with the formulation of a research hypothesis.

A. Kozłowska [12] tried to define the most important factors that determine the quality of financial reporting. She pointed out that it is an issue of the qualitative features of financial information in financial statements that are increasingly raised in the context of the harmonization and standardization of accounting. The author used the qualitative analysis of financial statements. This has been carried out in accordance with the Framework for Financial Reporting Concepts. It was also based on the results of scientific research in this field.

The study presented by G. Iatridis [13–15] indicates that the implementation of International Financial Reporting Standards (IFRS) generally reinforces quality of financial information. The author pointed out that the implementation of IFRS reduces the scope for earnings management. The good quality of accounting information, especially financial statements, is related to more timely loss recognition and provides relevant accounting measures which are also connected with accounting disclosures.

Similar studies were conducted by M. E. Barth, W. R. Landsman, and M. H. Lang, [16] who pointed to the impact of the use of accounting standards on its quality. They referred to the coupling taking place in the application of IFRS in the context of the interpretation of solutions and their practical application due to limited income manipulation, the faster recognition of losses, and the better adequacy of the information presented.

The confirmation of this opinion is also the result of the deliberations of many researchers, e.g., [17,18], analyzing the quality of accounting information in the context of unified solutions in its field.

Conclusions on the impact of the accounting quality on the quality of reports were also presented by Sumiyana et al. [19], who argued that the quality of accounting information is extremely important for the target readers of financial reporting. Consequently, it results in a lower cost of internal and external capital and affects the development of business entities.

In the context of the quality of financial information presented in financial statements, a particular resource to be reported is goodwill. Goodwill is defined in the literature in different ways; however, in most of the presented definitions, its essence is emphasized as "surplus value", referring to the expected profit or assets of the entity [20–23]. It represents the additional gain to be realized in the future and is therefore the present value of the additional benefits that an entity acquiring another entity expects to realize. Meanwhile, Scott [9] indicates that this value is still related to intangible assets. At the same time, Hendriksen [8], while characterizing goodwill both on the basis of economic theory and accounting theory, additionally defines it as a surplus of the value of an entity as a whole over the sum of its tangible and intangible net assets. As an asset, goodwill is recognized in the statement of financial position (consolidated statement of financial position) only when control is obtained over another entityin accordance with IFRS 3 "Business Combinations". It therefore represents the value acquired by the investor and verified by means of the actual transaction of acquiring a controlling interest. It is the result of the existence of other intangible assets that have not been identified, individually measured, and disclosed in the acquired entity's statement of financial position. In the context of the high expectations of users regarding financial reporting, the scope of the presented disclosures becomes of key importance [22–24]. In the research concerning the quality of information on goodwill presented in consolidated financial statements, the basis was the reference to the concept of true and fair view and the requirements of the International Accounting Standards (IAS) and IFRS [25] (Amel-Zadeh, Glaum, and Sellhorn, 2021). M. Gierusz [26] points out the inconsistency between the rules of showing and valuation of goodwill in financial statements and the definition of assets adopted in the IFRS. In the opinion of A. Amel-Zadeh, M. Glaum, and T. Sellhorn [25], the means of identification and valuation, as well as the presentation of a company's value, is crucial for the correct assessment of economic entities' conditions, but it is also strongly influenced by the area of managerial activities. In this area, we can also distinguish research focusing on the problem of the allocation of the purchase price to goodwill and other identifiable intangibles [27,28].

The quality of disclosures concerned with goodwill is a subject of many studies. Khairi et al. [29] conducted a study which answered several important questions regarding the quality of information disclosed on the goodwill impairment process under the requirements of the Financial Reporting Standard (FRS) 36. This study investigated the compliance level and disclosure quality of the financial statements of the top 20 Singaporean firms listed on the stock exchange-SGX in 2007 based on their market capitalization.

The results of an investigation by M. Boučková [30] have shown a generally low level of compliance with disclosure requirements of goodwill impairment within the selected entities. This caused little comparability between financial statements.

A critical review of the literature and research on the disclosure of goodwill were conducted by C. Carvalho, A.M. Rodrigues, and C Ferreira [31], who indicated the main contribution of the literature as well as its limitations. They also suggested new approaches to future research connected with goodwill and the main determinants of its disclosures in the literature and the market. The authors also discussed the need to implement enforcement mechanisms to improve the level of compliance regarding disclosures on goodwill and their impairment tests. They concluded that most of the analyzed literature shows that the information disclosed about goodwill is incomplete and largely heterogeneous. It also confirms a reduced level of compliance with the disclosures indicated by accounting standards.

In the light of the above-mentioned literature, the following hypothesis was formulated:

H1: *During the COVID-19 pandemic, there has been a slight improvement in the quality of goodwill disclosures in the consolidated financial statements of groups in the energy, fuel, and mining sectors.*

Goodwill, due to its uniqueness compared to other assets, is a subject of research concerning the valuation process. Part of the research concerns factors influencing the identification and valuation of company value (e.g., [10,32,33]).

Another group of studies addresses methods of accounting for goodwill, including assessing the effects of abandoning amortization in favor of impairment testing (e.g., [34–38]). The resignation from amortization in IAS/IFRS in favor of goodwill impairment testing caused discussion among researchers about the implications of this change and its compliance with the true and fair view concept. The results of conducted studies were not unequivocal in their assessment. The results of the study by P. Van Hulzen et al. [39] indicate a higher quality of financial statements when the amortization method is used. However other studies indicate that accounting for goodwill by utilizing impairment write-offs better reflects the economic value of goodwill than the method of systematic write-offs (write-downs) (e.g., [34,39]). Martinez et al. [40] indicate the need for changes in accounting regulation to combine depreciation with impairment testing. A SWOT analysis performed by M. Cieciura and H. Czaja-Cieszyńska [41] led to a conclusion about the advantage of impairment testing. The results of the research also point to the threats connected with this method, consisting of an intentional treatment of possible write-offs by managers [11,35,42–44].

There are some studies that verify whether companies disclose the information required by the IFRS on the impairment of goodwill [45,46].

What is important to the purpose of the article is the research on the goodwill impairment losses in terms of their determinants (e.g., [31,47,48]). M. Glaum et al. [49] investigated the main determinants of goodwill impairment decisions undertaken by entities applying the IFRS. The authors chose a comprehensive sample of stock-listed firms from 21 countries and indicated that goodwill impairment is negatively associated with their economic results. The authors concluded that firms in low-enforcement countries are less responsive to declines in the economic value of goodwill.

One of the factors analyzed included the global crisis of 2008–2010 [42]. Zemskova [50] researched the factors influencing goodwill and impairments of goodwill in the oil and gas sectors. Research in recent years has focused on the impact of the COVID-19 pandemic on goodwill write-downs [3,51]. G. Goswami and Kimmel [51] examined whether the deteriorating cash flow due to the pandemic affected goodwill write-downs.

A study of the energy sector in China found a strong negative impact of the pandemic on financial results in the sector [3]. The results indicate that the presentation of write-offs resulted in a moderating variable; companies with goodwill impairment were more strongly affected by the pandemic. The situation of entities that did not show such a write-off was significantly more favorable [3]. This section may be divided into subheadings. It should provide a concise and precise description of the experimental results, their interpretation, as well as the experimental conclusions that can be drawn.

In this context, the second hypothesis was formulated:

H2: *The COVID-19 pandemic caused a decrease in goodwill due to impairment losses in the consolidated financial statements of groups in the energy, fuel, and mining sectors.*

It is worth emphasizing that there has been very few studies on the impact of the COVID-19 pandemic on the valuation of goodwill, particularly impairment write-offs in selected economic sectors, including the energy, fuel, and mining sectors, which are the research areas of this paper. The small amount of research is due to the very short research period from the outbreak of the pandemic and thus the limited possibilities of observing its direct and indirect effects in selected sectors.

3. Material and Methods

In line with the adopted objective, the research was carried out in two stages. The first stage focused on the quality of disclosures regarding goodwill presented in consolidated financial statements. In the second stage, the relationship between the pandemic and write-offs (including write-downs) of goodwill was investigated.

The results of the study presented below are aimed at confirming or rejecting the hypotheses H1 and H2. Both hypotheses were verified on the basis of the consolidated financial statements of the same groups. The study covered the consolidated financial statements for the period 2018–2021 of 23 leading stock exchange groups listed in Warsaw, belonging to three sectors of the economy:

1. Energy (12 groups): CEZ, Kogenera, Polenerga, Będzin, Enea, PGE, Tauron, Zepak, Interaolt, Photon, MLSystem, Onde.
2. Mining (5 groups): Coal Energy, Greenx, JSW. KGHM, Bogdanka,
3. Fuel (6 groups): Lotos, MOL, PGNiG, Orlen, Serenius, Unimot.

The selected sample was targeted and covered the energy sector as well as related industries. On the other hand, the selection of exchange-traded entities was dictated by the fact that such entities, as leading players on the market, conclude the most transactions to acquire other business units. Therefore, it can be expected that the acquired goodwill will constitute a significant percentage of their balance sheet total. In addition, in the case of entities listed on European stock exchanges, all entities report in accordance with IFRS.

The verification of hypothesis H1 required the development of a unique research tool, which is the **self-developed index of disclosure quality**, assuming the perspective of a financial statement user who is interested in understanding the essence of goodwill in a concise way. It is necessary to explain why this method was chosen. There are many methods of assessing the quality of financial statements, including mathematical and statistical methods—for example, relating to earnings management. Different methods are used to evaluate the narrative part of financial statements. Two groups of methods can be distinguished, namely, subjective or semi-objective methods. V. Beattie, B. McInnes, and S. Fearnley [52] consider the disclosure index—for example, in the form of a checklist—to be the semi-objective method. The authors cited distinguish three types of disclosure indexes, namely, binary, weighted, and nested. Taking into account the narrow research area in this article, covering one item of assets (goodwill), a decision was made to use a binary index.

The study used its own disclosure quality index regarding goodwill, which was constructed based on the list of control questions. An affirmative answer results in awarding

"1", whereas a negative answer results in "0". For simplicity, no additional weights were assigned to the responses. Goodwill resulting from a merger or attributed to subsidiaries or associates was disclosed in 65 statements (80% of the audited entities). Companies that did not show goodwill (16 financial statements) included only general disclosures regarding goodwill in their accounting policies. The list of control questions is closely related to the scope of disclosures regarding goodwill specified in IAS/IFRS. The selection of questions is aimed at highlighting the information that should be disclosed and is important for the user of the consolidated financial statements to correctly interpret the goodwill. In our opinion, this list of questions can be used to test the quality of goodwill disclosures in the financial statements of entities of different sectors, not only those that we research in this study.

In order to verify research hypothesis H2, an analysis of numerical data was carried out, comparing the goodwill value with other items of the financial statements, such as:

- opening balance sheet, impairment losses, and closing balance sheet for goodwill,
- total assets,
- equity,
- revenue,
- operating profit,
- expenditure on fixed assets, intangible assets, and subsidiaries.

The research part, aimed at verifying the second hypothesis, was divided into the following stages of the analysis

Study 1—change in revenue;
Study 2—change in operating profits;
Study 3—share of goodwill in total assets;
Study 4—share of goodwill in total equity;
Study 5—amount of impairment losses in goodwill;
Study 6—frequency of impairment losses in goodwill;
Study 7—average useful life of fixed assets;
Study 8—the "Rollover Indicator" for goodwill

It should be noted that, despite the selection of a relatively small number of entities from related sectors, it is difficult to talk about full comparability in terms of the scale of business activity. The balance sheet totals and sales revenue indicate that the largest entities' operations amount to hundreds of billions of zlotys (Orlen). In contrast, other entities conduct business activity at the level of tens of millions of zlotys (Photon, Serenius).

We can assume that, due to the radically different business potential, the goodwill presented on the balance sheets of these entities may constitute a different percentage of the balance sheet total, and the transactions of business combinations have different importance and frequency.

Another element that should be noted at the beginning of the study is the number of entities in the research sample; in the analyzed period of 2018–2021, the acquired goodwill did appear on the balance sheet. In the sample, there were 16 entities from 23 companies initially covered by the study. In the further part of the study, results will be presented for each of the three sectors (energy, mining, and fuel) separately and collectively, which should increase the cognitive value of the analysis.

4. Results and Discussion

4.1. *Analysis of the Quality of Disclosures*

The recognition and measurement of goodwill raise many doubts. Some even claim that this balance sheet item is so questionable that it does not meet the definition of assets. The argument for this statement is the inability to sell goodwill separately. The occurrence of doubts as to the existence and measurement of goodwill justifies the expectation of disclosures being of high quality regarding the recognition and impairment of goodwill.

It can be assumed that a measure of the quality of financial statements is the quality of disclosures relating to "soft" assets, such as goodwill.

As already mentioned, the analysis covered 81 financial statements. The first hypothesis is verified based on 65 reports in which goodwill appeared. In the remaining statements, there are only general references to goodwill in the accounting policy. They usually repeat the provisions of accounting standards. The binary index consisted of seven questions. One point was assigned for an affirmative answer; hence, the maximum number of points was 65. The questions were deliberately not limited to the requirements imposed by the accounting standards. The collective responses are presented in Table 1.

Table 1. The number of points for the quality of disclosures by question.

Number	Question	Number Points (Max. 65)	%
1	Is the financial statement searchable?	60	92%
2	Have the means of creation of goodwill been disclosed?	27	42%
3	Has the date of creation of goodwill been disclosed?	19	29%
4	Has the interpretation of the goodwill been disclosed?	10	15%
5	Has the allocation of goodwill to the CGU been disclosed?	49	75%
6	Have the indications of impairment of goodwill been disclosed?	25	38%
7	Have the discount rates used to estimate the recoverable amount been disclosed?	36	55%

Question 1 concerns a technical issue. Some of the audited financial statements were made available as scans with the official signatures of management and accountants. This method of communication, appearing only in 2018–2019, was used in five reports. In recent years, it has been observed that the traditional form of presenting financial statements in pdf format has been replaced by a more friendly form: the electronic standard (ESEF).

Question 2 concerned the sources of goodwill. In some financial reports—for example, in Unimot SA—the history of each element of goodwill is presented in detail, specifying not only the means but also the moment of its creation. This allows users to assess whether the elements of goodwill are the result of several transactions or one single transaction. The opposite of this approach, applied in numerous reports, is limited to the recognition of goodwill in the notes concerning intangible assets without providing any explanations. This approach, in our opinion, has low informative value and does not allow users to assess the risks and future economic benefits of goodwill. Most of the investigated entities limited their reporting to the description of transactions generating goodwill in a reporting period. To learn more about the sources of goodwill, it would be necessary to study the financial statements from previous periods.

The answer to question 3 can only be found in 19 financial statements. The lower number of points awarded in response to this question is a consequence of indicating the transactions that are the reason for recognizing goodwill without indicating the date of these transactions. Such a solution leads to difficulties in analyzing changes in the goodwill over time.

Question 4 requires some clarification. By the interpretation of goodwill in this article, we mean the explanation of the meaning of a given component of goodwill. In many analyzed reports, there are general statements that repeat the definition of IFRS goodwill. Goodwill may include the following components [53]:

- Component 1—The excess of the fair value over the book value of the acquiree's net assets at the date of acquisition,
- Component 2—The fair value of other net assets that the acquiree had not previously recognized,

- Component 3—The fair value of the going concern element of the acquiree's existing business, reflecting on the ability of the acquiree as a standalone business to earn a higher rate of return on an organized collection of net assets than would be expected if those net assets had to be acquired separately,
- Component 4—The fair value of the expected synergies and other benefits from combining the acquirer's and acquiree's net assets and businesses,
- Component 5—Overvaluation of the consideration paid by the acquirer stemming from errors in valuing the consideration tendered,
- Component 6—Overpayment or underpayment by the acquirer.

From the point of view of the user of the financial statements, it is not important how goodwill was calculated. The economic substance of the item is more substantial. An example of the disclosure of economic content may be the disclosure in the reports of the Będzin company: "The recognized goodwill resulted primarily from the possessed experience and specialist knowledge in the financial sector, the reputation of the acquired company, established processes and business projects in the financial area, and access to markets". Another company (Orlen) posted the following content: "Goodwill created from the acquisition of RUCH results from the forecast synergies resulting from the merger of RUCH's operations with the ORLEN Group and presents the value of assets that could not be recognized separately in accordance with the requirements of International Accounting Standard (IAS) 38 (employees and their knowledge)". The presented disclosures, in conjunction with the transaction data (subject, date), assist users in assessing the potential of future cash flows related to a given component of goodwill. In accordance with the requirements of IAS 36 (paragraph 134), an entity is required to disclose goodwill allocated to cash-generating units. This is essential because goodwill does not generate economic benefits on its own but rather in combination with other components of the cash-generating units (CGU).

When analyzing the number of answers to question 5, it should be stated that 25% of the studied financial statements do not meet these requirements. The lack of an attribution of goodwill to the CGU deprives users of key information, reducing the predictive value of financial statements. Informing only about global amounts of goodwill, without their allocation to CGU, excludes assessment of the probability of the impact of economic benefits in a given segment of activity. It also makes it impossible to verify any assumptions accepted for the purpose of measuring the impairment of goodwill.

An entity's obligation under IAS 36 (paragraph 10) is to test goodwill for impairment, regardless of any impairment indicators. Therefore, it is assumed that goodwill is an item that is particularly vulnerable to value adjustments. An entity that recognizes impairment losses is required to disclose the events and circumstances that led to impairment losses (IAS 36 para 130a). Therefore, it is assumed that entities that have not recognized impairment losses of goodwill are not required to disclose the impairment's indicators. The study showed entities that disclosed impairment indicators despite not recognizing any impairment loss. For example, the Polenergia report for 2021 revealed that: "The conducted analysis of the indicators has shown that changes in the expected price levels of electricity, gas and CO2 emission allowances contribute to a decrease in the forecast margin on electricity production." The treatment of the determinants of the impairment of assets provides a background for evaluating the assumptions accepted by an entity for the purpose of measuring impairment.

The last question was about the disclosure of the discount rates applied to measure the recoverable amount of the CGUs. Despite the obligation to disclose key assumptions of the measurement, in almost half of the financial statements, this information was omitted. It is also worth noting that quite different discount rates were applied—from 4% to 12%. It cannot be concluded that higher rates were used in the 2020–2021 reports, which would be justified by expected inflation.

In summary, most entities comply with their obligations regarding disclosure imposed by the IFRS. However, this is not the same as high-quality disclosure, as judged from the perspective of a financial statement for users who expect clear and concise information about goodwill to enable assessment of the risk and the potential forfuture benefits generated by this questionable item of assets.

The following section presents the ranking of disclosure quality scores by year and entity (Table 2). This section may be divided into subheadings. It should provide a concise and precise description of the experimental results, their interpretation, as well as the experimental conclusions that can be drawn.

Table 2. The number of points for the quality of disclosures (by the company) (max. 7 points/year).

Place	Company	2018	2019	2020	2021	Total
1	CEZ	7	7	7	7	28
2	Unimot	6	6	6	6	24
3	Będzin	4	5	6	7	22
4	Lotos	4	4	4	4	16
4	Polenerg	4	4	4	4	16
4	Tauron	4	4	4	4	16
5	MOL	3	3	3	6	15
6	Inter	4	4	5	1	14
7	Orlen	1	1	5	6	13
8	Enea	3	3	3	3	12
8	Kogener	3	3	3	3	12
8	PGE	3	3	3	3	12
9	ML	2	2	2	2	8
9	PGNIG	2	2	2	2	8
10	Photon	1	1	1	1	4
10	Serinius	1	1	1	1	4
11	Onde	N/A	N/A	N/A	3	3
Total		52	53	59	63	227

Source: the authors' own study.

It can be concluded that the quality of disclosures regarding goodwill: (1) is not dependent on the affiliation of the capital group with a sector and (2) is gradually improving. The research carried out with the use of a binary index allows for a positive verification of hypothesis H1: During the COVID-19 pandemic, there has been a slight improvement in the quality of goodwill disclosures in the consolidated financial statements of groups in the energy, fuel, and mining sectors. However, it is difficult to link this improvement with the COVID-19 pandemic. It can only be assumed that, due to the increased risk, the examined entities were willing to present more information, increasing the chances of users assessing the ability to generate economic benefits by the CGU to which goodwill was allocated. This tendency should therefore be assessed positively.

4.2. Recognition and Measurement of Acquired Goodwill in Companies Listed on the Warsaw Stock Exchange

Study 1—Change in revenue

The purpose of this analysis is to evaluate how the revenues from the core business of the companies covered by the study behaved in the last four years—thus, the results for 2018/19 and 2020/21 will be compared with each other. This will allow for a determination of what impact the coronavirus pandemic had on the scale of business activity. At this stage, it can be presumed that a possible decrease in revenue, i.e., limited demand for the products and services offered by companies, should be a significant reason for testing goodwill for impairment and should consequently lead to the recognition of impairment losses in this respect. Table 3 presents the changes in revenue by sector and collectively.

Table 3. Change in revenue compared to the previous year.

Sector/Years	2018	2019	2020	2021
ENERGY	20.8%	15.7%	1.7%	47.0%
FUEL	22.6%	36.8%	−11.7%	64.5%
MINING	−15.4%	−2.7%	−22.6%	−5.7%
TOTAL	13.4%	17.2%	−7.1%	40.1%

Source: the authors' own study.

From the above analysis, several interesting conclusions can be drawn:

1. The effects of the coronavirus pandemic are very clearly visible in the financial results of all industries in 2020. In the case of the energy and fuel sectors, we are dealing with a recession. Although the energy sector recorded minimal growth in nominal terms, its scale is smaller than the inflation in Poland for this period, oscillating around 3.4%. Realistically, therefore, we were dealing with a decline. On the other hand, the second of the periods marked by the epidemic, i.e., 2021, is already experiencing a significant rebound and, to a large extent, apart from the mining industry, is making up for losses for the previous reporting period.
2. The mining sector has been struggling with recession and a decline in the scale of operations for many years. This phenomenon intensified in the pandemic-tainted year 2020, but the next year was also marked by declines from the already lower base.
3. The best of the analyzed sectors were companies from the fuel industry, recording solid growth in the pre-covid period and more than making up for losses in 2021 after the collapse a year earlier.
4. In more analytical terms, 4 out of 5 entities from the mining sector in 2020 recorded a decrease in revenue, while in the fuel sector, it was 4 companies out of 6, and in the energy sector, it was only 4 out of 11 units. In 2021, declines were limited to some units of the mining sector.

In conclusion, the first study showed that more than half (12 out of 23) of the entities felt the effects of the epidemic in the form of a decrease in their revenues from core activities in the particularly difficult year of 2020. It should be considered that the negative dynamics of sales meet the criterion set for the impairment of assets in accordance with IAS 36, and they therefore should be the basis for conducting formal tests for impairment.

Study 2—Change in operating profits

The purpose of this study was to see whether the negative trend noticed in sales revenue was also reflected in the dynamics of operating profits. Theoretically, one would expect that the time of economic downturn in the markets could be the perfect time to introduce various types of rationalization activities, which are easier to justify and carry out than in a period of prosperity. In addition, it should be emphasized that the operating profit—i.e., before operations recognized as a result of accounting convention (e.g., revaluation to fair value, impairment losses), one-off events, and financial income and expenses—has the strongest link to cash flows from operating activities, which are key determinants of the generation of added value by an enterprise. Table 4 shows the change in operating profits by sector:

Table 4. Change in operating profits [1].

Sector/Years	2018	2019	2020	2021
ENERGY	−43.8%	−41.2%	16.6%	58.8%
FUEL	−23.8%	−9.0%	−77.9%	191.2%
MINING	−59.1%	49.4%	−133.1%	122.8%
TOTAL	−41.9%	−13.1%	−40.6%	107.2%

Source: the authors' own study. [1] (The following results must be interpreted with a great deal of caution, as operating profit can be both positive and negative. For example, if a company makes a profit of 10 units in one year and a loss of 40 units in the next year, the decrease is −400%. In addition, if, in the third period, the loss is to be 20 units, we are dealing with an increase in the ratio by 50%. In the case of entities where the increase or decrease in profits exceeded 300%, it was decided to "stabilize" its level and adopt a value of 300%. The aim of this procedure is to ensure that the extremely high or low result of one company does not burden the average too much).

The conclusions to be drawn from the above-presented analysis are as follows:

1. All the analyzed sectors between 2018 and 2020 faced difficult business realities, and operating results deteriorated over this time. Even the rebound in the mining sector in 2019 could not compensate for the losses incurred a year earlier. Only 2021 saw a noticeable improvement in the economic performance in all three sectors.
2. In more detail, only 3 out of the 23 surveyed entities recorded a negative operating profit in 2018. A total of 7 companies did in 2019, 9 companies did in 2020, and 5 companies did in 2021. The above results, therefore, seem to coincide with the trend observed in the context of operating profit dynamics.
3. The above indicates a high degree of correlation between the negative dynamics of revenues and operating profits, so the entities were not able to significantly reduce operating costs and introduce savings and rationalization activities on a wider scale.

Study 3—Share of goodwill in total assets

The aim of this study is to examine what percentage of goodwill constitutes the total assets of the surveyed entities, which will determine the importance of tests for the impairment of goodwill. Guided by the principle of materiality, it can be assumed that the role of the procedures provided for in IAS 36 will be much more important in those entities where goodwill constitutes a significant percentage of the balance sheet total.

The results only for those entities for which goodwill has occurred are presented in Table 5:

Table 5. Share of goodwill in total assets—all entities with purchased goodwill.

Sector/Years	2018	2019	2020	2021
ENERGY	0.8%	0.8%	0.6%	0.4%
FUEL	0.9%	0.8%	1.1%	1.2%
MINING	0.0%	0.4%	0.4%	0.4%
TOTAL	0.8%	0.8%	0.7%	0.7%

Source: the authors' own study.

From the above data, the following conclusions can be drawn:

1. The goodwill for the entities in which it occurred oscillates around one percent of the balance sheet total. Although this figure may not seem impressive, it exceeds the level often accepted by auditors as the materiality threshold when working on the audit of financial statements. Goodwill in the analyzed entities is therefore an important balance sheet item.
2. The relatively low level of goodwill can be explained by the fact that the analyzed entities belong to traditional sectors of the economy, where the emphasis is primarily on tangible, non-current assets—intangible assets play a smaller role.

3. The highest share of goodwill can be seen in the fuel industry. The successive increase in this value is connected with the acquisitions of business units realized by the Hungarian MOL.

Behind these ratios, there are specific monetary values, as presented in Table 6.

Table 6. The nominal value of goodwill—all entities with purchased goodwill (PLN mln).

Sector/Years	2018	2019	2020	2021
ENERGY	2653	2788	2480	2923
FUEL	784	788	2221	4096
MINING	0	57	57	57
TOTAL	3437	3633	4758	7076

Source: the authors' own study.

On the basis of the above data, it can be concluded that:

1. Goodwill in nominal terms has been gradually growing in recent years—it doubled between 2018 and 2021 from PLN 3.4 billion to almost PLN 7.1 billion. A significant acceleration in the growth of nominal amounts took place in the years marked by the coronavirus pandemic. Despite the uncertainty associated with the epidemic, at least some entities were not afraid to invest in the acquisition of further subsidiaries.
2. Goodwill plays a particularly important role in the energy and fuel sectors, while in the mining industry, its level is negligible. This may be due to the fact that entities operating in this sector have been struggling with recession and a decrease in revenue for years (see Table 3), which does not create the right climate for further mergers and acquisitions.

Study 4—Share of goodwill in total equity

Equity is often interpreted by readers of financial statements as a "safety cushion", ensuring that an entity's resources are financed in an appropriate proportion—not only by debt but by funds at the disposal of the owners. Too low of a level of equity compared to liabilities leads to an increase in the marginal cost of lending and, as a result, the weighted average capital cost (WACC). The analysis will determine to what extent companies are sensitive to possible impairment losses. It will be possible to determine what impact a one-off impairment of the entire goodwill would have on the net assets of the analyzed entities. This can be an important indicator in the context of identifying entities or sectors that are particularly interested in protecting goodwill in order to avoid a sharp and significant deterioration in the financial situation. The results are presented in Table 7.

Table 7. Share of goodwill in total equity.

Sector/Years	2018	2019	2020	2021
ENERGY	2.0%	1.7%	1.3%	1.5%
FUEL	2.5%	2.0%	2.9%	3.2%
MINING	0.0%	0.7%	0.8%	0.7%
TOTAL	2.1%	1.7%	1.8%	2.0%

Source: the authors' own study.

From the above analysis, the following conclusions can be drawn:

1. In all three analyzed sectors, the average level of the share of goodwill in equity oscillated at the level of 2%. The highest share by far was characterized by the fuel industry (over 3%), and the lowest was characterized by mining (below 1%).
2. Values of 2% of equity shall be considered, as in the case of the share of goodwill in total assets, to be substantial and to affect the image presented by an enterprise. A

one-off reduction of equity by the impairment of goodwill would already be felt by the company and would have an impact on the deterioration of its financial ratios.

3. In more analytical terms, in the case of five entities, the share of goodwill in equity in at least one of the analyzed years and, in most cases, in the entire period of 2018–2021 exceeded 4%, or the average share exceeded 2.5%. These companies are: POLENERGA, CEZ, INTERAOLT, MOL, UNIMOT, and KOGENERA. In their case, the significance of a possible impairment loss in goodwill would be particularly high.

Study 5—Amount of impairment losses in goodwill

Goodwill is an intangible asset that, due to its indefinite useful life, is not subject to amortization. Its role is taken over by impairment tests, which must be carried out at least once a year. It is therefore possible to determine the rate of impairment losses, which will be the equivalent of an amortization rate. As explained in the theoretical part of the study, goodwill is a unique resource of an entity representing:

- the above-average ability to generate profits by the entity—this competitive advantage is ultimately eliminated by other entities that copy innovative solutions;
- processes taking place in the entity—they require constant redefinition and adaptation to the changing realities of management;
- synergy benefits—the impulse associated with the acquisition of another enterprise brings the greatest benefits in the periods immediately after the acquisition,
- resources not included in the company's balance sheet—if they do not meet the criteria allowing for individual recognition in the company's balance sheet, it is difficult to consider that they have a longer useful life than resources presented as assets in the financial statements;

Table 8 presents the share of impairment losses in goodwill in its opening balance for the years 2018–2021:

Table 8. Share of impairment losses in goodwill in its opening balance for a period.

Sector/Years	2018	2019	2020	2021
ENERGY	6.2%	11.1%	11.4%	0.0%
FUEL	0.6%	0.0%	0.0%	0.2%
MINING	0.0%	0.0%	0.0%	0.0%
TOTAL	4.1%	6.9%	7.1%	0.1%

Source: the authors' own study.

From the above data, the following conclusions can be drawn:

1. In the "COVID year" of 2020, the impairment losses in goodwill were not significantly higher than those in the previous two reporting periods. In the second of the years marked by the effects of the pandemic (2021), there were practically no write-offs recorded.
2. The only sector in which impairment losses in goodwill were accounted for was the fuel industry—in the case of energy and mining, write-offs were absent.
3. This approach of companies to goodwill may seem surprising, especially in the light of the previously analyzed data on the dynamics of revenues, profits, and cashflows from operating activities. It could be assumed that a significant deterioration in the financial results will be more clearly reflected in the level of impairment losses in goodwill.
4. Also in a more analytical approach, the write-offs do not represent a different trend. For the five companies selected previously, the impairment rate in subsequent years was at the level of: 5.5% (2018), 2.2% (2019), 22.8% (2020), and 0.2% in 2021, respectively. The high result of 2020 is influenced by the large impairment in goodwill made by ITERAOLT, which strongly affects the average.

It is worth noting that, on the basis of the results obtained, the average impairment rates of goodwill for the years 2018–2021 for individual industries can be determined, which are, respectively: energy—7.2% fuel—0.2% mining—0.0%, and total—4.6%. An amortization rate of 4.6% corresponds to an economic useful life of more than 21 years. It can be assumed that, during this period, the analyzed entities amortized their goodwill.

5. It is also worth disaggregating the results into two two-year periods—before and during the pandemic. The data are then as follows:
   - 2018/19: 5.5%
   - 2020/21: 3.6%

The above shows that, paradoxically, during the epidemic, despite worse economic results, the surveyed companies were less willing to make write-downs of goodwill. The reasons for this can be manifold. One can only speculate about the reasons for this, but the concerns of unit management about the additional deterioration of financial results in already difficult economic times may play an important role. In addition, the logic of the IAS 36 tests is that they are highly prospectively oriented and thus largely abstract from the weaker performance of the current period, emphasizing the optimistic forecasts of management for subsequent periods. An important element may also be subjective assumptions and estimates when constructing the forecasts of discounted cash flows of the cash-generating unit—this makes models difficult to verify by an external auditor. It should also be noted that making an impairment loss in goodwill can also be read as an admission of the company's management to the error of acquiring a subsidiary—if it is necessary to make a write-off, it means that the investment was a failure. Entities may therefore treat a write-off as a loss to their reputation.

Study 6—Frequency of impairment losses in goodwill

The purpose of this study is to see how often companies noticed the need to make impairment losses. Table 9 presents the frequency of impairment losses in goodwill for the analyzed years (2018–2021):

Table 9. Frequency of impairment losses in goodwill.

Sector/Years	2018	2019	2020	2021
ENERGY	2	2	2	0
FUEL	1	0	0	1
MINING	0	0	0	0
TOTAL	3	2	2	1

Source: the authors' own study.

The above data allow for the drawing the following conclusions:

1. The probability of making a write-off by the analyzed companies was about 15%, which results from multiplying the sum of write-offs over a period of 4 years (nine times) by the total number of financial statements containing purchased goodwill prepared by these entities at that time (58 reports).
2. This means that companies impair their goodwill, on average, once every seven years. If this result is compared with the previously determined average amortization rate of goodwill, which amounted to 4.6%, it can be concluded that companies make write-offs less often but for larger amounts.
3. If we multiply the aforementioned result of 4.6% by 7 (due to making write-offs once every seven years), we will get 32.2% as an approximate value of the write-off made once every seven years. As a result, such a write-off must be made three times, which gives 21 years, to impair goodwill in full.
4. It is also worth disaggregating the results into two two-year periods—before and during the pandemic. The data are then as follows:

- 2018/19: five write-offs
- 2020/21: three write-offs

It is confirmed that goodwill write-offs occurred more often in the period before the pandemic than during it. Even the particularly difficult year of 2020 was not marked by more frequent write-offs made potentially for smaller amounts.

Study 7—Average useful life of non-current assets

The aim of this study was to determine at what rate all fixed assets rotate in the entities covered by the study. Such a calculation will be very useful in order to compare the obtained results with the estimated useful life of goodwill, determined to be 21 years. According to the authors, the period of holding non-current assets by enterprises should be longer than goodwill due to the fact that non-current assets consist, to a large extent, of tangible resources, which, by their nature, seem to have, in principle, a greater durability than intangible assets, with a particular emphasis on such a volatile resource as goodwill. In order to determine the useful life of all non-current assets, the expenditure on the purchases of new assets as part of investment activities was divided by the sum of non-current assets. This will allow to determine the rate of their rotation. Table 10 presents the turnover of assets in the analyzed 16 entities for the years 2018–2021.

Table 10. Turnover of non-current assets.

Sector/Years	2018	2019	2020	2021
ENERGY	8.7%	7.7%	9.8%	9.8%
FUEL	9.6%	10.1%	9.7%	11.0%
MINING	12.2%	17.7%	17.7%	13.3%
TOTAL	9.8%	9.6%	10.9%	10.9%

Source: the authors' own study.

From the above data, the following conclusions can be drawn:

1. The analyzed rotation rates on an annual basis present a fairly stable level over a period of four years. With a few exceptions, the fluctuation does not exceed the level of 2–3 percentage points, which proves the consistent policy of the companies in the field of CAPEX investments and the systematic replacement of already used-up, non-current assets by new purchases. The averages of individual industries oscillate around the level of 10–11%—the total average for all industries over a period of four years is 10.3%, so the full exchange of non-current assets takes place in the analyzed entities, on average, after less than 10 years.
2. The turnover rate of non-current assets in each industry is significantly higher than the period of amortization of goodwill in the analyzed entities (4.6%). This proves the above-average longevity of goodwill in the analyzed financial statements but at the same time contradicts the economic interpretation of this balance sheet category. Goodwill, despite its volatility and constant redefinition, is kept in the balance sheet in reporting practice longer than seemingly more durable material resources. According to the authors, this leads to a distortion of the image presented in the balance sheets of companies.

Study 8—The goodwill rollover ratio analysis

The theoretical part of the study indicates that goodwill is a specific resource of an entity that is not able to generate economic benefits individually but only within cash-generating units. It should therefore be assumed that it cannot function independently and thus shows an economic useful life exceeding the life of CGU assets, with a particular emphasis on non-current assets. IAS 38 states that the useful life of goodwill cannot be determined, but the economic logic would suggest that it does not seem to be longer than the useful life of the non-current assets to which it is assigned. This claim is evidenced by the fact that IAS 36, as part of the tests for impairment when calculating the present value of discounted cash flows, expressly prohibits the inclusion of inflows and outflows for the upgrade of existing and purchases of new fixed assets. In other words, the CGU is being tested in its current form. If the period of maintaining goodwill on the balance sheet exceeds the rotation period of all non-current assets, there is a rollover of goodwill. In practice, this means that, when conducting impairment tests, goodwill is assigned to a different set of assets in subsequent reporting periods than those for its initial recognition. The rollover ratio is determined by dividing the fixed asset turnover rate by the goodwill impairment ratio. Table 11 shows the rollover rate for the years analyzed.

Table 11. Goodwill rollover ratio.

	2018	2019	2020	2021
TOTAL	2.4	1.4	1.5	182.7 [1]

Source: the authors' own study. [1] Result distorted due to almost no impairments taking place in 2021.

From the above data, the following conclusions can be drawn:

1. In all analyzed industries, the value of the ratio exceeds 1, which means that goodwill is kept on the balance sheet longer than the average for non-current assets, so there is a rollover of goodwill, which, according to the authors, is inconsistent with the economic content of the discussed balance sheet category as well as with the guidelines for the procedure for conducting IAS 36 tests prohibiting taking into account increases in CGU assets.
2. The average value of the rollover ratio of goodwill for all industries has been set at 2.23 since the exchange of non-current assets takes place, on average, after about 10 years (the turnover rate of non-current assets was 10.3%), and the average rate of the depreciation of goodwill is, as previously established, 4.6%.
3. The ratio of 2.23 indicates that goodwill lives more than twice as long as other seemingly more durable non-current assets and has a useful life of about 21 years. In an era of a dynamically changing economic environment, it is difficult to expect an entity's intangible assets to be held in the financial statements for more than two decades. In particular, this applies to a resource defined as:
    - the entity's above-average ability to generate profits;
    - processes occurring in the entity;
    - synergy benefits;
    - resources not individually included in the company's balance sheet;
4. The year most affected by the coronavirus was 2020, when both sales revenue and operating results deteriorated significantly.
5. The share of goodwill in both total assets and equity is above the materiality threshold adopted by the auditor during the audit of the financial statements.
6. Based on the amounts and frequency of goodwill write-downs, the average depreciation rate was determined at 4.6%, which corresponds to the useful life of 21 years.
7. The economic useful life of non-current assets was determined to be 10 years, on average.
8. The goodwill rollover rate was determined to be 2.23, which does not appear to be consistent with the economic substance of this balance sheet item.

The research shows that, in general, the pandemic period did not affect the frequency and level of write-downs of goodwill, although significantly deteriorated results for 2020 were recorded. The results of the second stage of research did not confirm hypothesis H2: The COVID-19 pandemic caused a decrease in goodwill through impairment losses in the consolidated financial statements of groups in the energy, fuel, and mining sectors.

The results obtained by us referred to other studies. Research conducted by M. Glaum, W.R. Landsman, and Wyrwa [49] indicated that goodwill impairment losses are negatively associated with the economic performance of entities. However, based on our study, it can be seen that deteriorated results do not always lead to a reduction in goodwill. Additional observations can be made at specific times, such as a global crisis or a pandemic.

The results of our study do not coincide with the research conducted by M. Glaum and S. Wyrwa [42] on the companies of the 12 largest European stock exchanges in the period 2007–2009, i.e., the period including the global crisis. The researchers noted an increased frequency of both write-offs and their higher level in relation to the period before the crisis. However, there was clear sectoral variation. The energy industry and related sectors showed average data. As can be seen above, the pandemic should be viewed from a different perspective than the man-made financial crisis of the time.

G. Goswami and Kimmel [51] indicate a trend of deteriorating performance and cash flow under the impact of the COVID-19 pandemic, which should result in goodwill write-downs. In light of the study of T. Kiestik et al. [54] on Slovak enterprises, which showed the main sources of goodwill, one would expect that the reduction of these figures should be combined with goodwill write-downs. Thus, in this context, Polish companies from the energy, mining, and fuel sectors should show goodwill write-downs. Their absence or low level may indicate a lack of economic adequacy of goodwill presented in the balance sheet as an asset with the current way of accounting for goodwill. Such conclusions are, to a large extent, in line with those derived by Martinez et al. [40], who point out the need to modify the accounting for "wear and tear" of this group asset in the direction of combining depreciation charges with impairment testing. On the other hand, the results of Fu and Shen [3] show a strong impact of the pandemic on the financial situation of entities in the energy sector. Companies with goodwill impairment were more strongly affected by the pandemic. The situation of entities that did not show such impairment was significantly more favorable. In the context of the cited research results, the dissimilarity of the obtained results can be noted.

C. Carvalho, A.M. Rodrigues, and C. Ferreira [31] point to the problems of incompleteness and wide variation in the form and scope of reported information on goodwill and its write-downs. The results of our research indicate that, during the analyzed period, the quality of reporting improved, but there is no basis to associate this improvement with the impact of the pandemic. The research suggests that the information was not complete. Similar conclusions regarding completeness were drawn by M. Glaum and S. Wyrwa [42], who also took into account an important external factor, namely, the global crisis in 2007–2009. Both studies pointed out the lack of information, among other factors, on the discount rates adopted. The research results indicated above are consistent with those obtained by M. Boučková [30].

5. Conclusions

In the literature, there are studies in which the impact of the pandemic on the financial situation of entities in the energy sector was investigated [55–58]. Such research concerned various countries, including Poland. However, little research has been focused on the importance of goodwill as a specific asset of energy sector groups in the context of the pandemic. Goodwill can significantly affect the financial situation of groups in this sector. The acquired goodwill can constitute a high percentage of their balance sheet total. The findings of our study partially fill the gap regarding the impact of the pandemic on the goodwill presented in the consolidated financial statements of the energy, fuel, and mining sector groups in Poland. Our research also contributes to the discussion on the quality of goodwill disclosures in consolidated financial statements.

Our study extends the literature on the determinants of the quality of goodwill disclosures, particularly related to goodwill write-offs in energy, mining, and fuel sector groups. Previous studies have not taken into account the unique impact of factors such as pandemics. Economic changes, their scale, and their dynamics, as well as managers' reactions to them, represent a new challenge for researchers. The research hypotheses we have formulated focus on strategic sectors, generally concentrating on the largest groups. In case of Poland, these sectors are strongly interconnected by ownership, which makes them an interesting object of research. The obtained results provide a new perspective on goodwill as an asset of capital groups presented in consolidated financial statements.

Although there is a slight improvement in the quality of disclosures regarding goodwill based on the disclosure quality index used, this cannot be linked to the impact of COVID-19. In the analyzed financial statements, the reasons for updating the goodwill do not include factors resulting from COVID-19. Thus, it is possible to speak of a general improvement in the quality of disclosures, which contributes to the increase in the prognostic value of financial statements.

The basic conclusion of the study is that even the year 2020, particularly marked by the coronavirus pandemic, did not cause a significant increase in impairment losses in goodwill in the analyzed entities. The year 2021 already marks a clear economic recovery despite the still difficult epidemiological situation. At the same time, the tendency not to make write-offs is a common phenomenon both in the industry and in the time dimension —the volume of write-offs in 2018 and 2019, before the pandemic, also remained at a relatively low level. According to the authors, the practices of the analyzed entities are not consistent with the economic essence of the analyzed balance sheet item, which, in the understanding of readers of financial statements, should be characterized by a relatively short useful life. In addition, the excessive long retention of goodwill in a balance sheet is inconsistent with the provisions of IAS/IFRS, which, despite stating that goodwill has an indefinite useful life (IAS 38), do not allow impairment tests to take into account increases and new purchases of fixed assets (IAS 36). The results of our research indicate the need for changes in the IAS/IFRS.

In summary, the years 2020 and 2021 and the events related to the global pandemic did not result in an increase in impairment losses, even in industries that seemed to be particularly vulnerable to the effects of this phenomenon. It is common to roll over goodwill, i.e., to keep it on the balance sheet for decades (at least twice as long as other non-current assets) and assign it as part of the impairment test to assets that were not in the CGU at the time of the acquisition of the subsidiary.

Our study has limitations due to the short time period that was considered. Due to the limited amount of research conducted on goodwill impairment in the energy, mining, and fuel sectors during this period, we were limited in our ability to compare the results with other research. We intend to continue our studies with a focus on the energy sector in EU countries in order to deepen our sectoral comparative analysis and identify further determinants of goodwill impairment.

Author Contributions: Conceptualization, M.G., S.H., M.S.-F. and K.Ś.; methodology, M.G. and S.H.; validation, M.G., S.H., M.S.-F. and K.Ś.; investigation, M.G., S.H., M.S.-F. and K.Ś.; resources, M.G., S.H., M.S.-F. and K.Ś.; data curation, M.G. and S.H.; writing—original draft preparation, M.G., S.H., M.S.-F. and K.Ś.; writing—review and editing, M.G., S.H., M.S.-F. and K.Ś.; visualization, M.G., S.H., M.S.-F. and K.Ś.; supervision, S.H.; project administration, M.G., S.H., M.S.-F. and K.Ś.; funding acquisition, M.G., S.H., M.S.-F. and K.Ś. All authors have read and agreed to the published version of the manuscript.

Funding: This research received no external funding.

Institutional Review Board Statement: Not applicable.

Informed Consent Statement: Not applicable.

Conflicts of Interest: The authors declare no conflict of interest.

References

1. Mukhtarov, S.; Humbatova, S.; Hajiyev, N.G.-O.; Aliyev, S. The Financial Development-Renewable Energy Consumption Nexus in the Case of Azerbaijan. *Energies* **2020**, *13*, 6265. [CrossRef]
2. Tu, K.J. *COVID-19 Pandemic's Impacts on China's Energy Sector: A Preliminary Analysis. Columbia SIPA, Center on Global Energy Policy*; Columbia SIPA Center on Global Energy Policy: New York, NY, USA, 2022.
3. Fu, M.; Shen, H. COVID-19 and Corporate Performance in the Energy Industry. *Energy Res. Lett.* **2020**, *1*, 1–4. [CrossRef]

4. Phan, D.H.B.; Narayan, P.K. Country Responses and the Reaction of the Stock Market to COVID-19-A Preliminary Exposition. Emerging Markets Finance and Trade 2020, Volume 56, Issue 10, Special Issue: Research on Pandemics. Available online: https://www.tandfonline.com/doi/full/10.1080/1540496X.2020.1784719 (accessed on 22 April 2022).
5. Platje, J.; Harvey, J.; Bacchus, L. COVID-19—Reflections on the Surprise of Both an Expected and Unexpected Event. *Cent. Eur. Rev. Econ. Manag.* **2020**, *4*, 149–162. [CrossRef]
6. Lisicki, B. Impairment of assets and market reaction during COVID-19 pandemic on the example of WSE. *Risks* **2021**, *9*, 183. [CrossRef]
7. Shaikh, I. On the relation between the crude oil market and pandemic COVID-19. *Eur. J. Manag. Bus. Econ.* **2021**, *30*, 331–356. [CrossRef]
8. Hendriksen, E.S. *Accounting Theory*, 2nd ed.; Richard, D., Ed.; Irvin, Inc.: Homewood, AL, USA; Illinois and Irwin-Dorsey Limited: Georgetown, ON, Canada, 1970; pp. 432–433.
9. Scott, W.R. *Financial Accounting Theory*, 5th ed.; Prentice Hall: Hoboken, NJ, USA; Pearson Canada Inc.: Toronto, OH, USA, 2009; p. 231.
10. Lycklama a Nijeholt, M.P.; Grift, Y.K.; Blommaert, J.M.J. *Goodwill Measuring Value Creation of Acquisitions: An Empirical Research. Working Papers from Utrecht School of Economics, Discussion Paper*; Series No 12–21; Tjalling, C., Ed.; Koopmans Research Institute: Utrecht, The Netherlands, 2022. Available online: https://dspace.library.uu.nl/bitstream/handle/1874/274895/12-21.pdf (accessed on 22 April 2022).
11. Strojek-Filus, M. *Determinanty Oraz Skutki Wynikowo-Bilansowe Identyfikacji i Rozliczania Wartości Firmy w Grupie Kapitałowej*; Wydawnictwo Uniwersytetu Ekonomicznego w Katowicach: Katowice, Poland, 2013; pp. 110–164.
12. Kozłowska, A. An attempt to identify major factors which determine a quality of financial statements. *Pr. Nauk. Uniw. Ekon. We Wrocławiu* **2018**, *503*, 246–258. [CrossRef]
13. Iatridis, G. International Financial Reporting Standards and the quality of financial statement information. *Int. Rev. Financ. Anal.* **2010**, *19*, 193–204. [CrossRef]
14. Iatridis, Accounting disclosures, accounting quality and conditional and unconditional conservatism. *Int. Rev. Financ. Anal.* **2011**, *20*, 88–102. [CrossRef]
15. Amiraslani, H.; Iatridis, G.; Pope, P.F. *Accounting for Asset Impairment: A Test for IFRS Compliance across Europe, A Research Report by the Centre for Financial Analysis and Reporting Research*; Cass Business School: London, UK, 2013.
16. Barth, M.E.; Landsman, W.R.; Lang, M.H. International Accounting Standards and Accounting Quality. *J. Account. Res. Wiley Online Libr.* **2008**, *46*, 467–498. [CrossRef]
17. Alnaas, A.; Rashid, A. Firm characteristics and compliance with IAS/IFRS. *J. Financ. Report. Account.* **2019**, *17*, 383–410. [CrossRef]
18. Nobes, C.; Stadler, C. How arbitrary are international accounting classifications? Lessons from centuries of classifying in many disciplines, and experiments with IFRS data. *Account. Organ. Soc.* **2013**, *38*, 573–595. [CrossRef]
19. Sumiyana, S.; Hendrian, H.; Effendi, R.; Fitrijati, K.; Sriwidharmanely, S. Accounting Quality and Its Challenges in 21st Century. Available online: https://www.intechopen.com/chapters/78825 (accessed on 22 April 2022).
20. Myers, S.C. Determinants of corporate borrowing. *J. Financ. Econ.* **1977**, *5*, 147–175. [CrossRef]
21. Johnson, J.; Tearney, M.G. Goodwill: A Eternal controversy. *CPA J.* **1993**, *63*, 58–62.
22. Zarzecki, D. Goodwill w rachunkowości i wycenie przedsiębiorstw. In *Inwestycje Finansowe i Ubezpieczeniowe–Tendencje Światowe a Polski Rynek*; Jajuga, K., Ed.; Ronka-Chmielowiec; Wydawnictwo Akademii Ekonomicznej: Wrocław, Poland, 2004; pp. 421–430.
23. Judkowiak, A. Disclosure Practices of Information in the Field of Financial Instruments: Evidence from Polish Companies Listed in the Warsaw Stock Exchange. *Eur. Res. Stud. J.* **2021**, *24*, 468–493. [CrossRef]
24. Khoram, A.; Abadi, A.C.; Janani, M.H. The role of disclosure quality in financial reporting. *Eur. Online J. Nat. Soc. Sci.* **2013**, *2*, 439–443.
25. Amel-Zadeh, A.; Glaum, M.; Sellhorn, T. Empirical Goodwill Research: Insights, Issues, and Implications for Standard Setting and Future Research. *Eur. Account. Rev.* **2021**, 1–32. [CrossRef]
26. Gierusz, M. *Ujęcie i Wycena Nabytej Wartości Firmy w Kontekście Użyteczności Sprawozdania Finansowego*; Wydawnictwo Uniwersytetu Gdańskiego: Gdańsk, Poland, 2020.
27. Beatty, A.L.; Harris, D.G. The effects of taxes, agency costs and information asymmetry on earnings management: A comparison of public and private firms. *Rev. Account. Stud.* **1998**, *326*, 299–326.
28. Paugam, L.; Astolfi, P.; Ramond, O. Accounting for business combinations: Do purchase price allocations matter? *J. Account. Public Policy* **2015**, *34*, 362–391. [CrossRef]
29. Khairi, K.F.; Laili, N.H.; Tran, D.M. Disclosure Quality of Goodwill Impairment Testing: A Disclosure Index. *J. Econ. Dev.* **2013**, *14*, 5–30.
30. Boučková, M. Quality of disclosed information with emphasis on goodwill impairment. *Eur. Financ. Account. J.* **2016**, *11*, 37–52. [CrossRef]
31. Carvalho, A.M.; Rodrigues, C.F. Goodwill and Mandatory Disclosure Compliance: A Critical Review of the Literature. *Aust. Account. Rev.* **2016**, *26*, 376–389. [CrossRef]
32. Gu, F.; Baruch, L. Overpriced Shares, 111–Advised Acquisitions, and Goodwill Impairment. *Account. Rev.* **2011**, *86*, 1995–2022. [CrossRef]

33. Moeller, S.B.; Schlingemann, F.P.; Stulz, R.M. Firm size and the gains from acquisitions. *J. Financ. Econ.* **2004**, *73*, 201–228. [CrossRef]
34. Bugeja, M.; Galery, N. Is older goodwill value relevant? *Account. Financ.* **2006**, *46*, 519–535. [CrossRef]
35. Verriest, A.; Gaeremynck, A. What determines goodwill impairments? *Rev. Bus. Econ.* **2009**, *54*, 106–128.
36. Chalmers, K.G.; Godfrey, J.M.; Webster, J.C. Does a goodwill impairment regime better reflect the underlying economic attributes of goodwill? *Account. Financ.* **2011**, *51*, 634–660. [CrossRef]
37. Godfrey, J.M.; Koh, P.-S. Goodwill impairment as a reflection of investment opportunities. *Account. Financ.* **2009**, *49*, 117–140. [CrossRef]
38. Bradbury, M.E.; Godfrey, J.M.; Koh, P.S. Investment opportunities et influence on goodwill amortization. *Asia Pac. J. Account. Econ.* **2011**, *10*, 57–79. [CrossRef]
39. Van Hulzen, P.; Alfonso, L.; Georgakopulos, G.; Sotiropoulos, I. Amortization versus impairment of goodwill and accounting quality. *Int. J. Econ. Sci. Appl. Res.* **2011**, *4*, 93–118.
40. Martinez, A.N.; Hernandez, M.; Cavero Rubio, J.A. The economic effects of IFRS goodwill reporting. *Aust. Account. Rev.* **2018**, *28*, 309–322. [CrossRef]
41. Cieciura, M.; Czaja-Cieszyńska, H. Rozliczenie wartości firmy w świetle polskich i międzynarodowych regulacji rachunkowości-analiza przyjętych rozwiązań. In *Wartości Niematerialne i Prawne–Ujmowanie i Prezentacja*; Cebrowska, T., Dotkuś, W., Eds.; Wydawnictwo Uniwersytetu Ekonomicznego we Wrocławiu: Wrocław, Poland, 2011; pp. 51–61.
42. Glaum, M.; Wyrwa, S. Making Acquisitions Transparent. Goodwill Accounting in Times of Crisis. 2011. Available online: www.pwc.com/ie/pubs/2011 (accessed on 22 April 2022).
43. Lemans, J. Goodwill Impairment as a Tool for Earning Management. 2010. Available online: https://www.researchgate.net/publication/254804379_Goodwill_Impairment_as_a_Tool_for_Earnings_Management (accessed on 22 April 2022).
44. AbuGhazaleh, N.M.; Al-Hares, O.M.; Roberts, C. Accounting Discretion in Goodwill Impairments: UK Evidence. *J. Int. Financ. Manag. Account.* **2011**, *22*, 165–204. [CrossRef]
45. Mazzi, F.; Andre, P.; Dionysiou, D.; Tsalavoutas, I. Compliance with goodwill-related mandatory disclosure requirements and the cost of equity capital. *Account. Bus. Res.* **2017**, *47*, 268–312. [CrossRef]
46. Mazzi, F.; Slack, R.; Tsalavoutas, I. The effect of corruption and culture on mandatory disclosure compliance levels: Goodwill reporting in Europe. *J. Int. Account. Audit. Tax.* **2018**, *31*, 52–73. [CrossRef]
47. Hirschey, M.; Richardson, V.J. Information content of accounting goodwill numbers. *J. Account. Public Policy* **2002**, *21*, 173–191. [CrossRef]
48. Ramanna, K. The implications of unverifiable fair-value accounting: Evidence from the political economy of goodwill accounting. *J. Account. Econ.* **2008**, *45*, 253–281. [CrossRef]
49. Glaum, M.; Landsman, W.R.; Wyrwa, S. Goodwill Impairment: The Effects of Public En-forcement and Monitoring by Institutional Investors. *Account. Rev.* **2018**, *93*, 149–180. [CrossRef]
50. Zemskova, I. Acquisition premiums and subsequent impairment of goodwill in 2010–2015 in oil and gas sector. *Rev. Bus. Econ. Stud.* **2016**, *2*, 65–82.
51. Goswami, G.; Kimmel, J. Assessing Goodwill Impairment Amid COVID-19. 2021. Available online: https://www.pwc.com/ph/en/deals-corporate-finance/ph-covid-19-impact-on-goodwill-impairment-testing.pdf (accessed on 22 April 2022).
52. Beattie, V.; McInnes, B.; Fearnley, S. A methodology for analysing and evaluating narratives in annual reports: A comprehensive descriptive profile and metrics for disclosure quality attributes. *Account. Forum* **2004**, *28*, 205–236. [CrossRef]
53. EFRAG. *Should Goodwill Still Not Be Amortised? Accounting and Disclosure for Goodwill.* Discussion Paper. 2014. Available online: https://www.accountancyeurope.eu/consultation-response/fee-comments-on-efrag-s-asbj-s-and-oic-s-discussion-paper-should-goodwill-still-not-be-amortised-accounting-and-disclosure-for-goodwill/ (accessed on 22 April 2022).
54. Kiestik, T.; Kovacova, M.; Podhorska, I.; Kiestikova, J. Searching for key sources of goodwill creation as new global managerial challenge. *Pol. J. Manag. Stud.* **2018**, *17*, 144–153.
55. Rutkowska-Tomaszewska, E.; Łakomiak, A.; Stanisławska, M. The Economic Effect of the Pandemic in the Energy Sector on the Example of Listed Energy Companies. *Energies* **2022**, *15*, 158. [CrossRef]
56. Shaikh, I. Impact of COVID-19 pandemic on the energy markets. *Econ. Change Restruct.* **2022**, *55*, 433–484. [CrossRef]
57. Bocca, R. As Coronavirus Shocks the Energy Sector and Economy, Is Now the Time for a New Energy Order? *World Economic Forum.* 2020. Available online: https://www.weforum.org/agenda/2020/04/energy-oil-gas-electricity-sustainability-economy-covid19-coronavirus-pandemic-market-stability/ (accessed on 22 April 2022).
58. Siksnelyte-Butkiene, I.; Karpavicius, T.; Streimikiene, D.; Balezentis, T. The achivements of climate change and energy policy in the European Union. *Energies* **2022**, *15*, 5128. [CrossRef]

energies

Article

Pro-Inflationary Impact of the Oil Market—A Study for Poland

Grzegorz Przekota [1,*] and Anna Szczepańska-Przekota [2,*]

1 Department of Economics, Faculty of Economic Sciences, Koszalin University of Technology, 75-453 Koszalin, Poland
2 Department of Finance, Faculty of Economic Sciences, Koszalin University of Technology, 75-453 Koszalin, Poland
* Correspondence: grzegorz.przekota@tu.koszalin.pl (G.P.); anna.szczepanska-przekota@tu.koszalin.pl (A.S.-P.)

Abstract: The economic activity of businesses and the living standards of the population are largely dependent on inflation. Here, energy prices are of particular importance. Energy is what offers a competitive edge to economies. Therefore, many energy sectors still remain under state control. However, the fuel market is free although highly concentrated. The primary objective of this study was to determine the impact of fuel price changes on inflation in Poland. The research was based on causality models and regression models including asymmetry correction. The flow path was analyzed of price impulses from the basic raw material (i.e., crude oil) through wholesale diesel prices to inflation. The study demonstrates that with each successive stage of raw material processing, price volatility proves to be weaker. However, the final effect is still significant: inflation is largely shaped by energy carriers and, here, specifically by fuel prices. Such results have serious implications for the state's economic policy. On one hand, they point to the limitations of this policy and, on the other hand, they raise questions about the legitimacy of the reforms that free up energy markets.

Keywords: oil prices; inflation; exchange rate; causality

Citation: Przekota, G.; Szczepańska-Przekota, A. Pro-Inflationary Impact of the Oil Market—A Study for Poland. *Energies* **2022**, *15*, 3045. https://doi.org/10.3390/en15093045

Academic Editor: David Borge-Diez

Received: 14 March 2022
Accepted: 18 April 2022
Published: 21 April 2022

1. Introduction

Energy is one of the fundamental concepts in physics, and it is also an important factor in overall economic growth, which is why it is given special importance in economics. In physics, energy refers to the ability of a system to do work or produce heat. In economics, energy refers to any raw material and resource containing significant amounts of physical energy, thus enabling work to be performed [1]. The economic analysis of production is not oriented toward energy flows and the performance of work in the physical sense, but toward the process of value creation.

When analyzing the relationships between the economy and energy, the following facts can be noted:

- cheap energy, and even more so, energy efficiency, are crucial for economic growth [2];
- from a historical perspective, the increase in energy consumption due to the supply of a relatively cheap source of energy such as coal is considered as a key factor in the industrial revolution in Great Britain, and later worldwide [3];
- there is a clear, long-term relationship between the volume of the domestic product and the volume of energy consumption in the economy. This relationship is significant, especially for countries with lower energy consumption levels and it clearly weakens for countries with medium and high levels of energy consumption. This is because highly developed countries use energy more efficiently. Due to technological progress, there has been faster growth in domestic product than in energy production. In the 20th century, the volume of the product per one unit of energy consumption roughly doubled [4]; and

- there is a clear short-term relationship between changes in energy consumption and changes in GDP. This relationship is almost proportional, being strictest on the level of the global economy. However, the positive income elasticity of energy demand is gradually declining.

The determinants of energy consumption and changes in energy consumption in the national economy may be represented using a simple identity:

$$E = L \cdot GDP/L \cdot E/GDP, \tag{1}$$

This leads to three simple conclusions:

- population growth (L) increases energy consumption E in the economy,
- an increase in product per capita (GDP/L) raises energy consumption in the economy, and
- a decrease in specific energy consumption (E/GDP) and, therefore an increase in energy efficiency, results in a reduction in energy consumption in the economy.

However, the formula presented above does not specify any quantitative relationships between variables and, above all, it does not rely on any theoretical premises that could provide grounds for modeling the relationship between the economy and energy consumption. Attempts may be taken to derive these from economic growth models. In this context, what remains controversial is whether to treat energy as a factor of production, just like human labor and capital. However, regardless of the differences between the different versions, economic growth models do not consider raw materials including energy as a separate variable. Mainstream economic models decouple economic growth from raw materials including energy [5], which in practice could lead to their depletion and welfare falling to zero [6,7].

A useful theoretical approach to account for the interaction between energy, environment, and economic growth is the models proposed by Cass [8] and Koopmans [9]. In these, the future-oriented behaviors of producers and consumers are combined with the past-derived linkages between investments and capital resources. Alternative growth paths do not directly depend on energy and environmental policies, but this is indirectly through their impact on medium-term changes [10].

Increasing importance is being attached to the issue of ecology. In particular, the use of non-renewable energy sources—coal, oil and gas—is being critically viewed from an ecological perspective. Differences between traditional and ecological approaches to energy and economic growth include the following issues, among others [11,12]:

- identifying the main source of productivity growth. Traditionally, this is assumed to be technical progress, while in the green approach, this is an increasing availability of high quality energy;
- possibilities for a substitution of inputs. In the traditional approach, these can be determined by a flexibility of substitution at the sectoral level, while in the green approach, the flexibilities estimated in this manner were assumed to be overestimated as they did not take energy into account; and
- the marginal productivity of energy inputs. In the traditional approach, this is proportional to the share of energy in the value of the product, while in the ecological approach, it is greater than this share.

In models proposed by ecological economists, increases in energy inputs are crucial in explaining economic growth [13,14].

The long-term connections between energy consumption and economic growth are two-way connections as a rule, and this was only the methodology adopted and the research sample as well as the degree of economic development of the country that determines which direction will prove to be stronger [15–18]. Apart from long-term studies, the analyses have also covered short-term relationships between the economy and energy, primarily in the context of cyclical fluctuations. The focus has primarily been on the effect of shocks that are taken into account in short-run equilibrium models [19]. The vulnerability of the economy to supply shocks has decreased markedly over the past 200 years as the economy shifted from coal to oil [20,21].

In the past few decades, the greatest number of analyses have been conducted to assess the impact of oil shocks on the economy [22–24]. Detailed analyses have focused on the impact of energy prices, especially oil prices, on the macro economy in the post-World War II period. Many studies have found a significant negative effect of oil price increases on GDP [25–27], although net positive effects were found for energy importers [28,29].

Several studies have found an asymmetric relationship between the domestic product and oil prices. The effect for price increases was stronger than for price decreases [30,31]. In some studies, the results obtained indicated that, after taking into account other macroeconomic variables, the impact of oil prices on changes in the domestic product was insignificant [32]. Much of the recessionary impact of oil price increases may be due not to oil price changes as such, but due to endogenously determined effects on the part of monetary policy [33,34]. The rise in oil prices caused inflation to rise, prompting central banks to tighten monetary policy.

An important area related to energy production and consumption is the problem of energy market regulation and the state energy policy. Energy policy focuses on three main areas [35]:

- energy security, understood as the ability to meet the current and future energy demand and to withstand any potential systemic shocks in relation to energy supply at the level of a national economy and/or a group of countries;
- energy justice, understood as an ability to ensure common access to energy at affordable and fair prices that ensure the competitiveness of the economy and its stable growth; and
- environmental sustainability, understood as a transition of the energy system toward mitigating and avoiding potential environmental damage and climate change impacts.

These goals can be formulated on the level of the national policies of individual countries, but also on the level of economic and political groupings and on the global level. In the conditions of a multiplicity of goals and participants in the decision-making process, a conflict between these is quite natural, and a mechanism for resolving these disputes and decision making needs to constitute a component of the energy policy. In the energy policy of the European Union, these are supplemented by the creation of an internal energy market.

Energy policy makes use of various instruments of influence. Among these, we can distinguish regulatory and market-based instruments. Examples of the former include quality standards, quotas, and prohibitions. The second group includes fiscal and non-fiscal instruments. A well-known proposal in economic theory to internalize external costs (the costs are borne by the issuer) is the Pigou tax. This tax should be equal to the full marginal cost/loss resulting from the emission. As a result, the volume of pollutant emissions is reduced to a level at which the marginal benefits of the emitter are equal to the marginal social costs of the emissions. The advantages of this tax, in addition to those outlined above, are that it offers producers a high degree of flexibility in their operations, relatively low administrative costs, and it stimulates the development of low-carbon technologies [36]. Nevertheless, the rationale and applicability of the Pigou tax are subject to theoretical and practical controversies. Theoretical arguments point to the assumed determinants of the efficiency of the tax. The interaction with other taxes causes the size of the optimal tax to be below the optimum based on the marginal cost criterion of emissions [37]. In turn, the cross elasticity between energy prices and leisure time may lead to the conclusion that the optimal tax should be higher in relation to this criterion [38]. Practical problems arise from the difficulty of estimating the marginal social costs of emissions and their variation depending on, among others, the type of emissions and the location of the issuer. What might be an alternative proposal under these conditions is a criterion based on environmental objectives [39]. Under such an approach, the objective could be to limit emissions to a certain level. Regulations based on such a principle could reduce the marginal costs of emissions and provide an incentive to develop innovations that reduce emissions. The efficiency of such regulations is enhanced by a system of tradable emission rights [40]. Distortions in the functioning of markets that are due, for example, to price and wage rigidity, undermine the effectiveness of general tax instruments. As a result, energy policy should also include other instruments (e.g., those that take into account differentiation of taxes according to products and raw materials, subsidies, etc.) [41].

Despite the high politicization of the energy market, economic instruments play a fundamental role in its regulation. According to the traditional classification, they can be divided into fiscal and non-fiscal instruments. In economic policy, regulation of the energy sector has largely been subordinated to the achievement of macroeconomic objectives including control of inflation, balance of payments, and technological development [42]. The apogee of these actions was during the oil crises of the 1970s. The ban on oil exports in the USA was a spectacular, but not the only example of such a policy. Energy companies were the main target of the policy, and they incurred costs, but also had some benefits [43]. The source of the latter was primarily regulations protecting domestic companies from international competition.

In many countries, what was an instrument for achieving the primary objectives had a direct influence on energy sector companies, often state-owned ones as well as direct interventions to limit competition in the industry [44]. The experience of the past decades shows that the hierarchy of objectives in the energy policy has changed and, consequently, the instruments preferred by states to influence the sector have also changed. First of all, energy is no longer perceived as a good of social necessity, which was used to justify state support and interference in the past; external costs undermine the idea of always available, cheap energy; they question the idea of the economies of scale underlying the preference for large, centralized electricity systems and energy policy.

At the same time, however, this policy continues to emphasize the importance of technological progress to counteract the scarcity of resources and to enable the supply of energy to meet the expected growth in demand. A review of energy policy goals is also evident in China and in the European Union [45]. There is a stronger emphasis on increasing energy efficiency and reducing the environmental impact of emissions.

In empirical studies carried out for various countries, a statistically significant impact of energy prices on inflation has most frequently been revealed. However, conclusions were often drawn on the basis of primary energy prices including oil prices [46–50], while from the point of view of households and businesses, it is petrol and diesel prices that they observe directly and make decisions on the basis of these prices [51,52] and not the prices of primary energy (e.g., oil). Therefore, the price path from primary energy to the final product needs to be included in the research. The significant influence of the petrol price rather than the oil price has been demonstrated in more recent studies [53–56].

The oil crises of the 1970s became the main cause of increased interest in the subject of the relationship between the energy commodity market and economic development and inflation. Oil prices were pointed out as those responsible for economic recessions [57], although modeling the relationship between oil prices and economic activity provided many problems, especially those related to the constancy of this relationship and linearity [58], which is related to an improvement in the efficiency of energy use.

From a theoretical perspective, oil price volatility affects major macroeconomic processes through supply and inflationary transmission channels [59].

Through the supply channel, changes in oil prices have a direct impact on production, where changes in marginal production costs are the cause. Decreases and corresponding increases in production costs are caused by lower and adequately higher raw material prices [60]. For the economy, uncertainty related to fluctuations in raw material prices is particularly dangerous, as it limits the amount of investment [61]. The inflation channel, on the other hand, indicates the effect of oil price changes on core inflation or inflation related expectations [62]. There is a fairly simple relationship between supply and inflation channels; changes in the production costs of a whole range of energy-intensive goods result in changes in their prices, which affects the prices of consumer goods, thus having a direct impact on inflation [63].

The economy is particularly stimulated by falling oil prices, as household budgets are relieved by lower energy bills, and overall consumption then rises [64]. On one hand, rising consumption triggers a demand inflation, while on the other, falling oil prices mitigate its effects [65]. Hence, further difficulties arise in modeling the impact of energy prices on economic activity and inflation.

The results of extensive research by Fuinhas et al. [66] prove that energy consumption drives economic growth, but only in the short-run. The ratio of oil production to oil consumption has exerted a positive impact on growth in both the short- and long-run. Oil prices only exert a positive effect on growth in the short-run. Oil rents depress growth, suggesting that oil is more of a curse than a blessing for economies.

The impact of oil prices on inflation occurs through several channels. On one hand, petroleum products constitute a component of consumption baskets, so changes in their prices directly affect inflation rates; on the other hand, these products are used in production and transport, so their price increase generates higher production costs and, consequently, higher prices of consumption goods.

Fluctuations in oil prices in world markets have particularly negative consequences for the functioning of the economies of those countries that import significant quantities of this raw material. However, in general, whether a country is an oil exporter or importer, economic activity depends on oil prices [67,68] and, even more importantly, a significant impact was found on exchange rates, interest rates, inflation, and unemployment [69–71].

In many countries including Poland, the years 2021 and 2022 brought a sharp rise in inflation. The causes of this phenomenon are seen in many social, political, and economic aspects. The most commonly cited are overly expansive fiscal and monetary policies, broken supply chains as a result of the pandemic, and the society's unwinding after the lockdown period; however, a lot of attention is paid to energy markets. In 2021, the prices of all primary energy sources: coal, oil, and gas rose sharply. Policy makers very often use the impact of energy prices on the economy to explain the general rise in inflation. The problem adopted in the study concerns the determination of the impact of energy prices and, more specifically, oil prices, on the overall price increase. Therefore, the aim of the study was to determine the direction, strength, and statistical significance of the relationship between oil prices and inflation in Poland.

Hypotheses have been put forward that:

Hypothesis 1 (H1). *Changes in oil prices in the world markets are an important pro-inflationary factor in Poland.*

Hypothesis 2 (H2). *The price impulses from oil world market indirectly passes through the prices of diesel and gasoline.*

In a practical assessment of the significance of the impact of oil prices on inflation, it is not only oil prices that may be important, but due to the fact that there are several processing stages between the primary energy source and the final product, there is an additional problem of determining the significance of the product flow chain.

The research is important because the possible confirmation of the hypotheses calls into question the effectiveness of classical methods of monetary policy in terms of price normalization. Rather, it will move toward fiscal policy. The model of the central bank's independence may be undermined. Since the causes of inflation are of a cost nature, it is easier to regulate prices with the tax system than with interest rates.

In this respect, the situation of Poland is a valuable research object because after the period of stable prices, recent years have brought increased inflation. On one hand, the Polish government has supported enterprises with anti-COVID shields, and on the other hand, energy prices have risen worldwide. Today's effect is inflation that has been unheard of for many years.

2. Materials and Methods

The research used monthly Brent crude oil prices, monthly USD-PLN exchange rate quotations, monthly wholesale prices of diesel oil offered by PKN Orlen, and the CPI inflation index. The data covered the years 2004–2021. The monthly frequency of data is dictated by the frequency of the calculations of the CPI inflation index reported by the Central Statistical Office. The data used makes it possible to trace price impulses directly from the Brent crude oil market on inflation in Poland, but also to trace price impulses in intermediate links and, hence, from Brent quotations through the exchange rate and wholesale diesel oil prices to the inflation index, according to Scheme 1.

Brent Crude Oil (in USD/bbl) → USD/PLN → Brent Crude Oil (converted in PLN/bbl) → diesel fuel (in PLN/tone) → CPI

Scheme 1. Tested price impulses.

What is very important in the context of the whole study is the presentation of world oil prices in the national currency. This is a procedure that is recommended by numerous authors [72–75].

The research was conducted in several stages:

1. Shaping the values of the variables under evaluation. The graphs illustrate the trends of the variables under evaluation in their original version. No transformation was made at this point. The relevant pairs shown in the graphs refer to the successive stages of diesel production and they end with the inflation rate.
2. Correlational study. The study allowed for an assessment of the long-term relationship between the time series examined. The data for the correlation study was logarithmized, additionally further modeling of the relationship was carried out on the basis of the data logarithmized. Due to the possible occurrence of apparent correlations, this study was not interpreted in a causal convention but only through the prism of trend consistency.
3. Study of variability. Simple statistics concerning the mean and standard deviation of the logarithmized values of the time series. The behavior of the standard deviation is crucial here, which shows the magnitude of the variability transmission. Although this is not a classic fiscal policy task, it can be used to manage the market and inflation. This is currently happening in Poland.
4. Tests of the stationarity of the time series. The study used the ADF test [76]; this allowed for an assessment of the fulfilment of the assumptions of the applicability and reliability of modeling the relationship between the time series evaluated.
5. Causality study. The Granger test for a two-variable VAR model with k lags of the form [77] and:

$$\begin{aligned} x_t &= a_1 + a_{1.1}x_{t-1} + \ldots + a_{1.k}x_{t-k} + b_{1.1}y_{t-1} + \ldots + b_{1.k}y_{t-k}, \\ y_t &= a_2 + a_{2.1}x_{t-1} + \ldots + a_{2.k}x_{t-k} + b_{2.1}y_{t-1} + \ldots + b_{2.k}y_{t-k}. \end{aligned} \quad (2)$$

The significance of the $a_{i.k}$ and $b_{i.k}$ parameters was tested with the F statistic.

6. Cointegration testing. Cointegration was tested on the basis of the following equations:

$$\ln(Y) = a_1 \cdot \ln(X) + a_0, \quad (3)$$

where the relationships between the X and Y variables were consistent with the course marked in Scheme 1. The residuals of these equations were subjected to the ADF stationarity test. The aforementioned equations determined the long-run equilibrium path (equation) around which the values of the economic processes analyzed were run. The differences between the value of the time series and the path determined of the long-run equilibrium were presented in the graphs and interpreted as short-run deviations.

7. Application of the Engel–Granger theorem [78]. According to the Engel–Granger theorem, if X and Y variables are integrated to the degree of (1.1), that is, the processes are non-stationary but their first differences are stationary, and it is possible to determine a long-run equilibrium path whose residuals will be stationary, then it is possible to represent, in a single equation, the short-run relationship between these variables and the process of reaching long-run equilibrium:

$$\Delta y_t = \alpha ECT^{+}_{t-1} + \beta ECT^{-}_{t-1} + \sum_{i=1}^{k-1} \theta_i \Delta y_{t-i} + \sum_{i=0}^{k-1} \gamma_i \Delta x_{t-i} + \varepsilon_t \quad (4)$$

where:

ECT_{t-1}—series of positive (+) and negative (−) residuals from the cointegrating equation;
α, β—the rate at which Y variable adjusts to the long-run equilibrium level with X variable after positive (α) or negative (β) precipitation; in order for the rebalancing mechanism to work properly, the value of this parameter needs to be negative;
θ_i—the impact of lagged values of the increment of Y variable on the current increment of this variable; and
γ_i—the effect of current and lagged values of the increment of X variables on the current increment of Y variable.

8. Graphical representation of the importance of oil price lags in shaping inflation.

The use of the Engel–Granger model allows for the simultaneous testing of short-term and long-term effects. This is an unquestionable advantage of this model, as its results may be an important implication for macroeconomic policy. Short-term and long-term reactions as well as time shifts in the transmission of price impulses are important for its effectiveness.

3. Results

The research results are summarized under three headings: (1) an evolution of the variables evaluated and the correlations between the variables; (2) a causality analysis; and (3) modeling of dependencies.

3.1. Evolution the Values of Variables Evaluated

The time series analyzed in their original form are presented in Figure 1. The order of the presentation is consistent with the importance of the volumes for the economy, starting from the most global ones and descending to domestic volumes. Therefore, the first graph presents Brent crude quotations (USD/bbl) and the USD/PLN exchange rate; the second graph converts Brent crude quotations into PLN and shows the wholesale prices of diesel oil (PLN/ton), while the third graph presents the CPI Y/Y inflation indices (month-to-month inflation in the corresponding month of the previous year) and consumer price levels in subsequent months relative to January 2004 prices (CPI 01.2004 = 100).

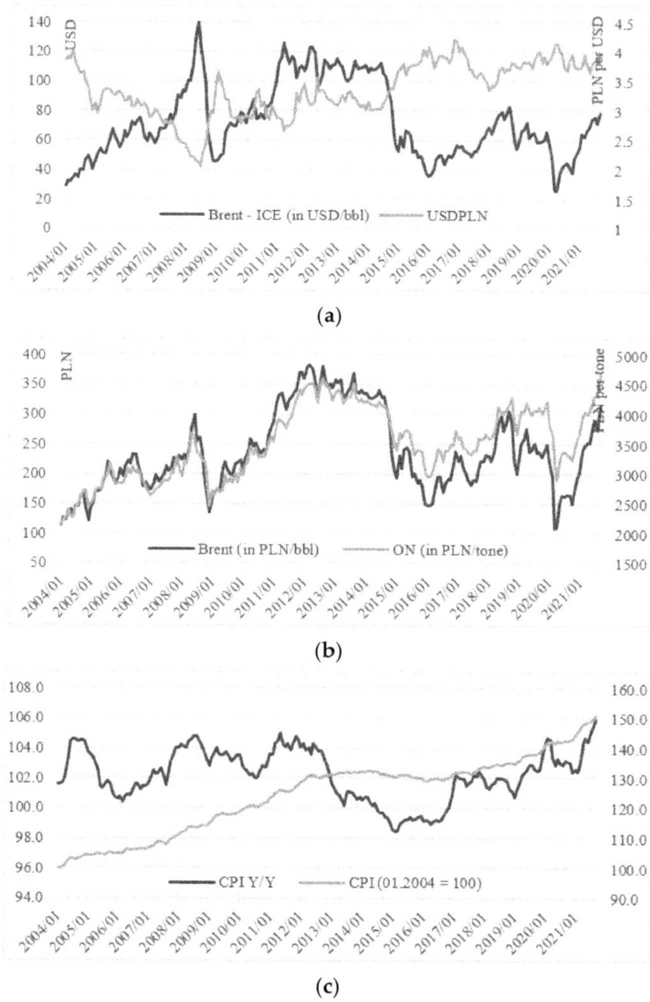

Figure 1. The values of the variables evaluated. (**a**) Stock exchange oil prices and exchange rate; (**b**) Oil and diesel oil prices in Poland; (**c**) Inflation indicators in Poland.

The logarithm values of the variables analyzed underwent a correlation study (Table 1). This study can be interpreted in the context of a long-term relationship.

Table 1. Correlation connections of the logarithmic variable levels.

Correlations	LN_BRENT_USD	LN_USDPLN	LN_BRENT_PLN	LN_ON (Orlen)
LN_BRENT_USD	1	−0.6360	0.9187	0.6395
LN_USDPLN	−0.6360	1	−0.2795	0.1164 *
LN_BRENT_PLN	0.9187	−0.2795	1	0.8552
LN_ON (Orlen)	0.6395	0.1164 *	0.8552	1
CPI Y/Y	0.1922	−0.3358	0.0672 *	−0.0372 *
CPI (01.2004 = 100)	0.0994 *	0.5139	0.3867	0.7661

* Statistically insignificant at $p = 0.05$.

Over the period of 2004–2021, the variables under study followed different trends. The period was long enough to include both significant sharp increases in quotations and spectacular decreases. Oil prices recorded historic highs in 2008, when oil cost around 140 USD/bbl, and a historic low during the uncertainty surrounding the coronavirus pandemic in March 2020, when prices fell below 30 USD/bbl. Local maxima also occurred in 2011 and 2018, and the years 2009 and 2016 saw the minima. At the end of the period under review, the oil price approached 80 USD/bbl, which was slightly above the period's average of 72 USD/bbl.

What happened in the world oil market had a very strong impact on the domestic oil and diesel market. The turning points of development trends fell in the same periods, and correlation links were very strong. The correlation between Brent oil quotations in USD/bbl and Brent oil quotations expressed in PLN (in PLN/bbl) was estimated at 0.9187. This result was possible to achieve despite the negative relationship between the Brent oil quotations and the exchange rate (−0.6360). Additionally, the wholesale diesel prices of PKN Orlen were strongly correlated with oil quotations (0.6395 and 0.8552, expressed in USD and PLN, respectively). The exchange rate was very weakly related to oil and diesel prices in the domestic market but, as indicated above, it proved to be quite strongly and inversely dependent on world oil quotations.

From the perspective of the objective of this study, however, it is important which of the oil market related parameters affects inflation in Poland. It becomes evident that the current price changes expressed by CPI Y/Y are weakly influenced by world oil prices, while the general price level is influenced by diesel prices. Here, the relationship of the CPI (01.2004 = 100) with wholesale oil prices was as high as 0.7661.

In assessing the evolution of oil prices, the issue of volatility looks interesting (Table 2).

Table 2. Volatility of crude oil and diesel oil prices.

Time Series	Statistics	
	Mean	**St. Dev.**
LN_BRENT_USD	4.23	0.36
LN_USDPLN	1.20	0.15
LN_BRENT_PLN	5.43	0.29
LN_ON (Orlen)	8.14	0.18

Due to the logarithmic transformation, it becomes possible to assess the scale of the variability of the time series under study. From the perspective of domestic economy stability, it is quite important to note that the scale of wholesale diesel price variability is twice lower than the variability scale of the basic raw material (i.e., Brent crude oil: 0.18 and 0.36, respectively). This is largely influenced by the exchange rate, as after converting Brent crude prices from USD to PLN, the price volatility decreased from 0.36 to 0.29. Another issue is the stability of other production costs; it is natural that the volatility of the price of the primary raw material is higher than that of the final product, but here the scale of the difference proved to be significant and in favor of the domestic market

3.2. Causality Testing

The time series studied were classical time series in which the levels are non-stationary, and the first differences are stationary (Table 3).

Table 3. Stationarity tests.

Time Series	I(0)		I(1)	
	t-Stat	Prob.	*t*-Stat	Prob.
LN_BRENT_USD	0.2196	0.7491	−11.4215	0.0000
LN_USDPLN	−0.1760	0.6217	−13.4881	0.0000
LN_BRENT_PLN	0.4209	0.8034	−12.5273	0.0000
LN_ON (Orlen)	1.2114	0.9422	−13.6231	0.0000
CPI_Y/Y	0.5256	0.8286	−10.1048	0.0000
CPI_01.2004 = 100	4.5079	1.0000	−7.6490	0.0000

The reason for the non-stationarity here is the trend, which is also a classical situation. This situation forces the modeling of the relationship using the first differences. The trend may be responsible for the occurrence of apparent dependencies. Although the nature of the study excludes apparent dependencies, the final model was nevertheless performed for the first differences.

Very important information in the context of the problem covered by the study is contained in Table 4.

Table 4. Causality tests.

Cause (X)	Effect (Y)	Lags: 1		Lags: 2		Lags: 3		Lags: 4	
		F-Stat	Prob.	F-Stat	Prob.	F-Stat	Prob.	F-Stat	Prob.
LN_BRENT_USD	LN_USDPLN	0.1285	0.7204	2.4685	0.0872	1.7158	0.1649	1.6451	0.1644
	LN_BRENT_PLN	0.5575	0.4561	2.0265	0.1344	1.8527	0.1388	2.1488	0.0762
	LN_ON (Orlen)	0.4356	0.5100	20.1150	0.0000	14.0981	0.0000	11.3545	0.0000
	CPI_Y/Y	0.1629	0.6869	12.4984	0.0000	8.8983	0.0000	6.6557	0.0001
	CPI_01.2004 = 100	0.1945	0.6597	6.2317	0.0024	5.4606	0.0013	5.1746	0.0005
LN_USDPLN	LN_BRENT_USD	1.4272	0.2336	1.0621	0.3476	1.2001	0.3108	0.9763	0.4216
	LN_BRENT_PLN	0.5575	0.4561	2.0265	0.1344	1.8527	0.1388	2.1488	0.0762
	LN_ON (Orlen)	0.0041	0.9490	5.1204	0.0068	3.4408	0.0178	2.8493	0.0250
	CPI_Y/Y	2.8843	0.0909	1.2572	0.2866	0.8265	0.4806	1.6642	0.1598
	CPI_01.2004 = 100	0.0001	0.9912	1.8474	0.1603	1.2306	0.2997	1.3027	0.2703
LN_BRENT_PLN	LN_BRENT_USD	1.4272	0.2336	1.0621	0.3476	1.2001	0.3108	0.9763	0.4216
	LN_USDPLN	0.1285	0.7204	2.4685	0.0872	1.7158	0.1649	1.6451	0.1644
	LN_ON (Orlen)	1.3376	0.2488	15.6520	0.0000	11.4826	0.0000	8.8874	0.0000
	CPI_Y/Y	0.0982	0.7543	18.5652	0.0000	13.8072	0.0000	10.7103	0.0000
	CPI_01.2004 = 100	0.3446	0.5578	11.4435	0.0000	9.8737	0.0000	8.8370	0.0000
LN_ON (Orlen)	LN_BRENT_USD	1.9707	0.1619	0.3966	0.6731	0.3607	0.7814	0.3559	0.8397
	LN_USDPLN	2.1439	0.1446	3.9852	0.0200	3.2669	0.0223	2.9233	0.0222
	LN_BRENT_PLN	0.0294	0.8641	0.1699	0.8439	0.0980	0.9610	0.2806	0.8903
	CPI_Y/Y	0.0011	0.9733	14.3009	0.0000	9.8345	0.0000	7.1069	0.0000
	CPI_01.2004 = 100	0.0763	0.7827	12.5271	0.0000	9.1428	0.0000	6.6758	0.0001
CPI_Y/Y	LN_BRENT_USD	0.3681	0.5447	2.6104	0.0759	2.1172	0.0992	1.7283	0.1451
	LN_USDPLN	0.6363	0.4260	0.3357	0.7152	0.2802	0.8396	0.7201	0.5791
	LN_BRENT_PLN	0.0432	0.8355	1.8151	0.1654	1.7809	0.1520	1.0640	0.3755
	LN_ON (Orlen)	0.2192	0.6401	0.5863	0.5573	0.9526	0.4162	1.0066	0.4052
	CPI_01.2004 = 100	17.3214	0.0001	2.0450	0.1320	1.5107	0.2129	2.0330	0.0912
CPI_01.2004 = 100	LN_BRENT_USD	0.6119	0.4350	0.3243	0.7234	0.2975	0.8272	1.0389	0.3882
	LN_USDPLN	6.4005	0.0121	3.9869	0.0200	3.0559	0.0294	3.9583	0.0041
	LN_BRENT_PLN	0.2607	0.6102	0.6853	0.5051	0.2843	0.8367	0.3934	0.8132
	LN_ON (Orlen)	3.8648	0.0506	2.3814	0.0950	2.1150	0.0995	1.2897	0.2753
	CPI_Y/Y	0.0286	0.8659	0.4815	0.6186	0.4969	0.6849	0.4000	0.8085

This study shows the results of the causality test. The direction of the impulse flows and the response latency can be read here. The most important findings include the following:

- Brent crude oil quotations are the cause of wholesale diesel prices in Poland and the CPI_Y/Y inflation index with a minimum lag of two months and the CPI_01.2004 = 100 index with a minimum lag of four months;
- the USD/PLN exchange rate is the cause of wholesale diesel prices in Poland with a minimum lag of two months;
- Brent crude quotations expressed in PLN (and thus the combined effect of Brent crude quotations and the exchange rate) are the cause of wholesale diesel prices in Poland, the CPI_Y/Y inflation index, and the CPI_01.2004 = 100 index with a minimum lag of two months;
- wholesale diesel prices in Poland are the cause of the CPI_Y/Y and CPI_01.2004 = 100 inflation indices with a minimum lag of two months, but also of the exchange rate with a minimum lag of two months;
- CPI_Y/Y inflation index is not a cause of any variables; and
- CPI_01.2004 = 100 is the cause of the exchange rate with a minimum lag of two months.

In general, as expected, all the causal relationships listed in Scheme 1 proved to be statistically significant. In addition, the relationship between the exchange rate and wholesale diesel prices appeared to be two-way but the direction indicated in Scheme 1 was stronger than the reverse direction. The effect of inflation on the exchange rate can also be revealed, but this is a side effect to the flow of impulses in Scheme 1.

3.3. Modeling of Dependencies

Modeling of the relationship began with the implementation of cointegrating models. The long-term relationship, however, one that is on the verge of statistical significance, concerns the impact of Brent oil quotations expressed in USD and after taking the USD/PLN exchange rate into account as well as the impact of wholesale diesel prices on the CPI_Y/Y inflation rate (Table 5). The impact of wholesale diesel prices was the strongest here.

Table 5. Cointegrations models and tests (A).

Independent Variable (X)	Dependent Variable (Y)							
	CPI_Y/Y				CPI_01.2004 = 100			
	Cointegration Model		Cointegration Test		Cointegration Model		Cointegration Test	
	Coeff.	Prob.	*t*-Stat	Prob.	Coeff.	Prob.	*t*-Stat	Prob.
LN_BRENT_USD C	0.9194 98.3251	0.0049 0.0000	−1.8018	0.0681	3.6084 109.1536	0.1482 0.0000	0.2070	0.7455
LN_BRENT_PLN C	0.4002 100.0396	0.3289 0.0000	−1.8393	0.0628	17.4616 29.6693	0.0000 0.0582	−0.5074	0.4957
LN_ON (Orlen) C	−0.3567 105.1156	0.5888 0.0000	−1.8853	0.0568	55.6659 −328.8170	0.0000 0.0000	−1.9773	0.0462

On the other hand, it was only the wholesale prices of diesel oil that had a significant impact on the CPI_01.2004 = 100 inflation index. No significant impact of Brent crude oil quotations was revealed here. This result was due to lagged responses, which were not examined here.

Within the time series of crude oil and diesel, a significant influence of Brent on wholesale crude oil prices was revealed, but without cointegration (Table 6). This is partly a result of economic and political decisions related to the fuel price formation in Poland. On the other hand, Brant crude prices expressed in PLN were strongly cointegrated with original Brent crude prices.

Table 6. Cointegrations models and tests (B).

Independent Variable (X)	Dependent Variable (Y)							
	LN_ON (Orlen)				LN_BRENT_PLN			
	Cointegration Model		Cointegration Test		Cointegration Model		Cointegration Test	
	Coeff.	Prob.	*t*-Stat	Prob.	Coeff.	Prob.	*t*-Stat	Prob.
LN_BRENT_USD	0.3195	0.0000	−1.0918	0.2487	0.7384	0.0000	−2.1243	0.0326
C	6.7918	0.0000			2.3048	0.0000		
LN_BRENT_PLN	0.5315	0.0000	−1.4495	0.1372	-	-	-	-
C	5.2581	0.0000			-	-		

Problems with cointegration are visible in the graphs of residuals (Figures 2 and 3). Generally, stationary graphs are expected, while the trend in question is visible. This may mean that the relationships examined are not long-term in reality, but are short-term only.

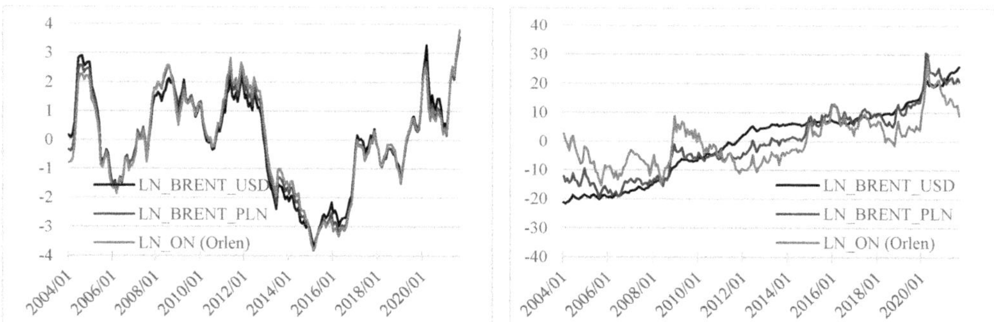

(**a**) (**b**)

Figure 2. Residuals from cointegrating models: (**a**) CPI_Y/Y; (**b**) CPI_01.2004 = 100.

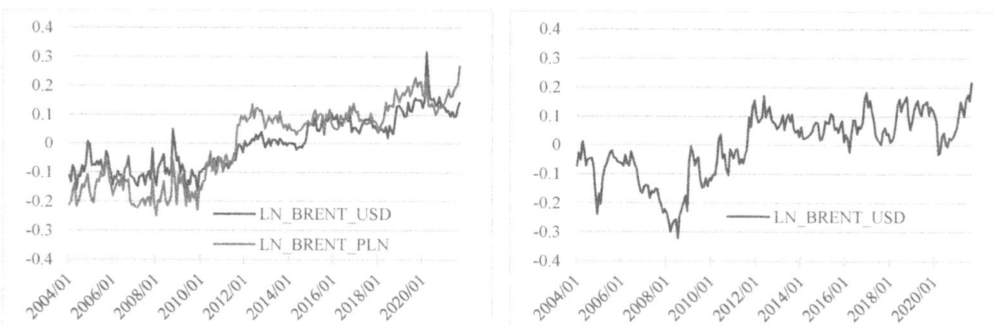

(**a**) (**b**)

Figure 3. Residuals from cointegrating models: (**a**) LN_ON (Orlen); (**b**) LN_BRENT_PLN.

Thus, inflation could be explained by changes in fuel prices only in the short-term, recognizing that in the long-term, these variables are independent. However, a more reasonable explanation of this phenomenon is to recognize the surging influence of fuel quotations on inflation. Thus, any upward jump in fuel prices will potentially increase inflation on a permanent basis, while temporary decreases in fuel prices will not be of any special significance. This effect can be attributed to entrepreneurs' reluctance to reduce the prices of their products, even if production costs are falling. It is natural that in such circumstances, they will opt for a higher margin. This phenomenon, if true, should be observed in the error correction model with asymmetry (Table 7). This explains the procedure followed in the study.

Table 7. ECT models with asymmetry.

Variable	Dependent Variable: d(LN_BRENT_PLN)		Variable	Dependent Variable: d(LN_ON (Orlen))	
	Coeff.	**Prob.**		**Coeff.**	**Prob.**
d(LN_BRENT_USD)	0.8432	0.0000	d(LN_BRENT_PLN)	0.3317	0.0000
d(LN_BRENT_USD_(-1))	−0.0177	0.7907	d(LN_BRENT_PLN_(-1))	0.2053	0.0000
d(LN_BRENT_USD_(-2))	0.0263	0.6883	d(LN_BRENT_PLN_(-2))	0.1170	0.0003
d(LN_BRENT_PLN_(-1))	−0.0053	0.9408	d(LN_ON (Orlen)_(-1))	−0.4055	0.0000
d(LN_BRENT_PLN_(-2))	−0.0533	0.4556	d(LN_ON (Orlen)_(-2))	−0.1747	0.0078
ect_plus_(-1)	−0.0455	0.4365	ect_plus_(-1)	0.0197	0.6235
ect_minus_(-1)	−0.0311	0.4433	ect_minus_(-1)	−0.0896	0.0458
C	0.0017	0.7252	C	−0.0017	0.6225
Variable	**Dependent Variable: d(CPI_Y/Y)**		**Variable**	**Dependent Variable: d(CPI_01.2004 = 100)**	
	Coeff.	**Prob.**		**Coeff.**	**Prob.**
d(LN_ON (Orlen))	0.6266	0.2605	d(LN_ON (Orlen))	1.1183	0.0697
d(LN_ON (Orlen)_(-1))	2.9098	0.0000	d(LN_ON (Orlen)_(-1))	3.0643	0.0000
d(LN_ON (Orlen)_(-2))	0.4816	0.4185	d(LN_ON (Orlen)_(-2))	0.9523	0.1479
d(CPI_Y/Y_(-1))	0.3206	0.0000	d(CPI_01.2004 = 100_(-1))	0.3384	0.0000
d(CPI_Y/Y_(-2))	0.0523	0.4387	d(CPI_01.2004 = 100_(-2))	0.0407	0.5417
ect_plus_(-1)	−0.0520	0.1112	ect_plus_(-1)	0.0062	0.2913
ect_minus_(-1)	−0.0196	0.4712	ect_minus_(-1)	−0.0032	0.6744
C	0.0225	0.6085	C	0.1011	0.0456

Four error correction models were determined. These models apply to successive price transmissions concerning Scheme 1.

The first model concerns the Brent_USD→Brent_PLN transition, hence, this is between the world oil price expressed in USD and the price expressed in PLN; in essence, it is an exchange rate effect. There were no time shifts in this relationship, the current changes in Brent_PLN depend directly on the current changes in Brant_USD, and the strength of this translation was estimated to be 0.8432. In this model, the ect parameters were insignificant, which is in line with the expectations, because in fact, the study concerns the same quantity, only expressed in a different currency. Thus, it is not possible to talk about any long-run equilibrium here, since it is the same variable. However, from a practical point of view, what is most important is a combination of the information that past oil price volatility does not affect the present one and that oil price volatility expressed in PLN is smaller than that expressed in USD (Table 2). This results in a greater stability of the oil price in the domestic market.

The next phase of the transition from oil prices to inflation is between world oil quotations and domestic wholesale prices. Here, this is after taking into account the exchange rate (i.e., the Brent_PLN→ON(Orlen) model). In this model, the outcome variable was increments in wholesale diesel prices, and these were dependent on the current increments in world prices, plus their first and second lags. It is thus a reaction up to three months back, which is a positive reaction. Thus, a rise in world prices significantly increases the domestic prices, but a fall in world prices also lowers domestic prices. There was also an opposite reaction to lagged price changes. This reaction is methodologically justified because it means that the series of increments does not have a trend. The results of the ect parameter are interesting. The ect(plus) parameter was insignificant but the ect(minus) parameter was significant. This means that if the price in the domestic market deviates downward from the equilibrium price with the world price for some reason, a process is quickly triggered to restore this equilibrium, but if the price deviates upward, there is no significance of such a process. The most important finding of this model was the significant positive response of changes in domestic prices to changes in world prices and the impossibility of a permanent reduction in domestic prices relative to the world price.

What is of key importance is what is contained in models 3 and 4. These models concern the impact of wholesale diesel prices on inflation. What is also important is that the conclusions only partly depend on the CPI_Y/Y or CPI_01.2004 = 100 inflation indices adopted; mostly, they were common. First, inflation was strongly and statistically significantly influenced by changes in the price of diesel fuel, and this was a reaction with a lag of one month. Furthermore, inflation was significantly fixed as it reacted positively to its lag. In contrast, there was no long-run relationship with oil prices. However, such a rapid short-term reaction to changes in wholesale diesel prices gives ground to consider the fuel market as a key pro-inflationary factor. All the more so since these changes are unidirectional from fuel prices to inflation.

Figure 4 provides a simplified visualization of relationships that occur in the models discussed. Thus, starting from Figure 4a, current changes in world oil prices expressed in PLN are directly dependent on original prices expressed in USD. In Figure 4b, current changes in wholesale diesel prices depend on current changes in world oil prices but, also on their lags. In Figure 4c,d, inflation appears to lag one month in relation to changes in wholesale diesel prices.

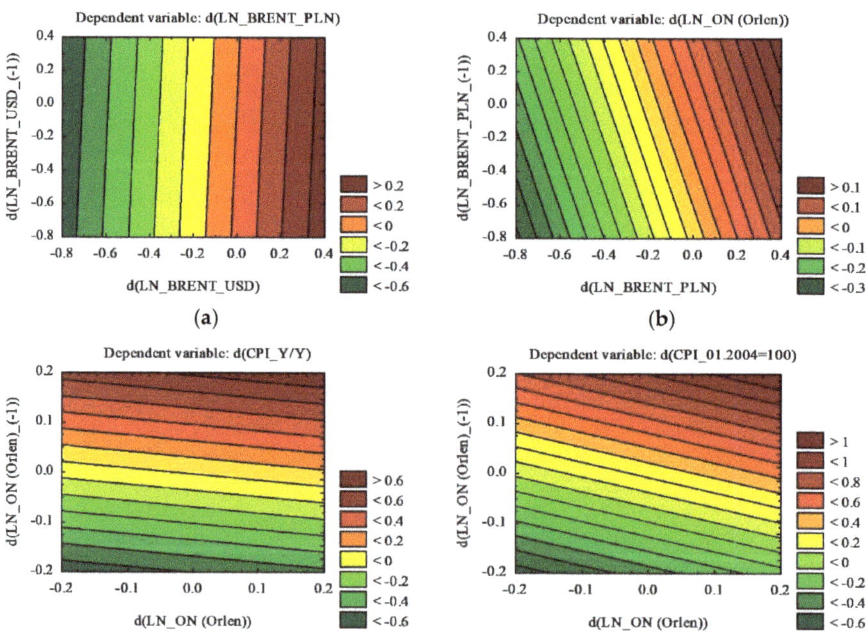

Figure 4. Dependencies taking into account time shifts. (**a**) Crude oil price response (in PLN) to changes in crude oil prices in the world markets; (**b**) Diesel oil prices response to changes in crude oil prices (in PLN); (**c**) Inflation (CPI_Y/Y) response to changes in diesel oil prices; (**d**) Inflation (CPI_01.2004 = 100) response to changes in diesel oil prices.

4. Discussion

The present study deals with a problem that is important from the economic perspective (i.e., the response of household inflation expectations (the CPI index) to fuel price shocks). This problem is still relevant, and it has especially gained in importance in the periods of increased inflation [79–81]. The issue of the transmission of price shocks from the fuel market to inflation is shown as an important cause of price increases [82–84].

The fact that the fuel market influences inflation is important not only from the perspective of the country's economy, fiscal, or monetary policy, or simply from the perspective of households. The fact that the most important CPI risk factor is the fuel market is also recognized in financial markets, where a popular strategy is to combine positions in the derivatives market for CPI swaps and RBOB futures. This strategy works in the same manner as an elimination of food price volatility risks by constructing an equivalent basket of agricultural futures [85].

One of the most serious problems of the impact of the fuel market on inflation is the controversy surrounding the short-term and long-term approach. Empirical studies are unable to unequivocally question or confirm whether the fuel market is responsible for inflation in the short- or long-term. The study finds evidence that the relationship between inflation and the fuel market is of a short-term nature, and that there is no statistically significant relationship in the long run. This conclusion is consistent with a number of empirical studies [86–88]. Generally, based on the research carried out, it can be concluded that inflation reacts quickly (up to three months) to increases in fuel prices. However, it does not react to decreases in fuel prices. This means that changes in fuel prices permanently increase inflation. Some authors have explained that during periods of falling fuel prices, inflation does not rise, and this is shown as a positive effect. The economy is particularly stimulated by falling oil prices as a result of the burden on the household budget being relieved by a reduction in energy bills; overall consumption then rises [89].

The research conducted has highlighted a unidirectional flow of price impulses: from the fuel market to inflation. Sometimes, the other direction (i.e., from inflation to the raw materials market) is discussed in the literature. It is frequently, however, that such studies treat the raw materials market as a whole and explain the increase in the prices of raw materials by running away from inflation. The oil market appears to be a good investment market against the loss of the value of money [90], which, however, does not seem to be true in light of most studies and the one carried out in this work. This is especially true if one takes into account the considerable volatility of the oil market. Recently, however, this approach has been recommended [91], but it may be the result of an excessive quantity of cash in the market and the need to look for any investment rather than a real and rational approach.

5. Conclusions and Policy Implications

The global oil market has proved to be a key pro-inflationary factor. This is not a direct influence, but an indirect one through the domestic fuel market. There are time lags in this relationship, generally up to three months. However, inflation does not take over all the volatility of the oil market. This is natural, however, as there are many more inflationary factors. Research in the context of the importance of the fuel market has important implications for economic policy, as all types of fiscal and monetary measures aimed at influencing inflation should take into account the current and projected situation in the fuel market.

The research shows important implications for macroeconomic policy. Several conclusions and suggestions can be drawn:

1. The current inflation is largely a cost-type inflation.
2. Taking into account the previous actions of the Polish government consisting of supporting enterprises during the COVID pandemic, cost-type inflation is overlapped by demand-type inflation, but additionally by more money in the market, with weak economic growth means throwing the economy out of balance. Inflation is also an effect of rebalancing, and this process has become a negative driving force for the current price increase.
3. The crude oil market, as sensitive to political conflicts, is a difficult one to control. However, the pro-inflationary effect may be mitigated by the pathway from primary energy to final product, as it turns out that at each subsequent stage of oil processing, price volatility decreases.
4. The government should reflect on full freedom to trade in energy. On one hand, economic considerations and market freedom speak for it, but on the other hand, it comes at the cost of price uncertainty.
5. Energy of the crude oil type has high marginal costs, and therefore, it is in the interest of the economy as a whole to use energy with low marginal costs. Such energy sources should no longer be pro-inflationary.

Author Contributions: Conceptualization, A.S.-P.; Methodology, G.P.; Validation, G.P.; Formal analysis, G.P.; Investigation, G.P.; Resources, G.P. and A.S.-P.; Data curation, G.P. and A.S.-P.; Writing—original draft preparation, A.S.-P.; Writing—review and editing, G.P. and A.S.-P.; Visualization, G.P.; Supervision, G.P. and A.S.-P.; Project administration, G.P. and A.S.-P. All authors have read and agreed to the published version of the manuscript.

Funding: This research received no external funding.

Institutional Review Board Statement: Not applicable.

Informed Consent Statement: Not applicable.

Data Availability Statement: Publicly available data.

Conflicts of Interest: The authors declare no conflict of interest.

References

1. Sweeney, J. *Economics of Energy*; Stanford University Press: Redwood City, CA, USA, 2004.
2. Fouquet, R. The role of energy technologies in long run economic growth. *Int. Assoc. Energy Econ.* **2014**, *8*, 11–13.
3. Allen, R.C. *The British Revolution in Global Perspective*; Cambridge University Press: Cambridge, UK, 2009.
4. Safa, H. The impact of energy on global economy. *Int. J. Energy Econ. Policy* **2017**, *7*, 287–295.
5. Stern, D.I. Economic growth and energy. In *Encyclopedia of Energy*; Elsevier Inc.: Amsterdam, The Netherlands, 2004; Volume 2, pp. 35–78.
6. Solow, R.M. Intergenerational equity and exhaustible resources. *Rev. Econ. Stud.* **1974**, *41*, 29–45. [CrossRef]
7. Dasgupta, P.; Heal, G. The optima depletion of exhaustible resources. *Rev. Econ. Stud.* **1974**, *41*, 3–28. [CrossRef]
8. Cass, D. Optimum growth in an aggregative model of capital accumulation. *Rev. Econ. Stud.* **1965**, *32*, 233–240. [CrossRef]
9. Koopmans, T.C. Objectives, constraints, and outcomes in optimal growth models. In *Economic Models, Estimation and Risk Programming: Essays in Honor of Gerhard Tintner*; Springer: Berlin, Germany, 1967; pp. 110–132.
10. Jorgenson, D.W.; Goettle, R.J.; Ho, M.S.; Wilcoxen, P.J. Energy, the environment and US economic growth. In *Handbook of Computable General Equilibrium Modeling*; Elsevier: Amsterdam, The Netherlands, 2013; Volume 1, pp. 477–552.
11. Cleveland, R.B.; Cleveland, W.S.; McRae, J.E.; Terpenning, I.J. STL: A seasonal-trend decomposition procedure based on loess. *J. Off. Stat.* **1990**, *6*, 3–33.
12. Ayres, R.U.; Warr, B. Accounting for growth: The role of physical work. *Struct. Chang. Econ. Dyn.* **2005**, *16*, 181–209. [CrossRef]
13. Ayres, R.U.; van den Bergh, J.C. A theory of economic growth with material/energy resources and dematerialization: Interaction of three growth mechanisms. *Ecol. Econ.* **2005**, *55*, 96–118. [CrossRef]
14. Beaudreau, B.C. The impact of electric power on productivity: A study of US manufacturing 1950–1984. *Energ. Econ.* **1995**, *17*, 231–236. [CrossRef]
15. Chontanawat, J.; Hunt, L.C.; Pierse, R. Does energy consumption cause economic growth? Evidence from a systematic study of over 100 countries. *J. Policy Model.* **2008**, *30*, 209–220. [CrossRef]
16. Ferguson, R.; Wilkinson, W.; Hill, R. Electricity use and economic development. *Energy Policy* **2004**, *28*, 923–934. [CrossRef]
17. Soytas, U.; Sari, R. Energy consumption and GDP. Causality relationships in G-7 countries and emerging markets. *Energy Econ.* **2003**, *25*, 33–37. [CrossRef]

18. Oh, W.; Lee, K. Causal relationship between energy consumption and GDP revisited: Evidence from Taiwan. *Energy Econ.* **2004**, *26*, 51–59. [CrossRef]
19. Romer, D. *Advanced Macroeconomics*; McGraw-Hill: New York, NY, USA, 2012.
20. Kilian, L. The economic effects of energy price shocks. *J. Econ. Lit.* **2008**, *46*, 871–909. [CrossRef]
21. Van de Ven, D.J.; Fouquet, R. *Historical Energy Prices Shocks and Their Changing Effect on the Economy*; Working Paper no. 171; Centre for Climate Change Economics and Policy: Leeds/London, UK, 2014.
22. Bernanke, B.; Gertler, M.; Watson, M. Systematic monetary policy and the effect of oil price shocks. *Brook. Pap. Econ. Act.* **1997**, *1*, 91–142. [CrossRef]
23. Kilian, L. Not All oil price shocks are alike.: Distinguishing demand and supply shocks in thee crude oil market. *Am. Econ. Rev.* **2009**, *99*, 1053–1069. [CrossRef]
24. Lee, K.; Ni, S.; Ratti, R. Oil shocks and the macroeconomy: The role of price variability. *Energy J.* **1995**, *16*, 39–56. [CrossRef]
25. Bildirici, M.; Alp, E.A.; Bakirtas, T. The great recession and the effects of oil price shocks and the U.S. recession: A Markov-switching and TAR-VEC analysis. *J. Energy Dev.* **2011**, *35*, 215–277.
26. Burbidge, J.; Harrison, A. Testing for the effects of oil-price using vector autoregressions. *Int. Econ. Rev.* **1984**, *25*, 459–484. [CrossRef]
27. Hamilton, J.D. Oil and the macroeconomy since World War II. *J. Political Econ.* **1983**, *9*, 228–248. [CrossRef]
28. Mork, K.A.; Olsen, O.; Mysen, H.T. Macroeconomic responses to oil price increases and decreases in seven OECD countries. *Energy J.* **1994**, *15*, 19–35.
29. Lafakis, C.; Kamins, A.; Friedman, E.; White, D. *The Economics of Lower Oil Prices*; Moody's Analytics: New York, NY, USA, 2015.
30. Federer, J.P. Oil price volatility and the macroeconomy: A solution to the asymmetry puzzle. *J. Macroecon.* **1996**, *18*, 1–16. [CrossRef]
31. Lardic, S.; Mignon, V. Oil prices and economic activity: An asymmetric cointegration approach. *Energy Econ.* **2008**, *30*, 847–855. [CrossRef]
32. Darby, M.R. The price of oil and World inflation and recession. *Am. Econ. Rev.* **1982**, *72*, 738–751.
33. Barsky, R.B.; Kilian, L. *Do We Really Know That Oil Caused the Great Stagflation? A Monetary Alternative*; NBER 2001 Working Paper no. 8389; National Bureau of Economic Research: Cambridge, MA, USA, 2001.
34. Hunt, B.; Isard, P.; Laxton, D. The macroeconomic effects of higher oil prices. In *National Institute Economic Review*; Cambridge University Press: Cambridge, UK, 2002; pp. 87–103.
35. Goldthau, A. *The Handbook of Global Energy Policy*; John Wiley & Sons: Hoboken, NJ, USA, 2013.
36. Goulder, L.H.; Perry, I.W.H. Instrument choice and environmental policy. *Rev. Environ. Econ. Policy* **2008**, *2*, 152–174. [CrossRef]
37. Parry, I.W.H.; Oates, W.E. *Police Analysis in the Second-Best World*; Resources for the Future: Washington, DC, USA, 1998; pp. 94–148.
38. West, S.E.; Williams, R.C., III. *Empirical Estimates for Environmental Policy Making in the Second-Best Setting*; NBER 2004 Working Paper 10330; National Bureau of Economic Research: Cambridge, MA, USA, 2004.
39. Baumol, W.J.; Oates, W.E. *The Theory of Environmental Taxation*; Cambridge University Press: Cambridge, UK, 1998.
40. Nielsen, L.; Jeppesen, T. Tradable green certificates in selected European countries: Over-view and assessment. *Energy Policy* **2003**, *31*, 3–14. [CrossRef]
41. Kosonen, K.; Nicodème, G. *The Role of Fiscal Instruments in Environmental Policy*; Taxation Papers 2009, Working Paper no. 19; European Union: Maastricht, The Netherlands, 2009.
42. Prontera, A. Energy policy: Concepts, actors, instruments and recent developments. *World Political Sci.* **2009**, *5*, 7. [CrossRef]
43. Clark, J.G. *The Political Economy of World Energy: A Twentieth-Century Perspective*; University of North Carolina Press: Chapel Hill, NC, USA, 1990.
44. Alkhuzam, A.F.; Arlet, J.; Rocha, S. Private versus Public Electricity Distribution Utilities: Are Outcomes Different for End-Users? 2018. Available online: https://blogs.worldbank.org/developmenttalk/private-versus-public-electricity-distribution-utilities-are-outcomes-different-end-users (accessed on 3 May 2018).
45. Bertoldi, P.; Mosconi, R. Do energy efficiency policies save energy? A new approach based on energy policy indicators (in the EU Member States). *Energy Policy* **2020**, *139*, 111320. [CrossRef]
46. Castro, C.; Poncela, P.; Senra, E. Oil Prices and Inflation in the Euro Area and Its Main Countries: Germany, France, Italy and Spain. Available online: https://www.forecasters.org/proceedings12/Castro_Cesar.pdf (accessed on 3 May 2018).
47. Chen, S.-S. Oil price pass-through into inflation. *Energy Econ.* **2009**, *31*, 126–133. [CrossRef]
48. Kiptui, M. *Oil Price Pass-Through into Inflation in Kenya*; Kenya School of Monetary Studies Research Centre: Kasarani, Kenia, 2009.
49. O'Brien, D.; Weymes, L. The impact of oil prices on Irish inflation. *Q. Bull.* **2010**, *3*, 66–82.
50. Shaari, M.S.; Hussain, N.E.; Abdullah, H. The effects of oil price shocks and exchange rate volatility on inflation: Evidence from Malaysia. *Int. Bus. Res.* **2012**, *5*, 106–112. [CrossRef]
51. Bobai, F.D. An analysis of the relationship between petroleum prices and inflation in Nigeria. *Int. J. Bus. Commer.* **2012**, *1*, 1–7.
52. Perera, W.T.K. The impact of petroleum prices on inflation in Sri Lanka. *Staff. Stud.* **2009**, *35*, 107–127. [CrossRef]
53. Coibion, O.; Gorodnichenko, Y. Is the Phillips curve alive and well after all? Inflation expectations and the missing disinflation. *Am. Econ. J. Macroecon.* **2015**, *7*, 197–232. [CrossRef]
54. Binder, C. Inflation expectations and the price at the pump. *J. Macroecon.* **2018**, *58*, 1–18. [CrossRef]

55. Kilian, L.; Zhou, X. Does Drawing Down the U.S. Strategic Petroleum Reserve Help Stabilize Oil Prices? *J. Appl. Econom.* **2020**, *35*, 673–691. [CrossRef]
56. Kilian, L.; Zhou, X. *The Impact of Rising Oil Prices on U.S. Inflation and Inflation Expectations in 2020-23*; Working Paper 2116; Federal Reserve Bank of Dallas: Dallas, TX, USA, 2021. [CrossRef]
57. Hamilton, J.D. Nonlinearities and the Macroeconomic Effects of Oil Prices. *Macroecon. Dyn.* **2011**, *15*, 364–378. [CrossRef]
58. Hooker, M. What happened to the oil price-macroeconomy relationship? *J. Monet. Econ.* **1996**, *38*, 195–213. [CrossRef]
59. Brown, S.P.A.; Yucel, M.K. Energy Prices and Aggregate Economic Activity and Interpretative Survey. *Q. Rev. Econ. Financ.* **2002**, *42*, 193–208. [CrossRef]
60. Finn, M.G. Perfects Competition and the Effects on Energy Price Increases on Economic Activity. *J. Money Credit. Bank.* **2000**, *32*, 400–416. [CrossRef]
61. Elder, J.; Serletis, A. Oil Price Uncertainty. *J. Money Credit. Bank.* **2010**, *42*, 1137–1159. [CrossRef]
62. Tang, W.; Wu, L.; Zhang, Z. Oil Prices Shocks and their Short- and Long-term effects on the Chinese Economy. *Energy Econ.* **2010**, *32*, 3–14. [CrossRef]
63. Blanchard, O.J.; Gali, J. *The Macroeconomic Effects of Oil Price Shocks: Why Are the 2000s So Different from the 1970s?* NBER Working Paper No. 13368; National Bureau of Economic Research: Cambridge, MA, USA, 2008.
64. Hamilton, J.D. The Causes and Consequences of the Oil Shock of 2007-08. In *Brookings Papers on Economic Activity*; National Bureau of Economic Research: Cambridge, MA, USA, 2009; pp. 215–261.
65. Alvarez, L.; Hurtado, S.; Sanchez, I.; Thomas, C. The Impact of Oil Price Changes on Spanish and Euro Area Consumer Price Inflation. *Econ. Model.* **2011**, *28*, 422–431. [CrossRef]
66. Fuinhas, J.A.; Marques, A.C.; Koengkan, M.; Santiago, R.; Couto, A.P. The energy-growth nexus within oil production and oil rents context. *Rev. Estud. Soc.* **2019**, *21*, 161–173. [CrossRef]
67. He, Y.; Wang, S.; Lai, K.K. Global economic activity and crude oil prices: A cointegration analysis. *Energy Econ.* **2010**, *32*, 868–876. [CrossRef]
68. Yeh, F.-Y.; Hu, J.-L.; Lin, C.-H. Asymmetric impacts of international energy shocks on macroeconomic activities. *Energy Policy* **2012**, *44*, 10–22. [CrossRef]
69. Castillo, P.; Montoro, C.; Tuesta, V. *Inflation, Oil Price Volatility and Monetary Policy*; Working Papers 2010-002; Banco Central de Reserva del Perú: Santa Rosa, CA, USA, 2010. Available online: https://www.bcrp.gob.pe/docs/Publicaciones/Documentos-de-Trabajo/2010/Documento-de-Trabajo-02-2010.pdf (accessed on 31 January 2010).
70. Doğrul, H.G.; Soytas, U. Relationship between oil prices, interest rate, and unemployment: Evidence from an emerging market. *Energy Econ.* **2010**, *32*, 1523–1528. [CrossRef]
71. Rahman, S.; Serletis, A. The asymmetric effects of oil price and monetary policy shocks: A nonlinear VAR approach. *Energy Econ.* **2010**, *32*, 1460–1466. [CrossRef]
72. De Gregorio, J.; Landerretche, O.; Neilson, C. Another Pass-Through Bites the Dust? Oil Prices and Inflation. *Economia* **2007**, *7*, 155–208. [CrossRef]
73. Van Den Noord, P.; André, C. *Why Has Core Inflation Remained So Muted in the Face of Oil Shock?* Working Paper 551; OECD Economics Department: Paris, France, 2007.
74. Oinonen, S.; Paloviita, M. *Updating the Euro Area Phillips Curve: The Slope has Increased*; Discussion Papers 31, 2014, Bank of Finland Research; Bank of Finland: Helsinki, The Netherlands, 2014. [CrossRef]
75. Baffes, M.; Kose, A.; Ohnsorge, F.; Stock, M. *The Great Plunge in Oil Prices: Causes, Consequences, and Policy Responses*; Policy Research Notes; World Bank Group: Washington, DC, USA, 2015. [CrossRef]
76. Dickey, D.A.; Fuller, W.A. Likelihood ratio statistics for autoregressive time series with a unit root. *Econometrica* **1981**, *49*, 1057–1072. [CrossRef]
77. Granger, C.W.J. Investigating Causal Relations by Econometric Models and Cross-Spectral Methods. *Econometrica* **1969**, *37*, 24–36. [CrossRef]
78. Engle, R.F.; Granger, C.W.J. Co-Integration and Error Correction: Representation, Estimation, and Testing. *Econometrica* **1987**, *55*, 251–276. [CrossRef]
79. Wong, B. Do inflation expectations propagate the inflationary impact of real oil price shocks? Evidence from the Michigan Survey. *J. Money Credit. Bank.* **2015**, *47*, 1673–1689. [CrossRef]
80. Conflitti, C.; Cristadoro, R. *Oil Prices and Inflation Expectations*; Banca d'Italia Occasional Papers No. 423; Bank of Italy: Rome, Italy, 2018.
81. Binder, C.; Makridis, C. Stuck in the seventies: Gas prices and macroeconomic expectations. *Rev. Econ. Stat.* **2022**, *104*, 293–305. [CrossRef]
82. Clark, T.E.; Terry, S.J. Time variation in the inflation passthrough of energy prices. *J. Money Credit. Bank.* **2010**, *42*, 1419–1433. [CrossRef]
83. Kilian, L.; Lewis, L.T. Does the Fed Respond to Oil Price Shocks? *Econ. J.* **2011**, *121*, 1047–1072. [CrossRef]
84. Conflitti, C.; Luciani, M. Oil price pass-through into core inflation. *Energy J.* **2019**, *40*, 221–247. [CrossRef]
85. Pond, M.; Weinblatt, V.; Skeoch, H.; Sooben, K. *Global Inflation-Linked Products: A User's Guide*; Barclays Capital: London, UK, 2019; pp. 61–69.

86. Badel, A.; McGillicuddy, J. *Oil Price and Inflation Expectations: Is There a Link?* Federal Reserve Bank of St. Louis: St. Louis, MO, USA, 2015. Available online: https://www.stlouisfed.org/~{}/media/Publications/Regional-Economist/2015/July/Oil.pdf (accessed on 31 July 2015).
87. Sussman, N.; Zohar, O. *Oil Prices, Inflation Expectations, and Monetary Policy*; Bank of Israel: Jerusalem, Israel, 2015.
88. Williams, P. *Inflation Expectations in U.S.: Linking Markets, Households, and Businesses*; IMF Working Paper, 2020, WP/20/240; IMF: Washington, DC, USA, 2020. Available online: https://www.imf.org/en/Publications/WP/Issues/2020/11/13/Inflation-Expectations-in-the-U-S-49815 (accessed on 13 November 2020).
89. Kilian, L. Oil Price Shocks: Causes and Consequences. *Annu. Rev. Resour. Econ.* **2014**, *6*, 133–154. [CrossRef]
90. Neville, H.; Draaisma, T.; Funnell, B.; Harvey, C.; Van Hammert, O. *The Best Strategies for Inflationary Times*; SSRN Working Paper; Elsevier: Amsterdam, The Netherlands, 2021. Available online: https://papers.ssrn.com/sol3/papers.cfm?abstract_id=3813202 (accessed on 25 May 2021).
91. Bouchouev, I. From risk bearing to propheteering. *Quant. Financ.* **2020**, *20*, 887–894. [CrossRef]

energies

MDPI

Article

The Heterogeneous Relationship between Pollution Charges and Enterprise Green Technology Innovation, Based on the Data of Chinese Industrial Enterprises

Mingyue Wang [1], Yingming Li [1,2,*], Zitong Wang [1,2] and Junqiang Li [3,*]

1 Institutes of Science and Development, Chinese Academy of Sciences, No. 15, Zhongguancun Beiyitiao, Haidian District, Beijing 100190, China; wangmingyue@casisd.cn (M.W.); zitongwangucas@126.com (Z.W.)
2 School of Public Policy and Management, Chinese Academy of Sciences, No. 19(A) Yuquan Road, Shijingshan District, Beijing 100049, China
3 School of Economics and Management, Tongji University, Tongji Building A, Siping Road 1500, Yangpu District, Shanghai 200092, China
* Correspondence: liyingming@casisd.cn (Y.L.); ljq@tongji.edu.cn (J.L.)

Abstract: Enterprises' green technology innovation is critical to achieving the "win-win" of enterprise competitiveness and environmental protection. The impact of environmental regulation on green technology innovation by enterprises has been widely considered, but the conclusion has not yet been determined, and needs to be studied in detail. To this end, we studied the impact of pollution charge policy on different types of green technology innovation by industrial enterprises in China. We found that (1) the impact of pollution charges on most types of green technology innovation by enterprises has increased significantly over time; (2) the pollution charge policy has a certain inhibition effect on the end-of-pipe technology innovation, but can promote the process improvement of reducing industrial wastewater emissions; (3) there is a *U*-shaped relationship between the pollution charges and some green technological innovation (e.g., emission intensity of SO_2, industrial wastewater emission intensity, and industrial wastewater removal intensity), which is dynamically adjusted over time; and (4) the larger the enterprise's solid assets, the faster the asset depreciation will inhibit the enterprise from adopting the green process innovation strategy.

Keywords: pollution charges; green technology innovation; industrial enterprise; heterogeneity analysis

Citation: Wang, M.; Li, Y.; Wang, Z.; Li, J. The Heterogeneous Relationship between Pollution Charges and Enterprise Green Technology Innovation, Based on the Data of Chinese Industrial Enterprises. *Energies* **2022**, *15*, 1663. https://doi.org/10.3390/en15051663

Academic Editors: Grzegorz Mentel and Sebastian Majewski

Received: 20 January 2022
Accepted: 17 February 2022
Published: 23 February 2022

1. Introduction

In 2019, China's GDP reached CNY 99,086.5 billion. With the rapid economic development, this surge was accompanied by a similarly prodigious growth in energy consumption, air pollution, and air-pollution-related deaths [1,2]. According to the *Bulletin of China's Ecological Environment*, in 2018, 121 of the 338 cities in China achieved ambient air quality standards, accounting for only 35.8%. Resolving the uncoordinated relationship between economy and environment is an important factor in high-quality development and the construction of ecological civilization [3–5]. We can no longer rely on investment-driven innovation, and must turn to environmentally friendly technological innovation [6–8]. Enterprises are not only the core drivers of social and economic wealth creation, but also the most critical factors in the coordination of economic development and environmental protection [9,10]. Unfortunately, most of the existing studies on the economic effectiveness of environmental regulation are from the perspective of government and society, while there is little literature investigating micro-enterprises and macro-policy from the system perspective [11,12]. Therefore, in-depth exploration of how enterprises deal with environmental regulation is of great significance for the implementation of environmental regulation policies and the improvement of the green competitiveness of enterprises.

The key to encouraging enterprises to participate in environmental management is to internalize externalities, including charging enterprises for polluting the environment and subsidizing enterprises for reducing pollution [13,14]. Pollution charges are an important policy tool of China's current environmental regulation system [15,16]. The policy basis is the *Administrative Regulations on Pollution Discharge Fee Levy* (http://www.gov.cn/zhengce/content/2008-03/28/content_5152.htm) (accessed on 12 May 2021), the *Environmental Protection Law of the People's Republic of China* (http://mee.gov.cn/ywgz/fgbz/fl/201811/t20181114_673632.shtml) (accessed on 12 May 2021), and the *Several Provisions on Strengthening the Management of Environmental Protection Subsidies* (http://www.csrcare.com/Law/Show?id=11782) (accessed on 12 May 2021). The purpose of pollution charges and environmental protection subsidies is to bring environmental factors into the production decision-making function of enterprise managers, so as to realize the internalization of the externality of environmental pollution [17,18]. While these environmental governance activities increase the additional costs faced by the enterprise and reduce the financial return of the shareholders [19], more and more studies have come to the conclusion that green technology innovation helps enterprises to reduce environmental costs, gain the trust of suppliers and customers, and seize the green competitive advantage [20–22]. Unlike direct participation in environmental governance and environmental protection investment, green technology innovation can not only reduce environmental pollution and improve environmental performance, but more importantly, green technology innovation enables enterprises to produce green-differentiated products, stimulate new market demand, and effectively improve their green competitiveness, so as to truly realize the "win-win" of economic efficiency and environmental protection [23,24].

Based on the above analysis, it is reasonable to believe that pollution charges have a certain influence on enterprises' green technology innovation, which is affected by factors such as charge intensity, types of green technology innovation, etc. The purpose of this article is to further reveal the multiple heterogeneity between pollution charges and green technology innovation from the level of Chinese industrial enterprises, including testing the impact of pollution charges on different types of green technology innovation, whether or not the relationship between them is, and the manifestations of the above relationship in different years. The rest of the paper is organized as follows: In Section 2, we review the literature on the relationship between pollution charges and green technology innovation. In Section 3, we describe our methodology in detail. Section 4 presents and discusses the empirical results. Section 5 concludes the paper and provides management implications.

2. Literature Review

The Nobel laureate in Economics Paul Romer pointed out that input factors and technological progress are the two main factors restricting economic growth [25]. Green technology innovation can reduce environmental pollution, save energy, and realize green, sustainable development, which coordinated between environmental protection and enterprise competitiveness [26–29]. No matter the general technological innovation or green technological innovation, innovation activities have the characteristics of long cycle, high investment, and high risk [30,31]. Therefore, the intensity of enterprises engaging in innovation activities depends on the managers' judgment of the risks and expected benefits of enterprise innovation activities [32]. However, the technological uncertainty, market uncertainty, and imperfect management system lead to the lack of motivation and ability in terms of enterprise innovation [10,33]. In order to minimize the negative impact of enterprise production on the environment, resources, and society, the government has adopted a series of environmental regulation measures, including administrative means [11,34], legal measures [35,36], and economic measures [37,38] (Cohen et al., 2017; Wang et al. 2020).

Pollution charges are a system wherein the state collects charges from organizations or individuals that discharge pollutants [39]. The neoclassical school believes that pollution charges increase the cost of enterprise compliance, as enterprises need to pay fees for their behavior of polluting the environment in the production process, increasing the capital burden on enterprises and occupying the resources of their green technology innovation [40]. Green technology innovation depends on a large amount of resource investment, and it takes a long time to show its positive impact on enterprise energy conservation, emissions reduction, and performance improvement [41]. Under the pressure of short-term performance and cash flow caused by pollution charges, managers are forced to give up green technology innovation with high investment, high risk, and high uncertainty [42].

Pollution charges as a regulation can force enterprises to adopt the strategy of green technology innovation, and provide compensatory benefits exceeding the cost of environmental regulation [43]. Enterprises can reduce their dependence on the original polluting production mode and effectively avoid the cost of pollution charges by applying green technology innovation to their production process [44]. The forced effect of pollution charges on enterprises' green technology innovation is reflected in the external pressure of stakeholders and the incentive factors within the enterprise.

In terms of external pressure, green development is the realistic appeal of external stakeholders to heavily polluting enterprises [45]. The *Administrative Regulations on Pollution Discharge Fee Levy* implemented in 2003 and the *Environmental Protection Law of the People's Republic of China* implemented in 2018 clearly stipulate that pollution charges shall be levied on the pollutant emissions of enterprises according to the pollution equivalent. Henriques and Sadorsky [46] found that stakeholders' pressure forces managers to weigh the consequences of environmental pollution, affecting the ways managers respond to pollution charges. Therefore, green technology innovation can enhance stakeholders' confidence in enterprises' green development, reduce stakeholders' negative expectations of enterprises' environmental pollution, and encourage managers to adopt a green innovation strategy in response to the demands of external stakeholders [47].

Although pollution charges constitute a cost for enterprises, and directly reduce their attainable profits, they can also encourage managers to actively reflect on the shortcomings of enterprises' own green development [48], effectively making up for the inherent defects of enterprises' governance mechanisms and overcoming the inertia of organizations not thinking about change [49]. Through green technology innovation, enterprises can not only realize the social benefits of energy conservation and emissions reduction, but also produce more green-differentiated products than their competitors, so as to obtain an increased market share and cultivate unique green competitive advantages [50,51]. In view of the above analysis, it can be assumed that pollution charges will encourage enterprises' managers to carry out green technology innovation.

It can be seen from the above that the research on environmental regulation and green technology innovation covers a wide range, and it is very important to clarify the attributes of the research object, or else it may reach biased conclusions. At present, more and more quantitative studies are describing the relationship between environmental regulation and green technology innovation more accurately, by subdividing the types of environmental regulation, green technology innovation, and enterprise. Under the theoretical framework of the Porter hypothesis, the unique advantages of green technology innovation encourage enterprise managers to incorporate green and sustainable innovation schemes into their business decisions and strategic plans [52]. The pressure of pollution charges and the demands of stakeholders more actively encourage managers to carry out green technology innovation.

3. Methodology

3.1. Data

The basic data used in this research come from China's Industrial Enterprise Database and China's Industrial Enterprise Pollution Emission Database, both of which are maintained by the National Bureau of Statistics of China. The databases include all state-owned enterprises and non-state-owned enterprises above a designated size (with an annual main business income of more than CNY 5 million). In view of the availability and reliability of China's Industrial Enterprise Database and China's Industrial Enterprise Pollution Emission Database, we mainly use the data from 2007 and 2012, because the latest data from China's Industrial Enterprise Database are for 2013, but the quality of the 2013 data is poor. China's Industrial Enterprise Database mainly includes basic information on enterprises and their main financial statistical indicators. China's Industrial Enterprise Pollution Emission Database mainly includes basic information on enterprises, consumption data for various energy resources, emission information on various pollutants, etc. Firstly, we matched China's Industrial Enterprise Database and China's Industrial Enterprise Pollution Emission Database of the corresponding year with the organization code and enterprise name, respectively. Then, the data obtained by matching the organization code and the data obtained by matching the enterprise name were integrated to delete duplicate samples, so as to obtain the original analysis data for the corresponding year. The original database with successful matching contains information on 47,172 enterprises in 2007 and 51,385 enterprises in 2012.

In view of the absence, omission, and error of some enterprises' observed values in the two databases used, before the data regression analysis, the sample data were processed as follows by referring to the practices of Brandt et al. [53] and Zhang et al. [54]: (1) exclude enterprise samples seriously lacking important financial indicators, including the net value of fixed assets and the year-end employment of enterprises; (2) exclude enterprise samples with less than 8 employees, because small enterprises with too few employees are more likely to misreport data; (3) exclude enterprise samples whose financial indicators are inconsistent with generally accepted accounting standards; (4) exclude enterprise samples established before 1949, because the enterprises established before 1949 are mainly Chinese medicine manufacturing enterprises, and it is difficult to know the age of these enterprises; (5) exclude enterprise samples whose intermediate input and wages payable to employees this year are less than 0; (6) exclude enterprise samples with fixed assets greater than their total assets, current assets greater than their total assets, current assets greater than their fixed assets, net value of fixed assets less than 0, depreciation in this year greater than accumulated depreciation, and industrial added value greater than total output value; (7) exclude enterprise samples whose assets are not equal to corporate liabilities and owners' equity; (8) exclude enterprise samples whose water and various energy consumption are less than 0; (9) exclude enterprise samples whose discharge and removal of waste water, SO_2, NOx, smoke dust, and industrial solid waste are missing values; (10) exclude enterprise samples above a designated size whose annual main business income is less than RMB 5 million. In addition, in view of the essential differences between enterprises in the waste resources and materials recovery processing industry, mining industry, power, heat, gas, and water production and supply industry, and ordinary industrial enterprises, refer to the practices of Lu and Tao [55], deleting the enterprise samples in the matching database with the industry code value of enterprises.

3.2. Variables

(1) Enterprises' green technology innovation

Considering the differences in industrial production and the availability of data, the measurement of the innovation behavior of enterprises' end-of-pipe technology innovation can be achieved via the removal rate of industrial wastewater, SO_2, smoke, and dust and the comprehensive treatment rate of industrial solid waste. Industrial wastewater removal rate, SO_2 removal rate, smoke dust removal rate, and the comprehensive treatment rate of industrial solid waste can be calculated by dividing the corresponding waste removal amount by the waste generation amount. The measurement of green process innovation is mainly based on energy use intensity, water use intensity, industrial wastewater discharge intensity, SO_2 discharge intensity, smoke dust discharge intensity, and industrial solid waste discharge intensity. Due to the inherent differences in production processes of enterprises, the impact of the same market regulation on different green technology innovations may be quite different. Because of the data, this study does not consider green product innovation;

(2) Pollution charges

Among many environmental regulation measures, pollution charges are a widely used and flexible method used for domestic environmental protection management at present, and these are also the main reasons for studying environmental regulations. The following mainly measures the intensity of environmental regulation by calculating the logarithm of the proportion of pollution charges to the total industrial output value. Figures 1–4 show the collection of pollution charges in China in 1998, 2002, 2007, and 2012, respectively. From the figures, the following conclusions can be roughly drawn: (1) the overall pollution charges show an upward trend in the time dimension; (2) in the spatial dimension, there are great differences in regional pollution charges—the pollution charges in the east are greater than those in the west, and those in the coastal areas are greater than those in the inland areas;

Figure 1. China's pollution charges in 1997.

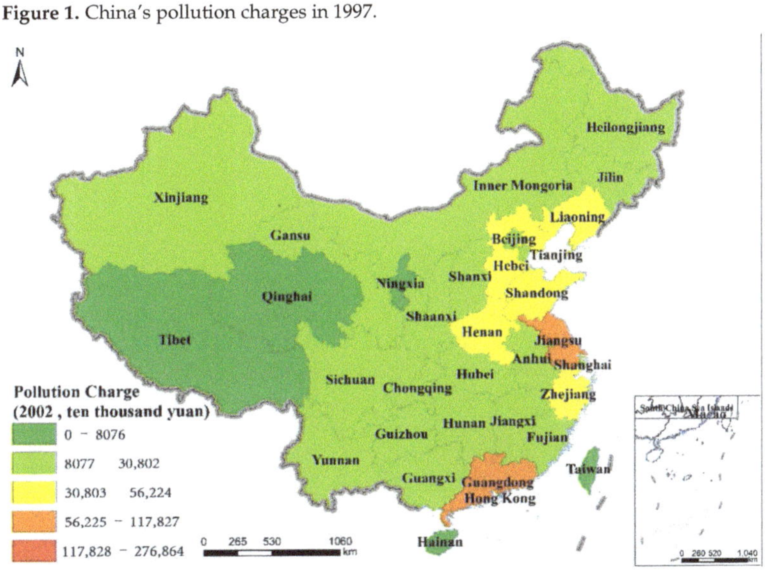

Figure 2. China's pollution charges in 2002.

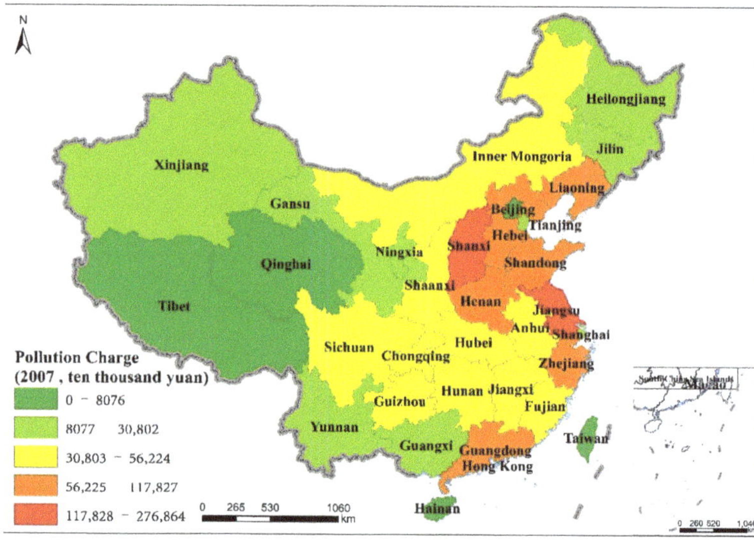

Figure 3. China's pollution charges in 2007.

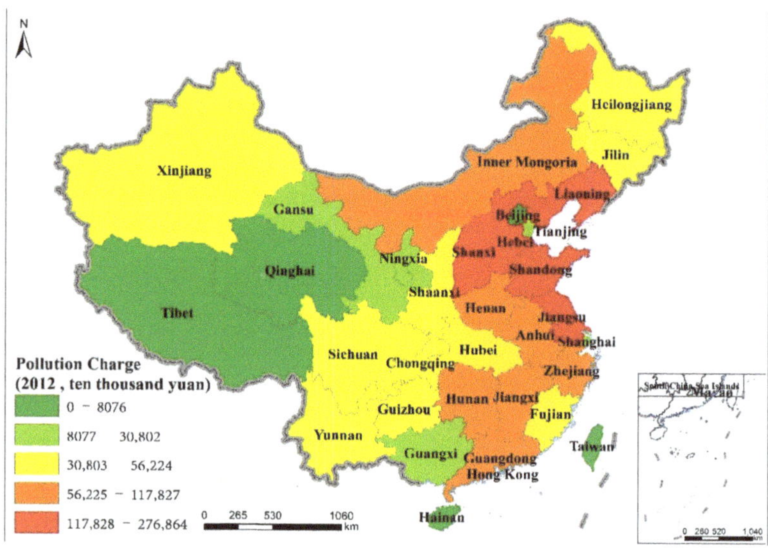

Figure 4. China's pollution charges in 2012.

(3) Control variables

In order to obtain estimation results as robust as possible, the control variables selected in this paper were as follows:

i. Industrial structure of the region where the enterprise is located: This article takes the proportion of regional industrial added value to total output value as the characteristic variable of industrial structure;
ii. ndustry that the enterprise belongs to: Considering that there are great differences in green technology innovation in different industries, it is necessary to consider the industrial characteristics of enterprises. Therefore, the industries to which

the enterprises belong are divided into primary production industries, labor- and resource-intensive industries, and technology-intensive industries;

iii. Region where the enterprise is located: Enterprises located in different regions bear different regulation intensity, so it is necessary to control for the regions of enterprises in the model. We can also see from Figures 1–4 that there is a significant difference in pollution charges between the east and the west;

iv. In addition to the variables listed above, the estimation model in 2007 also controls the value-added tax payable, total assets, asset structure, number of employees, total capital, total profit, intermediate investment, enterprise age, annual normal production time, and employee structure. Due to the change in statistical caliber, the total industrial output value, VAT payable, income tax payable, accumulated depreciation, total assets, asset structure, total liabilities, total profits, enterprise scale, enterprise age, number of employees, and annual normal production time were controlled in the estimation model for 2012;

v. Among all of the variables, the intensity of emissions of SO_2, smoke dust, industrial wastewater, and solid waste is measured by the logarithms of waste emission and total industrial output value quotient, Resource and energy consumption intensity are measured by the logarithms of resource and energy consumption and total industrial output value quotient, respectively. The waste removal rate is the logarithm of the actual removal rate of various waste products (SO_2, smoke dust, industrial wastewater, and solid waste), and the total industrial output value is controlled in the econometric model. Intermediate input, total industrial output value, interest expense, value-added tax payable, income tax payable, accumulated depreciation, total assets, total liabilities, total profits, enterprise age, number of employees, and annual normal production time are all logarithms of actual value. Enterprise age refers to the logarithm of survival time from opening to statistical node. The employee structure is the proportion of female employees to the total number of employees. Asset structure is the ratio of fixed assets to total assets.

3.3. Model

Because the current data are insufficient to construct panel data for empirical research, in order to achieve the research goal, this article mainly uses robust ordinary least squares (ROLS) to analyze the impact of pollution charges on different green technology innovations of enterprises. On the basis of previous studies, the quadratic term of pollution charge is introduced into the econometric model to further demonstrate the nonlinear relationship between pollution charges and enterprise green technology innovation, as shown in Equation (1).

$$y_i = \alpha + \beta_1 PC_i + \beta_2 PC_i{}^2 + \sum \gamma_i Cont_i + \varepsilon_i \quad (1)$$

where y_i is the green technology innovation of the enterprise, including end-pipe green technology innovation (measured by industrial wastewater removal intensity, SO_2 removal intensity, smoke and dust removal intensity, and comprehensive utilization intensity of industrial solid waste) and green process innovation (measured by energy use intensity, water resource use intensity, industrial wastewater discharge intensity, SO_2 emission intensity, smoke and dust emission intensity, and industrial solid waste emission intensity). $PC_i{}^2$ and PC_i are the square terms of the intensity of pollution charge and the intensity of pollution charge, respectively. $Cont_i$ represents control variables, ε_i is a random error term, and α is an intercept item.

4. Results

4.1. Descriptive Statistics

The mean value, standard deviation, maximum value, and minimum value of the main variables in 2012 are shown in Table 1, and the results in 2007 are similar, so they are not listed.

Table 1. Descriptive statistical results of the main variables for 2012.

Variables	Samples	Means	SD	Min	Max
SO_2 emission intensity	10,001	−9.311	2.035	−21.379	−1.786
Smoke and dust emission intensity	10,001	−8.356	2.352	−20.700	0.736
Solid waste discharge intensity	10,001	−5.894	2.019	−14.851	1.920
Wastewater discharge intensity	10,001	−1.232	2.139	−13.226	4.888
SO_2 removal intensity	10,001	−9.014	0.599	−9.210	−1.771
Smoke and dust removal intensity	10,001	−7.771	1.505	−9.210	0.735
Solid waste removal intensity	10,001	−5.821	1.790	−9.210	1.272
Wastewater removal intensity	10,001	−1.346	2.597	−9.210	5.420
Gross industrial output value	10,001	12.327	1.343	9.916	16.778
Pollution charges	10,030	−8.051	0.444	−9.270	−6.787
VAT payable	9462	8.418	1.746	2.944	13.352
Income tax payable	6602	7.310	2.041	1.386	12.276
Accumulated depreciation	9833	9.713	1.895	4.575	15.222
Total Assets	10,029	11.700	1.563	6.738	18.925
Asset structure	9943	0.378	0.219	0.011	0.954
Total liabilities	10,003	10.828	1.811	5.553	16.036
Total profit	9999	9.138	1.940	3.296	13.838
Enterprise scale	10,030	2.493	0.647	1.000	4.000
Enterprise age	10,011	12.872	10.616	1.000	58.000
Annual production time	10,030	8.363	0.618	5.704	9.078
Technology-intensive enterprise	10,030	0.487	0.500	0.000	1.000
Enterprise in the eastern provinces	10,030	0.606	0.489	0.000	1.000

By analyzing the correlation between pollution charges and enterprise green technology innovation in 2012, in Table 2, it can be seen that (1) the pollution charges in 2012 had a significant positive correlation with SO_2 emission intensity, smoke dust emission intensity, and SO_2 removal intensity at the 1% confidence level, and the correlation coefficients were 0.040, 0.030, and 0.073, respectively; (2) the pollution charges were negatively correlated with the discharge intensity and removal intensity of industrial wastewater at the confidence level of 1%, and the correlation coefficients are −0.045 and −0.054, respectively; and (3) pollution charges are negatively correlated with solid waste emission intensity, positively correlated with smoke and dust removal intensity, and negatively correlated with solid waste removal intensity, but the correlation coefficient is not significant at the 10% confidence level. The above results show that, except for smoke dust and solid waste, the pollution charges promote the end-treatment of SO_2, but inhibit the improvement of the SO_2 emission reduction process. However, the effect of pollution charges on industrial wastewater is the complete opposite—that is, pollution charges inhibit the end-treatment of industrial wastewater, but promote the improvement of the industrial wastewater removal process.

4.2. Empirical Test Results for 2012

Pollution charges at a given time can have a certain influence on different types of green technology innovation of enterprises. (1) From Table 3, it can be seen that pollution charges are positively correlated with SO_2 emission intensity, smoke dust emission intensity, and solid waste emission intensity at the 1% confidence level, with regression coefficients of 0.357, 0.293, and 0.162, respectively; that is, when increasing pollution charges in total industrial output value, SO_2 emission intensity, smoke dust emission intensity, and solid waste emission intensity increase correspondingly. There is a significant negative correlation between pollution charges and industrial wastewater discharge intensity at the confidence level of 5%, and the regression coefficient is −0.481. (2) From Table 4, it can be seen that the pollution charges are significantly positively correlated with the SO_2 removal intensity and solid waste removal intensity at the confidence levels of 1% and 10%, respectively, and the regression coefficients are 0.099 and 0.085, respectively. There is a significant negative correlation between pollution charges and industrial wastewater removal intensity at the confidence level of 1%, and the regression coefficient is −0.304. The regression coefficient between industrial pollution cost and smoke and dust removal intensity is not significant. (3) The results given in Table 5 show that the industrial pollution charges are negatively correlated with the total energy intensity, power consumption intensity, and fuel oil consumption intensity at the confidence levels of 1%, 1%, and 5% respectively, and the regression coefficients are −2.22, −0.428, and −0.339, respectively. Pollution charges are positively correlated with coal consumption intensity and natural gas consumption intensity at the confidence level of 1%, with regression coefficients of 0.555 and 0.378, respectively.

Table 2. Correlation coefficient matrix of major variables in 2012.

Variables	1	2	3	4	5	6	7	8	9
SO_2 emission intensity	1.000								
Smoke and dust emission intensity	0.740 ***	1.000							
Solid waste discharge intensity	0.675 ***	0.659 ***	1.000						
Wastewater discharge intensity	0.473 ***	0.368 ***	0.432 ***	1.000					
SO_2 removal intensity	0.225 ***	0.277 ***	0.273 ***	0.156 ***	1.000				
Smoke and dust removal intensity	0.575 ***	0.860 ***	0.584 ***	0.299 ***	0.322 ***	1.000			
Solid waste removal intensity	0.663 ***	0.653 ***	0.957 ***	0.426 ***	0.275 ***	0.604 ***	1.000		
Wastewater removal intensity	0.402 ***	0.347 ***	0.400 ***	0.761 ***	0.168 ***	0.337 ***	0.396 ***	1.000	
Pollution charges	0.040 ***	0.030 ***	−0.001	−0.045 ***	0.073 ***	0.008	−0.010	−0.054 ***	1.000

*** $p < 0.01$.

Table 3. Regression results of the impact of pollution charges on enterprises' process improvement and technological innovation in 2012.

	SO_2 Emission Intensity M1	Smoke and Dust Emission Intensity M2	Solid Waste Discharge Intensity M3	Industrial Water Intensity M4
Pollution charges	0.357 ***	0.293	0.162 *	−0481
	(6.67)	(4.43)	(2.99)	(−2.15)
Square of pollution charge	0.005	0.006	−0.001	0.196 *
	(0.26)	(0.35)	(−1.02)	(1.85)
Regional industrial structure	−1.561 ***	−0.665	−3.529 ***	1.811 ***
	(−2.01)	(−0.69)	(−4.49)	(−2.26)
VAT payable	0.0511 **	0.0789 ***	0.0305	0.06113 ***
	(2.56)	(3.20)	(1.51)	(3.05)
Income tax payable	−0.0222	−0.0377	−0.0374 *	−0.0446 **
	(−1.09)	(−1.51)	(−1.82)	(−2.18)
Accumulated depreciation	0.0947 ***	0.143 ***	0.178 ***	0.116 ***
	(4.52)	(5.52)	(8.40)	(5.51)
Gross industrial output value	−0.959 ***	−1.006 ***	−0.909 ***	−0.965 ***
	(−26.27)	(−22.32)	(−24.58)	(−26.22)
Asset structure	0.985 ***	1.233 ***	0.908 ***	0.520 ***
	(8.65)	(8.78)	(7.88)	(4.53)
Total liabilities	0.141 ***	0.210 ***	0.185 ***	0.131 ***
	(6.44)	(7.79)	(8.35)	(5.95)
Total profit	−0.0344	−0.0294	−0.0275	0.00257
	(−1.37)	(−0.95)	(−1.08)	(0.10)
Enterprise scale	−0.207	−1.202 ***	−1.379 ***	−0.585 **
	(−0.99)	(−4.64)	(−6.48)	(−2.76)
Enterprise age	−0.00465	−0.0106	−0.00147	0.0283 ***
	(−0.77)	(−1.42)	(−0.24)	(4.63)
Enterprise employment	0.0662 ***	0.0699 **	0.0875 ***	0.0695 ***
	(2.65)	(2.27)	(3.46)	(2.76)
Annual normal production time	1.199 ***	1.272 ***	1.076 ***	0.956 ***
	(31.91)	(27.41)	(28.25)	(25.22)
Technology-intensive enterprise	−0.388 ***	−0.304 ***	−0.323 ***	−1.457 ***
	(−8.83)	(−5.61)	(−7.27)	(−32.92)
Enterprise in the eastern provinces	−0.0692	−0.352 ***	−0.284 ***	−0.0532
	(−1.45)	(−5.98)	(−5.86)	(−1.10)
N	3115	3119	3680	3963
R^2	0.265	0.250	0.225	0.189

t-Statistics in parentheses; * $p < 0.1$, ** $p < 0.05$, *** $p < 0.01$.

Table 4. Regression results of the impact of pollution charges on enterprises' end-treatment technological innovation in 2012.

	SO_2 Removal Intensity M5	Smoke and Dust Removal Intensity M6	Solid Waste Removal Intensity M7	Wastewater Removal Intensity M8
Pollution charges	0.0991 ***	0.0658	0.0852 *	−0.304 ***
	(5.51)	(1.52)	(1.76)	(−4.13)
Square of pollution charge	−0.075	−0.398	−1.562	1.023
	(−1.09)	(−0.20)	(−0.82)	(0.59)
Regional industrial structure	−1.024 ***	−0.0156	−2.917 ***	5.449 ***
	(−3.93)	(−0.02)	(−4.15)	(5.11)
VAT payable	−0.00235	0.0531 **	0.0381 **	0.0716 ***
	(−0.35)	(3.28)	(2.11)	(2.61)
Income tax payable	−0.00354	−0.0124	−0.0390 **	0.0281
	(−0.52)	(−0.75)	(−2.12)	(1.01)
Accumulated depreciation	0.0364 ***	0.131 ***	0.146 ***	0.156 ***
	(5.19)	(7.71)	(7.69)	(5.42)
Gross industrial output value	−0.0460 ***	−0.565 ***	−0.797 ***	−0.882 ***
	(−3.75)	(−19.11)	(−24.10)	(−17.59)
Asset structure	0.188 ***	0.921 ***	0.950 ***	0.784 ***
	(4.92)	(10.00)	(9.22)	(5.02)
Total liabilities	0.0421 ***	0.141 ***	0.173 ***	0.125 ***
	(5.74)	(7.96)	(8.76)	(4.17)
Total profit	−0.0149 *	−0.0143	−0.0193	0.0824 **
	(−1.76)	(−0.70)	(−0.85)	(−2.38)
Enterprise scale	−0.207 ***	−1.016 ***	−1.262 ***	−0.342
	(−2.94)	(−5.97)	(−6.63)	(−1.19)
Enterprise age	−0.000678	−0.0106	−0.00147	0.0283 ***
	(−0.77)	(−1.42)	(−0.24)	(4.63)
Enterprise employment	0.00303	0.0475 **	0.0905 ***	0.143 ***
	(0.36)	(2.35)	(4.00)	(4.17)
Annual normal production time	0.145 ***	0.754 ***	0.946 ***	1.090 ***
	(11.50)	(24.77)	(27.79)	(21.11)
Technology-intensive enterprise	−0.0358 **	−0.147 ***	−0.305 ***	−1.387 ***
	(−2.43)	(−4.13)	(−7.68)	(−22.99)
Enterprise in the eastern provinces	−0.0358 **	−0.227 ***	−0.279 ***	−0.208 ***
	(−2.24)	(−5.86)	(−6.45)	(−3.17)
N	6283	6283	6283	6283
R^2	0.282	0.197	0.262	0.232

t-Statistics in parentheses; * $p < 0.1$, ** $p < 0.05$, *** $p < 0.01$.

Table 5. Regression results of the impact of pollution charges on enterprises' energy-saving technological innovation in 2012.

	Energy Consumption Intensity M9	Enterprise Power Intensity M10	Enterprise Coal Intensity M11	Enterprise Natural Gas Intensity M12	Enterprise Fuel Oil Intensity M13
Pollution charges	−2.222 ***	−0.428 ***	0.555 ***	0.378 ***	−0.339 **
	(−3.06)	(−3.72)	(5.57)	(3.46)	(−2.07)
Square of pollution charge	1.198 ***	0.235 ***	−1.036	−0.956	0.159
	(3.40)	(4.22)	(−0.38)	(−1.01)	(0.36)
Regional industrial structure	−1.558	−2.621 ***	−1.692	−1.798	2.650
	(−1.36)	(−6.18)	(−1.59)	(−1.58)	(1.48)
VAT payable	0.0507	0.0137	0.0506	0.0913 *	0.0539
	(1.32)	(1.24)	(1.35)	(2.48)	(1.02)
Income tax payable	−0.332 *	−0.122 *	−0.356 **	−0.0444	0.567
	(−1.80)	(−1.93)	(−2.06)	(−0.16)	(1.42)
Accumulated depreciation	0.193 ***	0.220 ***	0.220 ***	0.0409	0.161 **
	(4.99)	(18.80)	(6.13)	(0.86)	(2.40)
Gross industrial output value	−0.909 ***	−0.947 ***	−0.692 ***	−0.861 ***	−0.666 ***
	(−12.91)	(−45.71)	(−10.51)	(−11.57)	(−5.89)
Asset structure	1.318 ***	0.521 ***	0.895 ***	1.152 ***	−0.231
	(6.38)	(8.14)	(4.69)	(4.37)	(−0.59)
Total liabilities	0.191 ***	0.258 ***	0.195 ***	0.0873 *	−0.0578
	(4.50)	(21.14)	(4.98)	(1.73)	(−0.76)
Total profit	−0.101 ***	−0.0556 ***	−0.157 ***	−0.0613	−0.0666
	(−2.41)	(−3.97)	(−4.12)	(−1.18)	(−0.85)
Enterprise scale	−1.623 ***	−0.350 ***	−1.261 ***	−0.319	0.109
	(−4.14)	(−3.01)	(−3.57)	(−0.90)	(−0.21)
Enterprise age	−0.0246 **	−0.00845 **	−0.0212 *	0.00258	−0.0361 *
	(−2.18)	(−2.35)	(1.99)	(0.21)	(−1.93)
Enterprise employment	0.0681	0.116 ***	0.120 **	0.0233	−0.0331
	(1.41)	(7.88)	(2.70)	(0.56)	(−0.51)
Annual normal production time	1.186 ***	0.984 ***	1.187 ***	1.328 ***	0.708 ***
	(17.10)	(47.48)	(17.88)	(17.26)	(6.00)
Technology-intensive enterprise	−0.00822	0.330 ***	0.234 **	0.337 ***	0.173
	(−0.10)	(12.64)	(2.95)	(2.96)	(1.11)
Enterprise in the eastern provinces	−0.327 ***	0.295 ***	−0.210 **	−0.110	0.336 **
	(−3.78)	(10.76)	(−2.62)	(−1.11)	(2.01)
N	1674	18334	1739	1720	870
R^2	0.310	0.304	0.315	0.314	0.170

t-Statistics in parentheses; * $p < 0.1$, ** $p < 0.05$, *** $p < 0.01$.

The influence of pollution charges on different types of green technology innovation at a given time is inconsistent. (1) Pollution charges can improve the emission intensity of SO_2, smoke and dust, and solid waste, but reduce the emission intensity of industrial wastewater. Pollution charges can promote the intensity of SO_2 removal and solid waste discharge, but reduce the intensity of industrial wastewater removal. Pollution charges have a restraining effect on total energy intensity, power consumption intensity, and fuel oil consumption intensity, but they significantly increase coal consumption intensity and natural gas consumption intensity. (2) The impact of pollution charges on the end-of-pipe technology innovation of enterprises is linear (the square term of pollution charge is not significant). The influence of pollution charges on industrial wastewater discharge intensity, total energy intensity, and power consumption intensity is nonlinear, and there is a *U*-shaped relationship between them (the square coefficient of pollution charge is positive and the coefficient of pollution charge is negative). At the same time, the impact of pollution charges on the removal intensity of smoke and dust is not significant. (3) Although there is a *U*-shaped relationship between pollution charges and industrial wastewater discharge intensity, total energy intensity, and power consumption intensity, there are differences in the inflection points of the impact of pollution charges on the three types of green technology innovation. The curve inflection point of industrial wastewater discharge intensity is greater than that of the total energy consumption intensity or of the power consumption intensity.

From Tables 3 and 4, it can be seen that (1) pollution charges have pushed up the intensity of SO_2 emissions to a certain extent (the regression coefficient between pollution charges and SO_2 emission intensity is 0.357), while they have also improved the intensity of SO_2 removal (the regression coefficient between pollution charges and SO_2 removal intensity is 0.099), with a greater impact on the former than the latter; (2) pollution charges have pushed up the emission intensity of solid waste to a certain extent (the regression coefficient between industrial pollution charges and solid waste emission intensity is 0.162), and at the same time, they have also improved the removal intensity of solid waste (the regression coefficient between pollution charges and solid waste removal intensity is 0.066), with a greater impact on the former than the latter; (3) in a certain range, pollution charge reduces the intensity of industrial wastewater discharge (the regression coefficient between pollution charge sand industrial wastewater discharge intensity is -0.481), and at the same time, pollution charges also reduce the intensity of industrial wastewater removal (the regression coefficient between pollution charge and industrial wastewater removal intensity is -0.304); (4) electricity and fuel oil are the main energy sources produced by enterprises at the given time point, and they are also the main factors that determine the energy intensity, while coal and natural gas are not the main energy sources produced by enterprises at present. The above results show that pollution charges can promote the end-treatment of SO_2, but inhibit the process improvement of SO_2 emission reduction. However, the mechanism of pollution charges on industrial wastewater is the complete opposite; that is, pollution charges can inhibit the end-treatment of industrial wastewater, but promote the process improvement of industrial wastewater removal.

4.3. Empirical Test Results in 2007

Pollution charges at a given time have a certain influence on some types of green technology innovation by enterprises. (1) From Table 6, it can be seen that the pollution charges are only significantly correlated with the SO_2 emission intensity and industrial wastewater emission intensity at the confidence levels of 1% and 10%, respectively, and the regression coefficients are -3.583 and -3.447, respectively. The regression coefficient of pollution charges, smoke and dust emission intensity, and water use intensity of enterprises is not significant. (2) We can see from Table 7 that there is a significant negative correlation between pollution charges and industrial wastewater removal intensity at the confidence level of 5%, and the regression coefficient is -3.608; in addition, the regression coefficients of pollution charges with SO_2 removal intensity and smoke dust removal intensity are not significant. (3) The econometric results given in Table 8 show that pollution charges are positively correlated with total energy consumption intensity, coal consumption intensity, and natural gas consumption intensity at the 1% confidence level, with regression coefficients of 0.416, 0.340, and 0.953, respectively. There is a significant negative correlation between pollution charges and fuel oil consumption intensity at the confidence level of 1%, and the regression coefficient is −0.853.

Table 6. Regression results of the impact of pollution charges on enterprises' process improvement and technological innovation in 2007.

	SO_2 Emission Intensity M14	Smoke and Dust Emission Intensity M15	Wastewater Discharge Intensity M16	Industrial Water Intensity M17
Pollution charges	−3.538 ***	1.174	−3.447 *	−0.880
	(−2.60)	(0.66)	(−2.44)	(−0.65)
Square of pollution charge	2.145 ***	−0.088	1.851 **	0.610
	(3.14)	(−0.10)	(2.61)	(1.09)
Regional industrial structure	−7.649 ***	−1.614	−6.632 ***	−4.797 ***
	(−7.67)	(−1.26)	(−6.27)	(−4.88)
VAT payable	0.274 ***	0.347 ***	0.130 ***	0.157 ***
	(8.41)	(8.09)	(4.08)	(5.22)
Gross industrial output value	−1.088 ***	−1.297 ***	−1.172 ***	−1.251 ***
	(−6.34)	(−5.75)	(−6.61)	(−7.62)
Asset structure	1.536 ***	2.525 ***	0.849 ***	1.156 ***
	(9.41)	(11.94)	(5.00)	(7.41)
Enterprise employment	−1.686 **	−3.500 ***	−2.235 **	−2.232 ***
	(−2.46)	(−3.93)	(−3.17)	(−3.32)
Total asset	−0.0270	−0.0124	0.0822 **	0.119 ***
	(−1.03)	(−0.36)	(3.12)	(4.84)
Total profit	−0.0267	−0.00916	−0.0406 *	−0.0349 **
	(−1.63)	(−0.43)	(−2.39)	(−2.21)
Enterprise intermediate investment	0.167	0.234	0.361 *	0.532 ***
	(1.02)	(1.09)	(2.15)	(3.42)
Enterprise age	−0.00120	−0.0102	−0.0175	−0.0102
	(−0.13)	(−0.82)	(−1.80)	(−1.12)
Annual normal production time	0.675 ***	0.786 ***	0.541 ***	0.583 ***
	(13.08)	(11.81)	(10.24)	(12.16)
Employment structure	−1.486 ***	−2.114 ***	−1.016 ***	−1.127 ***
	(−8.50)	(−9.32)	(−5.89)	(−6.96)
Technology-intensive enterprise	−0.128 *	−0.0694	−1.317 ***	−1.031 ***
	(−1.72)	(−0.71)	(−17.91)	(−14.88)
Enterprise in the eastern provinces	0.731 ***	−0.0935	0.830 ***	0.484 ***
	(6.97)	(−0.69)	(7.38)	(4.64)
N	3115	3119	3680	3963
R^2	0.265	0.250	0.225	0.189

t-Statistics in parentheses; * $p < 0.1$, ** $p < 0.05$, *** $p < 0.01$.

Table 7. Regression results of the impact of pollution charges on enterprises' end-treatment technological innovation in 2007.

	SO_2 Removal Intensity M18	Smoke and Dust Removal Intensity M19	Wastewater Removal Intensity M20
Pollution charges	−1.555	−1.257	−3.608 **
	(−0.96)	(−0.71)	(−2.55)
Square of pollution charge	1.119	0.937	1.889 ***
	(1.37)	(1.06)	(2.65)
Regional industrial structure	−4.001 **	−2.967 **	−5.764 ***
	(−2.55)	(−2.28)	(−5.25)
VAT payable	0.208 ***	0.369 ***	0.133 ***
	(3.68)	(7.96)	(4.00)
Gross industrial output value	−0.948 ***	−1.303 ***	−1.113 ***
	(−3.27)	(−5.52)	(−5.73)
Asset structure	1.584 ***	2.211 ***	1.015 ***
	(5.76)	(9.91)	(5.63)
Enterprise employment	−0.914	−3.406 ***	−2.581 ***
	(−1.13)	(−3.93)	(−3.75)
Total asset	−0.0106	0.0538	0.0501 *
	(−0.24)	(1.47)	(1.84)
Total profit	0.0256	0.0226	−0.0517 ***
	(0.86)	(1.04)	(−2.90)
Enterprise intermediate investment	0.0468	0.217	0.277
	(0.17)	(0.96)	(1.51)
Enterprise age	−0.0142	−0.00545	−0.0143
	(−0.93)	(−0.42)	(−1.37)
Annual normal production time	0.540 ***	0.714 ***	0.500 ***
	(6.35)	(10.07)	(9.07)
Employment structure	−1.245 ***	−1.620 ***	−1.075 ***
	(−4.25)	(−6.79)	(−5.90)
Technology-intensive enterprise	−0.256 **	0.146	−1.261 ***
	(2.10)	(1.41)	(−16.33)
Enterprise in the eastern provinces	0.308 *	0.115	0.611 ***
	(1.73)	(0.83)	(4.96)
N	1452	2460	2971
R^2	0.196	0.226	0.259

t-Statistics in parentheses; * $p < 0.1$, ** $p < 0.05$, *** $p < 0.01$.

The influence of pollution charges on different types of green technology innovation at a given time is inconsistent. (1) From Tables 6–8, we can see that on the whole, pollution charges can promote some process improvement and end-of-pipe technology innovation, and can have a negative impact on most energy-saving green technology innovation. (2) The impact of pollution charges on energy-saving green technology innovation is linear (the square term of pollution charges is not significant). The influence of pollution charges on SO_2 emission intensity, industrial wastewater emission intensity, and industrial wastewater removal intensity is nonlinear, and there is a U-shaped relationship between them (the square coefficient of pollution charges is positive and the coefficient of pollution charges is negative). (3) Although there is a U-shaped relationship between pollution charges and

SO_2 emission intensity, industrial wastewater emission intensity, and industrial wastewater removal intensity, the inflexion points of pollution charges on the three kinds of green technological innovations are different. SO_2 emission intensity is less than industrial wastewater emission intensity and industrial wastewater removal intensity.

Table 8. Regression results of the impact of pollution charges on enterprises' energy-saving technological innovation in 2007.

	Energy Consumption Intensity M21	Enterprise Coal Intensity M22	Enterprise Fuel Oil Intensity M23	Enterprise Natural Gas Intensity M24
Pollution charges	0.416 ***	0.340 ***	−0.853 ***	0.953 ***
	(4.85)	(14.32)	(−8.05)	(6.16)
Square of pollution charge	−0.008	−0.001	2.361	−0.0652
	(−1.01)	(−1.09)	(0.02)	(−0.09)
Regional industrial structure	−2.326 ***	−1.675 ***	−0.760	−0.934
	(−2.71)	(−6.99)	(−0.83)	(−0.81)
VAT payable	0.248 ***	0.269 ***	0.124 ***	0.300 ***
	(8.71)	(26.10)	(3.91)	(5.42)
Gross industrial output value	−0.926 ***	−1.106 ***	−0.0530	−1.124 ***
	(−6.45)	(−22.36)	(−0.32)	(−5.00)
Asset structure	1.817 ***	1.362 ***	0.736 ***	1.399 ***
	(12.99)	(27.00)	(3.93)	(4.22)
Enterprise employment	−1.655 ***	−0.504 **	0.102	−1.903 ***
	(−5.51)	(−2.97)	(0.31)	(−5.17)
Total asset	0.00301	0.0867 ***	−0.0119	−0.0367
	(0.13)	(10.09)	(−0.42)	(−0.71)
Total profit	−0.0224	−0.0246 ***	−0.0629 ***	−0.0366
	(−1.59)	(−4.94)	(−3.32)	(−1.05)
Enterprise intermediate investment	0.0425	0.0909 *	−0.484 ***	0.242
	(0.31)	(1.92)	(−3.07)	(1.17)
Enterprise age	0.0119	−0.000373	−0.00287	−0.0820 ***
	(1.42)	(−0.13)	(−0.27)	(−4.45)
Annual normal production time	0.736 ***	0.970 ***	0.660 ***	0.837 ***
	(16.33)	(60.98)	(11.68)	(9.06)
Employment structure	−1.615 ***	−1.263 ***	−0.441 **	−0.443
	(−10.91)	(−23.81)	(−2.39)	(−1.28)
Technology-intensive enterprise	−0.0963	−0.151 ***	0.0863	0.210
	(−1.50)	(−6.57)	(1.03)	(1.25)
Enterprise in the eastern provinces	0.348 ***	−0.0120	0.861 ***	0.283 *
	(3.84)	(−0.44)	(6.88)	(1.82)
N	3385	22,011	3192	1343
R^2	0.296	0.369	0.207	0.238

t-Statistics in parentheses; * $p < 0.1$, ** $p < 0.05$, *** $p < 0.01$.

There are some common laws in the influence of pollution charges on different types of green technology innovation at a given time. From Tables 6 and 7, we can see that (1) within a certain range, pollution charges reduce industrial wastewater discharge intensity to a certain extent (the regression coefficient between pollution charges and industrial wastewater discharge intensity is −3.447), while pollution charges also reduce industrial wastewater removal intensity (the regression coefficient between pollution charges and industrial wastewater removal intensity is −3.608). (2) Pollution charges have no significant impact on the removal intensity and emission intensity of smoke and dust; that is, as far as smoke and dust are concerned, the collection of pollution charges does not significantly promote the technological innovation of emissions reduction at the source, nor does it significantly promote the technological innovation of emissions reduction at the end. (3) Coal and natural gas are the main energy sources produced by enterprises at the given time point, and they are also the main factors that determine the energy intensity, while fuel oil is not the main energy source produced by enterprises at present—that is, pollution charges can reduce the fuel oil consumption intensity, but can improve the total energy consumption intensity, so it is very likely that the intensity of coal and natural gas use is increased. The above results show that the pollution charges can inhibit the end-treatment of industrial wastewater, but can promote improvement of the process of industrial wastewater emission reduction.

4.4. Discussion

(1) The effectiveness of pollution charges on the green technology innovation of enterprises has been significantly improved. From the above empirical results, it can be seen that in 2012, the pollution charges had no significant effect on the intensity of smoke and dust removal, but had significant effects on other end-of-pipe technology innovation, green process innovation, and energy-saving green technology innovation of enterprises; that is, the government has imposed pollution discharge fees on enterprises, causing a certain impact on various green technology innovations of enterprises. At the same time, observing the empirical results for 2007, we can see that pollution charges had no significant effect on the intensity of smoke and dust removal, but had significant effects on the intensity of water production, SO_2 removal, and smoke and dust emissions; we can find similar rules for the degree of significance. Therefore, we can confirm that with the passage of time, the efficiency and validity of pollution charges in promoting green technology innovation by enterprises have been improved to a certain extent. At the same time, with the development of the economy and society, the eastern provinces and technology-intensive industries have an increasingly significant influence on the green technology innovation of process improvement;

(2) The regulation of pollution charges reduces the discharge intensity and removal intensity of industrial wastewater. From Tables 3–5 and Table 7, we can see that the regression coefficients of pollution charges and industrial wastewater discharge intensity and removal intensity are all negative; that is, pollution charges reduce industrial wastewater discharge intensity to a certain extent (the regression coefficients of pollution charges and industrial wastewater discharge intensity for 2007 and 2012 are −0.481 and −3.447, respectively), and pollution charges also reduce industrial wastewater removal intensity. The above results show that, for industrial wastewater, pollution charges can inhibit the end-of-pipe technology innovation, but can promote the process improvement of industrial wastewater emission reduction;

(3) There is a nonlinear dynamic relationship between pollution charges and green technology innovation. In the empirical results for 2007 and 2012, the influence of pollution charges on the emission intensity of smoke and dust was not significant, and because of statistical data, there was no record of solid waste discharge intensity in 2007, so this was not the object of follow-up analysis. From Tables 3–8, we can see that there is a *U*-shaped relationship between pollution charges and enterprises' green technology innovation. Specifically, pollution charges have a *U*-shaped influence on industrial wastewater

discharge intensity and industrial wastewater removal intensity; that is, in terms of industrial wastewater removal intensity and discharge intensity, the current regulation has not reached the inflection point, and moderately improving the regulation of wastewater removal will further improve the technological innovation ability of enterprises in this respect. In 2007, the relationship between pollution charges and SO_2 emission intensity was *U*-shaped; that is, the intensity of SO_2 emissions could be reduced by increasing the intensity of pollution charges within a certain range. However, the regression results for 2012 show that there was a linear increase between pollution charges and SO_2 emission intensity; that is, pollution charges pushed up the intensity of SO_2 emissions, largely because the regulation intensity had exceeded the inflection point (pollution charges squeezed the investment in green technology innovation of enterprises), so it was necessary to appropriately adjust the regulation of SO_2 emission intensity;

(4) The impacts of enterprise asset structure and depreciation on different types of green technology innovation are quite different. Although there are some differences in the selection of control variables between the empirical models in 2007 and 2012, we found that there was a significant positive correlation between asset structure and waste emission intensity (e.g., SO_2 emission intensity, smoke and dust emission intensity, solid waste emission intensity, industrial wastewater emission intensity), energy intensity, and enterprise production water intensity; that is, the higher the proportion of fixed assets, the higher the cost of process improvement for enterprises, because process improvement involves modification and replacement of existing production processes and production equipment. Therefore, the excessive proportion of fixed assets will reduce the enthusiasm of enterprises for green technology innovation; instead, they will choose end-of-pipe technology innovation, and the empirical results have been well verified. The greater the accumulated depreciation, the greater the risk of equipment investment; therefore, enterprises with greater accumulated depreciation are less motivated to improve production technology and update production equipment to reduce emissions. On the other hand, enterprises with greater depreciation tend to choose end-of-pipe technological innovation to achieve the legitimacy of enterprise production and reduce the environmental cost of enterprise production. We can see that larger solid assets of enterprises and faster depreciation of assets will inhibit enterprises from adopting the green process innovation, and promote enterprises to adopt the end-of-pipe technological innovation.

5. Conclusions

This article obtained basic data by matching China's Industrial Enterprise Database, China's Industrial Enterprise Pollution Emission Database, and regional pollution charges. Based on these data, 24 empirical models were constructed to analyze the impacts of pollution charges on different types green technology innovation by enterprises in 2007 and 2012. The results are consistent with our preliminary assumption, indicating that the pollution charges will indeed have a certain impact on the green technology innovation of enterprises. Such impact is not invariable; it is reflected in the type of green technology innovation and the intensity of the charges, as well as the time. At the same time, it enables the moderation of the internal characteristics of enterprises, including solid assets, asset depreciation, location, and industry. From the above research, the following conclusions can be obtained:

(1) The effectiveness of pollution charges on all kinds of green technology innovation by enterprises has significantly improved with the passage of time. This is consistent with the findings of Sellitto et al. and Porter [50,52], and will not change due to the difference in research subjects and methods. At the same time, with economic and social development, the impact of eastern provinces and technology-intensive industries on green process innovation is becoming more and more significant;

(2) Pollution charges can inhibit the end-of-pipe technology innovation for end-treatment of industrial wastewater, but can promote the process improvement of industrial wastewater emission reduction. This shows that the role of pollution charges in indus-

trial sewage treatment comes from process improvement. The "inhibition hypothesis" holds that pollution charges will increase costs to enterprises, reflected mainly in the increase in external pollution control costs and the increased innovation investment to adapt to the current policies [42,56];

(3) There is a nonlinear relationship between pollution charges and enterprises' green technology innovation, and it is constantly adjusted with the change in time. There is a *U*-shaped relationship between pollution charges and enterprises' green technology innovation (i.e., SO_2 emission intensity, industrial wastewater emission intensity, and industrial wastewater removal intensity). Sanchez-Vargas et al. [57] asserted that the relationship between environmental regulation and productivity is actually nonlinear. One explanation of the "uncertainty hypothesis" is that, in the short term, investment related to pollution control will crowd out green technology R&D investment, but in the long term, the compensatory effect of innovation can generate additional profits to promote technological progress [58];

(4) The larger the enterprise's solid assets, the faster the asset depreciation will inhibit the enterprise from adopting a green process innovation strategy. On the contrary, it will encourage the enterprise to adopt an end-pipe technology innovation strategy, meaning that the internal characteristics of the enterprise have a moderate effect on the role of pollution charges.

Therefore, we should not reduce the intensity of environmental regulations because of the current downward trajectory of the economy. Some local governments are trying to stimulate economic recovery by reducing the intensity of environmental regulations or starting high-carbon projects, which will have a negative impact on China's green economic recovery. The Porter hypothesis highlights how appropriate environmental regulation will stimulate enterprises' green technology innovation, so as to offset the environmental cost of production and, finally, improve enterprises' market competitiveness. The conclusion of this article further confirms this theoretical hypothesis. Therefore, at present, we cannot reduce the intensity of environmental regulations; instead, we should focus on the design and implementation of regulatory measures to drive enterprises' green technology innovation.

Nevertheless, this article has some limitations, which need to be improved in future research. Firstly, the data used in this article were cross-sectional data, which cannot capture the influence of time changes. Secondly, the model used in this article is the ROLS; although enough factors are controlled, the endogeneity still may not be ignored. Finally, the measurement variables of green technology innovation need to be further unified and standardized.

Author Contributions: All authors contributed to the study conception and design. Material preparation, data collection, and analysis were performed by M.W., Y.L., J.L. and Z.W. The first draft of the manuscript was written by M.W., and all authors commented on previous versions of the manuscript. All authors have read and agreed to the published version of the manuscript.

Funding: This work was supported by the China Postdoctoral Science Foundation (2020M670470).

Institutional Review Board Statement: Not applicable.

Informed Consent Statement: Informed consent was obtained from all subjects involved in the study.

Data Availability Statement: The datasets used and/or analyzed during the current study are available from the corresponding author on reasonable request.

Conflicts of Interest: The authors have no relevant financial or non-financial interests to disclose.

References

1. Geng, G.; Zheng, Y.; Zhang, Q.; Xue, T.; Zhao, H.; Tong, D.; Zheng, B.; Li, M.; Liu, F.; Hong, C.; et al. Drivers of PM2.5 air pollution deaths in China 2002–2017. *Nat. Geosci.* **2021**, *14*, 645–650. [CrossRef]
2. Zhang, Q.; He, K.; Hong, H. Cleaning China's air. *Nature* **2012**, *484*, 161–162. [CrossRef] [PubMed]
3. Wang, M.; Li, Y.; Li, M.; Wan, L.; Miao, L.; Wang, X. A comparative study on recycling amount and rate of used products under different regulatory scenarios. *J. Clean. Prod.* **2019**, *235*, 1153–1169. [CrossRef]

4. Zhan, Y.; Tan, K.H.; Ji, G.; Chung, L.; Chiu, A.S.F. Green and lean sustainable development path in China: Guanxi, practices and performance. *Resour. Conserv. Recycl.* **2018**, *128*, 240–249. [CrossRef]
5. Zhang, J.; Zhang, N.; Bai, S. Assessing the carbon emission changing for sustainability and high-quality economic development. *Environ. Technol. Innov.* **2021**, *22*, 101464. [CrossRef]
6. Lin, B.; Zhu, J. The role of renewable energy technological innovation on climate change: Empirical evidence from China. *Sci. Total Environ.* **2019**, *659*, 1505–1512. [CrossRef]
7. Wang, M.Y.; Li, Y.; Cheng, Z.; Zhong, C.; Ma, W. Evolution and equilibrium of a green technological innovation system: Simulation of a tripartite game model. *J. Clean. Prod.* **2021**, *278*, 123944. [CrossRef]
8. Wang, Q.; Qu, J.; Wang, B.; Wang, P.; Yang, T. Green technology innovation development in China in 1990–2015. *Sci. Total Environ.* **2019**, *696*, 134008. [CrossRef]
9. Wang, M.; Li, Y.; Li, J.; Wang, Z. Green process innovation, green product innovation and its economic performance improvement paths: A survey and structural model. *J. Environ. Manag.* **2021**, *297*, 113282. [CrossRef]
10. Wang, Z.; Wang, Q.; Chen, B.; Wang, Y. Evolutionary game analysis on behavioral strategies of multiple stakeholders in E-waste recycling industry. *Resour. Conserv. Recycl.* **2020**, *155*, 104618. [CrossRef]
11. Da Silva Rocha, A.B.; Salomão, G.M. Environmental policy regulation and corporate compliance in evolutionary game models with well-mixed and structured populations. *Eur. J. Oper. Res.* **2019**, *279*, 486–501. [CrossRef]
12. Dewick, P.; Maytorena-Sanchez, E.; Winch, G. Regulation and regenerative eco-innovation: The case of extracted materials in the UK. *Ecol. Econ.* **2019**, *160*, 38–51. [CrossRef]
13. Harford, J.D. Firm behavior under imperfectly enforceable pollution standards and taxes. *J. Environ. Econ. Manag.* **1978**, *5*, 26–43. [CrossRef]
14. Horbach, J. Determinants of environmental innovation—New evidence from German panel data sources. *Res. Policy* **2008**, *37*, 163–173. [CrossRef]
15. Fu, S.; Ma, Z.; Ni, B.; Peng, J.; Zhang, L.; Fu, Q. Research on the spatial differences of pollution-intensive industry transfer under the environmental regulation in China. *Ecol. Indic.* **2021**, *129*, 107921. [CrossRef]
16. Yang, Y.; Zhao, L.; Wang, C.; Xue, J. Towards more effective air pollution governance strategies in China: A systematic review of the literature. *J. Clean. Prod.* **2021**, *297*, 126724. [CrossRef]
17. Streimikiene, D. Review of internalization of externalities and dynamics of atmospheric emissions in energy sector of Baltic States. *Renew. Sustain. Energy Rev.* **2017**, *70*, 1131–1141. [CrossRef]
18. Nguyen, T.L.T.; Laratte, B.; Guillaume, B.; Hua, A. Quantifying environmental externalities with a view to internalizing them in the price of products, using different monetization models. *Resour. Conserv. Recycl.* **2016**, *109*, 13–23. [CrossRef]
19. Clarkson, P.M.; Li, Y.; Richardson, G.D. The Market Valuation of Environmental Capital Expenditures by Pulp and Paper Companies. *Account. Rev.* **2004**, *79*, 329–353. [CrossRef]
20. Brunel, C. Green innovation and green Imports:Links between environmental policies, innovation, and production. *J. Environ. Manag.* **2019**, *248*, 109290. [CrossRef]
21. Hazarika, N.; Zhang, X. Factors that drive and sustain eco-innovation in the construction industry: The case of Hong Kong. *J. Clean. Prod.* **2019**, *238*, 117816. [CrossRef]
22. Yusoff, Y.M.; Omar, M.K.; Kamarul Zaman, M.D.; Samad, S. Do all elements of green intellectual capital contribute toward business sustainability? Evidence from the Malaysian context using the Partial Least Squares method. *J. Clean. Prod.* **2019**, *234*, 626–637. [CrossRef]
23. Norberg-Bohm, V. Stimulating 'green' technological innovation: An analysis of alternative policy mechanisms. *Policy Sci.* **1999**, *32*, 13–38. [CrossRef]
24. Xu, L.; Zhang, Q.; Shi, X. Stakeholders strategies in poverty alleviation and clean energy access: A case study of China's PV poverty alleviation program. *Energy Policy* **2019**, *135*, 111011. [CrossRef]
25. Romer, P.M. Endogenous Technological Change. *J. Polit. Econ.* **1990**, *98*, S71–S102. [CrossRef]
26. Abbas, J.; Sağsan, M. Impact of knowledge management practices on green innovation and corporate sustainable development: A structural analysis. *J. Clean. Prod.* **2019**, *229*, 611–620. [CrossRef]
27. Luo, Q.; Miao, C.; Sun, L.; Meng, X.; Duan, M. Efficiency evaluation of green technology innovation of China's strategic emerging industries: An empirical analysis based on Malmquist-data envelopment analysis index. *J. Clean. Prod.* **2019**, *238*, 117782. [CrossRef]
28. Sun, H.; Edziah, B.K.; Sun, C.; Kporsu, A.K. Institutional quality, green innovation and energy efficiency. *Energy Policy* **2019**, *135*, 111002. [CrossRef]
29. Tariq, A.; Badir, Y.F.; Tariq, W.; Bhutta, U.S. Drivers and consequences of green product and process innovation: A systematic review, conceptual framework, and future outlook. *Technol. Soc.* **2017**, *51*, 8–23. [CrossRef]
30. Roper, S.; Tapinos, E. Taking risks in the face of uncertainty: An exploratory analysis of green innovation. *Technol. Forecast. Soc. Chang.* **2016**, *112*, 357–363. [CrossRef]
31. Wu, D.; Kefan, X.; Hua, L.; Shi, Z.; Olson, D.L. Modeling technological innovation risks of an entrepreneurial team using system dynamics: An agent-based perspective. *Technol. Forecast. Soc. Chang.* **2010**, *77*, 857–869. [CrossRef]
32. Brav, A.; Jiang, W.; Ma, S.; Tian, X. How does hedge fund activism reshape corporate innovation? *J. Financ. Econ.* **2018**, *130*, 237–264. [CrossRef]

33. Fernández Fernández, Y.; Fernández López, M.A.; Olmedillas Blanco, B. Innovation for sustainability: The impact of R&D spending on CO2 emissions. *J. Clean. Prod.* **2018**, *172*, 3459–3467. [CrossRef]
34. Zhao, S.; Jiang, Y.; Wang, S. Innovation stages, knowledge spillover, and green economy development: Moderating role of absorptive capacity and environmental regulation. *Environ. Sci. Pollut. Res.* **2019**, *26*, 25312–25325. [CrossRef]
35. Marino, M.; Parrotta, P.; Valletta, G. Electricity (de)regulation and innovation. *Res. Policy* **2019**, *48*, 748–758. [CrossRef]
36. Sgroi, F. Territorial development models: A new strategic vision to analyze the relationship between the environment, public goods and geographical indications. *Sci. Total Environ.* **2021**, *787*, 147585. [CrossRef]
37. Cohen, F.; Glachant, M.; Söderberg, M. The impact of energy prices on product innovation: Evidence from the UK refrigerator market. *Energy Econ.* **2017**, *68*, 81–88. [CrossRef]
38. Wang, M.; Cheng, Z.; Li, Y.; Li, J.; Guan, K. Impact of market regulation on economic and environmental performance: A game model of endogenous green technological innovation. *J. Clean. Prod.* **2020**, *277*, 123969. [CrossRef]
39. Beria, P. The role of transport supply in the acceptability of pollution charge extension. The case f Milan: A rejoinder. *Transp. Res. Part A Policy Pract.* **2020**, *133*, 381–382. [CrossRef]
40. Petroni, G.; Bigliardi, B.; Galati, F. Rethinking the Porter Hypothesis: The Underappreciated Importance of Value Appropriation and Pollution Intensity. *Rev. Policy Res.* **2018**, *36*, 121–140. [CrossRef]
41. Saunila, M.; Rantala, T.; Ukko, J.; Havukainen, J. Why invest in green technologies? Sustainability engagement among small businesses. *Technol. Anal. Strat. Manag.* **2019**, *31*, 653–666. [CrossRef]
42. Wagner, M. On the relationship between environmental management, environmental innovation and patenting: Evidence from German manufacturing firms. *Res. Policy* **2007**, *36*, 1587–1602. [CrossRef]
43. Yang, G.; Zha, D.; Wang, X.; Chen, Q. Exploring the nonlinear association between environmental regulation and carbon intensity in China: The mediating effect of green technology. *Ecol. Indic.* **2020**, *114*, 106309. [CrossRef]
44. Berrone, P.; Fosfuri, A.; Gelabert, L.; Gomez-Mejia, L. Necessity as the mother of "green" inventions: Institutional pressures and environmental innovations. *Strateg. Manag. J.* **2013**, *34*, 891–909. [CrossRef]
45. Qi, G.; Zeng, S.; Tam, C.; Yin, H.; Zou, H. Stakeholders' influences on corporate green innovation strategy: A case study of manufacturing firms in China. *Corp. Soc. Responsib. Environ. Manag.* **2013**, *20*, 1–14. [CrossRef]
46. Henriques, I.; Sadorsky, P. The Determinants of an Environmentally Responsive Firm: An Empirical Approach. *J. Environ. Econ. Manag.* **1996**, *30*, 381–395. [CrossRef]
47. Pansera, M.; Owen, R. Framing inclusive innovation within the discourse of development: Insights from case studies in India. *Res. Policy* **2018**, *47*, 23–34. [CrossRef]
48. Grossman, G. Growth, Trade, and Inequality. *Econometrica* **2018**, *86*, 37–83. [CrossRef]
49. Ambec, S.; Barla, P. A theoretical foundation of the Porter hypothesis. *Econ. Lett.* **2002**, *75*, 355–360. [CrossRef]
50. Sellitto, M.A.; Camfield, C.G.; Buzuku, S. Green innovation and competitive advantages in a furniture industrial cluster: A survey and structural model. *Sustain. Prod. Consum.* **2020**, *23*, 94–104. [CrossRef]
51. Tu, Y.; Wu, W. How does green innovation improve enterprises' competitive advantage? The role of organizational learning. *Sustain. Prod. Consum.* **2021**, *26*, 504–516. [CrossRef]
52. Porter, M.E. America's green strategy. *Sci. Am.* **1991**, *264*, 168. [CrossRef]
53. Brandt, L.; Van Biesebroeck, J.; Zhang, Y. Creative accounting or creative destruction? Firm-level productivity growth in Chinese manufacturing. *J. Dev. Econ.* **2012**, *97*, 339–351. [CrossRef]
54. Zhang, H.; Zhang, J.; He, Y. Are There Inter-enterprises Total Factor Productivity Spillover? An Investigation Based on China's Industrial Enterprises Data. *Nankai Econ. Stud.* **2018**, *202*, 102–119. [CrossRef]
55. Lu, J.; Tao, Z. Trends and determinants of China's industrial agglomeration. *J. Urban Econ.* **2009**, *65*, 167–180. [CrossRef]
56. Chintrakarn, P. Environmental regulation and U.S. states' technical inefficiency. *Econ. Lett.* **2008**, *100*, 363–365. [CrossRef]
57. Sanchez-Vargas, A.; Mansilla-Sanchez, R.; Aguilar-Ibarra, A. An Empirical Analysis of the Nonlinear Relationship Between Environmental Regulation and Manufacturing Productivity. *J. Appl. Econ.* **2013**, *16*, 357–372. [CrossRef]
58. Lanoie, P.; Patry, M.; Lajeunesse, R. Environmental regulation and productivity: Testing the porter hypothesis. *J. Product. Anal.* **2008**, *30*, 121–128. [CrossRef]

energies MDPI

Article

Industrialization and CO_2 Emissions in Sub-Saharan Africa: The Mitigating Role of Renewable Electricity

Urszula Mentel [1,*], Elżbieta Wolanin [2], Mansur Eshov [3] and Raufhon Salahodjaev [3,*]

[1] Department of Projects Management and Security Policy, Faculty of Management, Rzeszow University of Technology, 35-959 Rzeszow, Poland
[2] Department of Quantitative Methods, Faculty of Management, Rzeszow University of Technology, 35-959 Rzeszow, Poland; e.wolanin@prz.edu.pl
[3] Department of Academic Affairs, Tashkent State University of Economics, Islam Karimov Street 49, Tashkent 100066, Uzbekistan; m.eshov@tsue.uz
* Correspondence: u.mentel@prz.edu.pl (U.M.); r.salahodjaev@tsue.com or salahodjaev@gmail.com (R.S.)

Abstract: This study aims to explore the relationship between industry value added, renewable energy, and CO_2 emissions in a sample of 44 Sub-Saharan African countries over the period 2000–2015. This study makes several important contributions to extant research. While existing research was focused on the renewable energy-CO_2 emissions nexus, the current study assesses the moderating role of the renewables sector in the industrialization-CO2 emissions relationship. In addition, this study considers whether EKC relationships will hold after accounting for structural transformations (including industrial contributions to GDPs). Moreover, we are revising the existence of the EKC framework for the Sub-Saharan African countries. Using a two-step system GMM estimator, we found that the share of industry in GDP has a significant positive impact on CO_2 emissions, while renewable electricity output reduces CO_2 emissions. If causal, a one percentage point increase in renewable electricity output reduces carbon emissions by 0.22%. Moreover, the renewable energy sector then mediates the positive effect of industry value added on CO_2 emissions. We also find evidence for the statistical significance of the inverted U-shaped relationship between GDP per capita and CO_2 emissions.

Keywords: industry; renewable energy; CO_2 emissions; Sub-Saharan Africa

Citation: Mentel, U.; Wolanin, E.; Eshov, M.; Salahodjaev, R. Industrialization and CO_2 Emissions in Sub-Saharan Africa: The Mitigating Role of Renewable Electricity. *Energies* **2022**, *15*, 946. https://doi.org/10.3390/en15030946

Academic Editor: Štefan Bojnec

Received: 13 December 2021
Accepted: 23 January 2022
Published: 27 January 2022

1. Introduction

Research on the causes of CO_2 emission has proliferated in recent years [1–5]. One of the most important frameworks explored in this context was the existence of a non-linear (inverted U-shaped) relationship between GDP per capita and CO_2 emissions across countries, the so-called environmental Kuznets curve (EKC) phenomena. For example, the EKC framework was explored for Malaysia [6], China [7], Croatia [8], Turkey [9], Algeria [10], and Sub-Saharan Africa [11]. At the same time, another strand of studies suggested that economic growth, urbanization, trade, and renewable energy use are also important predictors of CO2 emissions across countries [12–14]. These studies have relied on the STRIPAT econometric framework [15,16].

While the global level of renewable energy consumption has been relatively stable over the past decade, Sub-Saharan African countries are among the top performers using renewables. At the same time, Figure 1 suggests significant differences in the levels of CO_2 emissions in this region, ranging from 0.04 tCO2 per capita in the Democratic Republic of Congo to 8.15 tCO_2 in South Africa. Therefore, the goal of this study is to explore the relationship between renewable energy use and CO_2 emissions in 44 Sub-Saharan Africa countries over the period 2000–2015. Our results make several important contributions to extant research. First, while existing research focused on the renewable energy-CO_2 emissions nexus, the current study assesses the moderating role of the renewables sector in the industrialization-CO_2 emissions relationship. Second, this study considers whether the EKC relationship holds after accounting for structural transformations (including industrial contribution to GDP). Third, in this study, we relied on a two-step system generalized method of moment (GMM) to explore the impact of renewable electricity output and industrialization on CO_2 emissions. Fourth, we suggest possible revisions to existing EKC frameworks in Sub-Saharan African countries.

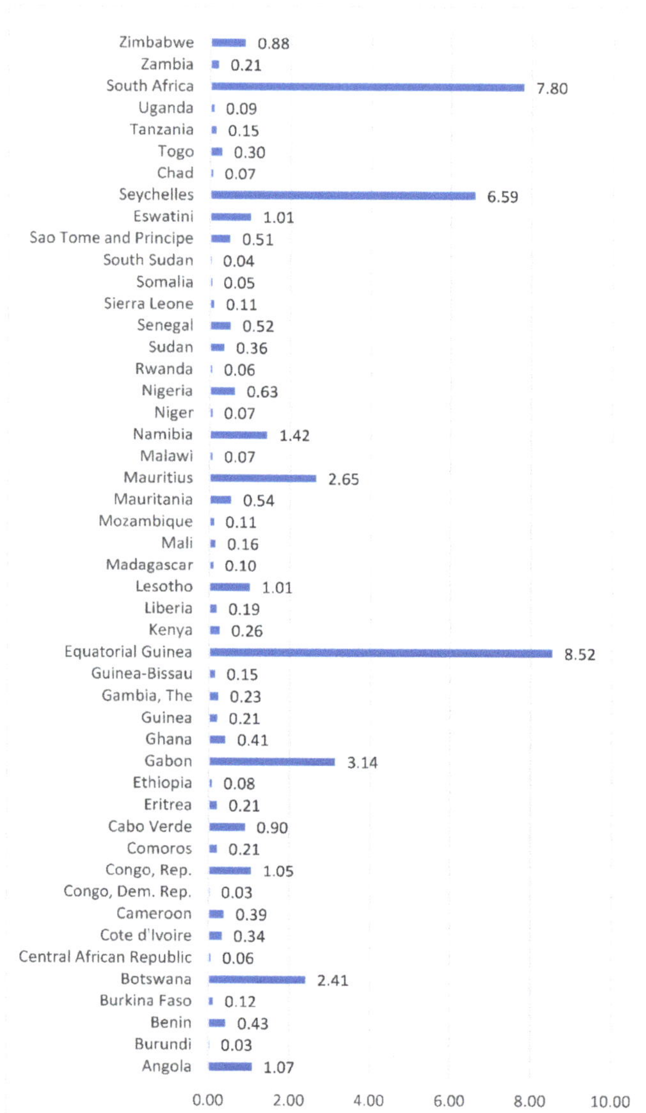

Figure 1. CO_2 emissions per person in Sub-Saharan Africa, 2000–2015.

The rest of the study is structured as follows: Section 2 reviews related literature, Section 3 presents data and methodology, Section 4 provides the empirical results, and Section 5 concludes the study.

2. Review of Related Literature

The role of renewable energy in explaining CO_2 emissions has been investigated in the context of the Environmental Kuznets Curve framework to assess the influence on GDP. For example, Zoundi [17], using the cointegration method for a sample of 25 African countries over the period 1980–2012, found that GDP increased CO_2 emissions, while renewable energy reduced air pollution in the long run. In a similar vein, Shafiei and Salim [18], analyzing data for the OECD countries for the years 1980–2011, documented that using renewable energy reduced CO2 emissions. Moreover, GDP per capita also had a positive impact on CO_2 emissions. Dogan and Seker [19] used the EKC theory to model the relationship between renewables and CO_2 emissions in the European Union from 1980–2012. Their dynamic ordinary least squares estimator results showed that trade openness and renewable energy reduced CO_2 emissions. Moreover, the Dumitrescu–Hurlin non-causality tests show a bi-directional relationship between renewable energy and CO2 emissions.

Saidi and Omri [20] further revisit the link between renewable energy and CO_2 emissions in a sample of 15 major energy-consuming countries. Results from the Granger causality test show the presence of bi-directional causality between renewable energy and CO_2 emissions in the long run and the absence of causality in the short run. Salahuddin et al. [21] found that renewable energy decreased CO_2 emissions and increased aggregate national savings in a sample of 34 Sub-Saharan Africa countries over the period 1984–2016. Sadorsky [22] also explored the relationship between renewable energy consumption and CO_2 emissions in a sample of G7 countries. The study used the panel cointegration method to find that causality runs from GDP per capita and CO_2 emission to renewable energy consumption in the long run. Therefore, renewable energy is not an instrumental variable to curb emissions in G7 countries.

Sebri and Ben-Salha [23] also did not report a significant causal influence of renewables on CO_2 emissions in BRICS over 1971–2010, using an ARDL estimator. The study found that economic growth and renewable energy are interrelated. Baloch et al. [24] also explored the relationship between renewable energy, GDP growth, and CO_2 emissions in BRICS over 1990–2015, using an augmented mean group estimator. In contrast, the study found that renewable energy use led to decreased CO_2 emissions for all BRICS countries except South Africa. Tiwari [25] explored the relationship between economic growth, renewable energy consumption, and carbon emissions in India from 1960–2009 using the vector autoregression method. The findings show, an impulse leading to a rise in renewable energy use will also increase economic growth and reduce CO_2 emissions.

Moreover, economic growth has led to a rise in air pollution. Boontome et al. [26] assessed the relationship between renewable energy use, economic growth, and carbon emissions in Thailand from 1971 to 2013. The panel cointegration results suggest that non-renewable energy use and GDP growth increase CO_2 emissions. The authors suggested that shifting to green energy sources will decrease environmental degradation without hampering economic growth prospects. Dong et al. [27] also assessed the relationship between renewable energy, GDP growth, and CO_2 emissions in a sample of 128 nations for the years 1990–2014 using the common correlated effects mean group method. The results suggested that renewable energy was instrumental in reducing CO_2 emissions across each geographic region. The observed effects were strongest in South America and Eurasia.

Mahmoodi [28] revisited the renewable energy-CO_2 emissions nexus for a sample of eleven developing countries over the period 2000–2014. Using panel cointegration estimation and VECM models, the study found bidirectional causality between renewable energy consumption and carbon emissions. Moreover, the alternative estimation methods demonstrated that renewables decrease emissions in general.

Abbasi et al. [29] explored the role of renewable energy within the framework of decreasing CO_2 emissions in Thailand by 25% by 2030. The ARDL simulation model for the years 1980–2018 showed that depletion of fossil fuels increased CO_2 emissions, while renewable energy consumption negatively affected CO_2 emissions in the short run. The authors highlighted a need for rapid energy sector transformation towards green energy consumption to achieve carbon mitigation targets.

Jebli and Youssef [30] assessed the links between renewable energy and CO2 emissions in North Africa over the period 1980–2011. The long-run estimates show a unidirectional causality from renewable energy to CO_2 emissions. In a similar vein, but for Pakistan, Waheed et al. [31], using ARDL estimator, find that greater renewable energy consumption leads to a decrease in carbon emissions.

Bhattacharya et al. [32] explored the role of renewable energy in reducing CO_2 emissions in 85 countries over the period 1991–2012. The study used a GMM estimator to find that rapid deployment of renewable energy technologies should lead to a decline in CO2 emissions. Nathaniel and Iheonu [33] also explored the effect of renewable and non-renewable demands on CO_2 emissions in a sample of 19 countries in Africa for the period 1990–2014 using the AMG method. The results showed that renewable energy use had no significant impact on environmental degradation while fossil fuel consumption led to a rise in CO_2 emissions.

While energy is considered one of the most important predictors of CO_2 emissions, industrialization is another factor of environmental degradation that has received attention in empirical literature [34]. For example, consider BRI countries such as China: "despite the economic benefits accrued from rapid industrialization, [China] has strained resource sources as labor, materials, and investment, and has incurred significant environmental degradation" [35] (p. 178). Li and Lin [36] argue that at earlier stages of economic development, industrialization was associated with greater energy demand and altered energy consumption models, increasing CO_2 emissions. The negative impact of industrialization on CO_2 emissions may be offset by the efficient use of infrastructure and agglomeration. However, many other factors should be considered when exploring the industrialization and CO_2 emissions nexus. For example, industrialization has led to urbanization and greater trade openness, which has also affected CO_2 emissions [37].

Other studies have explored the direct effect of industrialization on CO_2 emissions. For example, Shahbaz et al. [38] explored the relationship between industrialization, energy use, and CO_2 emissions in Bangladesh over the period 1975–2010. Using the ARDL bounds testing approach, the study found that energy use increased environmental degradation, while there was a non-linear, inverted U-shaped relationship between industrialization and CO_2 emissions. Ullah et al. [39] examined the relationship between industrialization and CO_2 emission in Pakistan over the period 1980–2018 using the ARDL estimator. Results suggest that an increase in the share of industry contributing to GDP led to a rise in CO_2 emissions, both in the short- and long-run. In addition, the study confirmed that urbanization and economic growth exerted a positive effect on environmental degradation. Mahmood et al. [40] further relied on the ARDL model to explore the industrialization-CO_2 emissions nexus in Saudi Arabia over the period 1968–2014. The results show that industrialization has had a significantly positive impact on environmental degradation (CO_2 emissions). The authors have suggested that it is important to enact more stringent industrial policies to reduce CO_2 emissions. Other studies also confirmed the significant effect of industrialization on CO_2 emissions in Korea, China, and the UAE [41–43].

Based on the abovementioned discussion we formulate the following hypothesis:

Hypothesis 1 (H1). *Industrialization leads to a rise in CO_2 emissions in Sub-Saharan Africa.*

Hypothesis 2 (H2). *Renewable energy enhances environmental quality in Sub-Saharan Africa.*

Hypothesis 3 (H3). *Renewable energy sector development offsets the negative effects of industrialization on CO_2 emissions in Sub-Saharan Africa.*

3. Data and Methods

In order to reach the goals of this study following extant research, we specified CO_2 emissions as a function of economic development (GDP), trade openness (T), urbanization (U), industrialization (I), and renewable energy (R). Thus, the econometric model can be specified as:

$$CO_{2i,t} = \alpha_0 + \alpha_1 CO_{2i,t-1} + \alpha_2 GDP_{i,t} + \alpha_3 GDP^2_{i,t} + \alpha_4 T_{i,t} + \alpha_5 U_{i,t} + \alpha_6 I_{i,t} + \alpha_7 R_{i,t} + \varepsilon_{i,t} \quad (1)$$

where i is the country, t denotes time (year), $\alpha_1 \ldots \ldots \ldots \alpha_7$ are parameters to be calculated, and ε is an error term. We also include the GDP per capita squared term to account for the EKC hypothesis in Sub-Saharan Africa [44,45]. Equation (1) is an estimated two-step system generalized method of moments (GMM) estimator. The two-step GMM estimator is used when (1) the number of panels (countries) is above the number of time periods (in years); (2) the empirical model includes lagged dependent variables; and (3) it is important to account for the problem of endogeneity and simultaneity. For example, if the inclusion of lagged CO_2 emissions leads to an emergence of this issue. For these reasons, many studies use the two-step system GMM to model the drivers of CO_2 emissions across countries [46–50].

Our data spanned the years 2000–2015 and covered 44 Sub-Saharan African countries. CO_2 emissions were measured as tCo2 emissions per person (Figure 1). GDP per capita was measured in constant international USD. As a proxy for FDI, we used net FDI inflows as percentage of GDP. Trade was the sum of exports and imports relative to GDP. Urbanization was the share of the urban population. Renewable energy was proxied by renewable electricity output as percentage of total electricity output, while industrialization was industry (including construction) value added as percentage of GDP. The descriptive statistics are presented in Table 1. The correlation matrix is reported in Table 2.

Table 1. Summary statistics.

Variable	Description	Mean	Std. Dev.	Min	Max
CO_2	tCO_2 emissions per person Source: Global Carbon Atlas	0.96	1.92	0.02	10.49
Industry	Industry (including construction), value added (percentage of GDP) Source: World Bank	25.05	13.87	2.07	84.35
Renewable energy	Renewable electricity output (percentage of total electricity output) Source: World Bank	43.09	37.82	0.00	100.00
GDP	GDP per capita, PPP (constant 2017 international USD) Source: World Bank	4.71	6.08	0.63	41.25
Trade	Trade as percentage of GDP Source: World Bank	73.39	38.86	19.10	311.35
Urbanization	Urbanization rate, percentage Source: World Bank	38.15	15.91	8.25	88.12

Table 2. Correlation matrix.

	CO_2	Industry	Renewable Energy	GDP	Urbanization	Trade
CO_2	1					
Industry	0.4476	1				
Renewable Energy	−0.3279	0.1454	1			
GDP	0.7983	0.5035	−0.2719	1		
Trade	0.5165	0.3366	−0.1027	0.4666	1	
Urbanization	0.6846	0.5121	−0.2625	0.5717	0.4076	1

Table 2 shows that correlations between main variables do not exceed 0.8; thus, multicollinearity should not be a problem in our study. The correlations matrix also shows that industry, GDP, trade openness, and urbanization are positively correlated with CO_2 emissions, while renewable energy has a negative correlation coefficient with CO_2 emissions. Figures 2 and 3 provide the visual associations between industry, renewable energy, and CO_2 emissions.

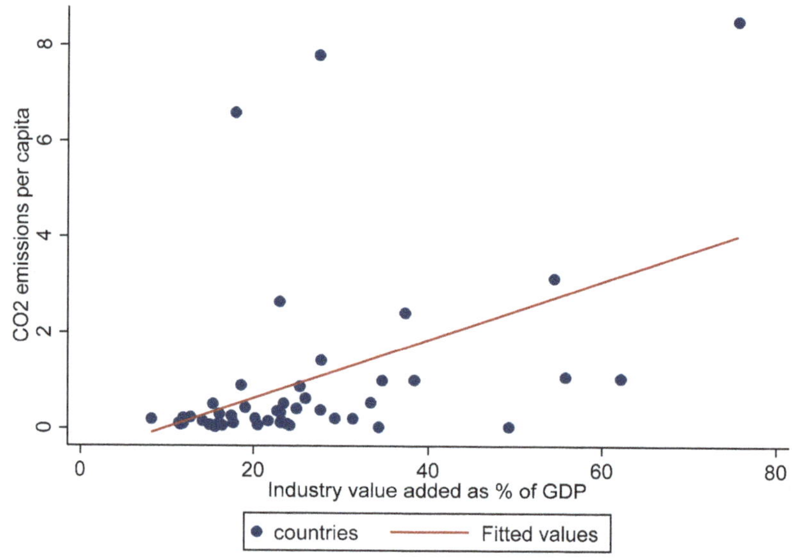

Figure 2. CO_2 emissions and industrialization, 2000–2015.

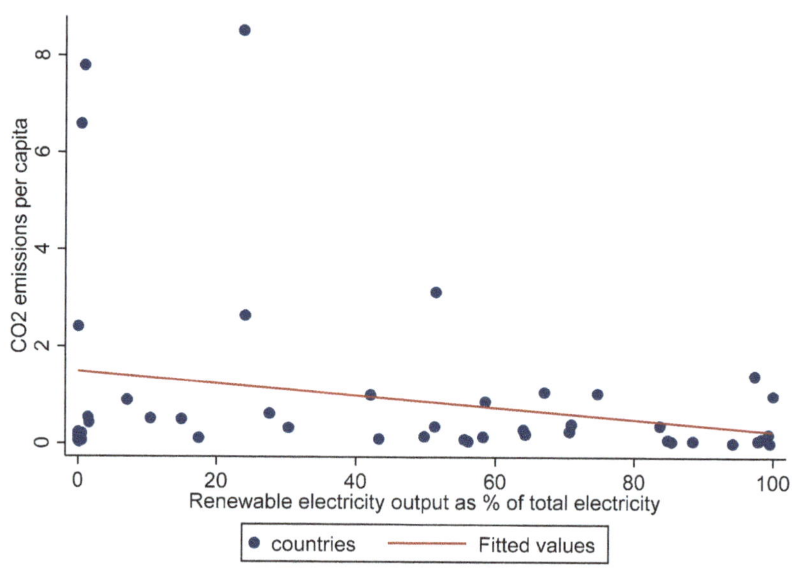

Figure 3. CO_2 emissions and renewable energy output, 2000–2015.

4. Results

The main results are reported in Table 3. Column 1 estimates the relationship between industry, control variables, and CO_2 emissions. First, we found that there was a positive relationship between industrialization and CO_2 emissions in Sub-Saharan Africa: a 1 percentage point increase in the share of industry in GDP led to a 0.3% increase in CO_2 emissions per person. We also documented an inverted U-shaped link between GDPs per capita and CO_2 emissions, confirming the statistical presence of the EKC in our sample with a turning point at approximately international constant USD 27,000. However, Figure 4 shows that only the GDP per capita of Equatorial Guinea was above the turning point in 2015. Therefore, the EKC did not have an economic implication in our study [30,45] we also failed to discover the EKC for African countries. Turning to other variables, we found that trade openness had a positive impact on CO_2 emissions in Sub-Saharan African countries. For example, a one percentage point increase in trade led to a 0.11% increase in CO_2 emissions. These results are in line with existing cross-country research [51]. Moreover, the positive effect of trade openness on environmental degradation was also documented [44]. Our results imply that trade liberalization has not improved the region's environmental conditions, suggesting that trade structure should change from energy-intensive products to knowledge-intensive goods and services. Indeed, Ncanywa et al. [52] found that the economic complexity of products produced in Sub-Saharan Africa is low, and this has had a negative impact on trade diversification in the region. Urbanization is insignificantly related to CO_2 emissions.

Table 3. Main results.

	I	II	III
$CO_{2_{t-1}}$	0.864609	0.856769	0.877561
	(38.08) ***	(33.13) ***	(45.63) ***
GDP	0.002957	0.003745	0.003537
	(3.59) ***	(4.05) ***	(4.68) ***
GDP squared	0.052024	0.050493	0.043806
	(4.76) ***	(3.82) ***	(5.70) ***
Trade	−0.096047	−0.099811	−0.101737
	(5.03) ***	(4.17) ***	(6.33) ***
Urbanization	0.001055	0.000886	0.001142
	(7.34) ***	(4.59) ***	(8.50) ***
Industry	−0.000016	0.000750	0.001124
	(0.01)	(0.38)	(0.74)
Renewable		−0.002203	0.000019
		(4.84) ***	(0.03)
Renewable * Industry			−0.000036
			(2.80) ***
Constant	−0.472627	−0.403860	−0.484623
	(5.14) ***	(3.55) ***	(5.68) ***
AR(1)	0.000	0.000	0.000
AR(2)	0.325	0.297	0.348
Hansen *p*-value	0.231	0.165	0.367
F-stat	51,985.22	407,952.02	794,403.81
N	628	628	628

* $p < 0.1$; *** $p < 0.01$.

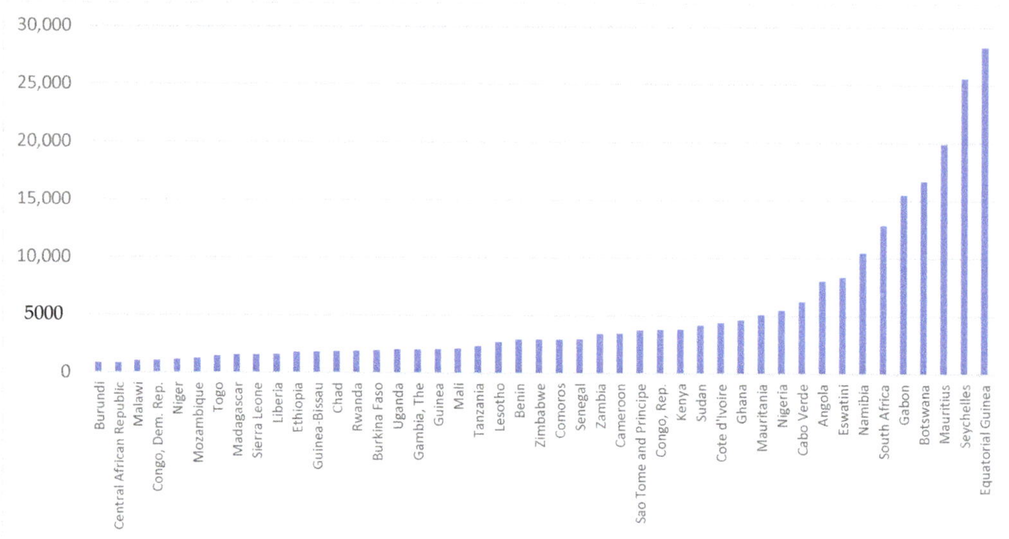

Figure 4. GDP per capita, 2000–2015.

In column 2, we included renewable electricity production. As expected, the coefficient for renewable energy was negative and significant at the 1% level. If causal, a one percentage point increase in renewable electricity output reduces carbon emissions by 0.22%. These results align with existing cross-country evidence [18] highlighting the importance of switching from fossil fuel energy to renewable energy consumption. We further include an interaction term between industry and renewable energy in Column 3. The interaction term is negative and significant, suggesting that the renewable energy sector is important to offset the negative effects of industrialization on CO_2 emissions. The coefficients in columns 1–3 change as we include additional variables and an interaction term between renewable energy and industrialization. The AR (2) and Hansen *p*-values confirm that our instruments are valid and reliable. The F-statistics exceed the threshold value of 10 confirming that overall; the econometric specification is significant in our analysis.

We also assess the robustness of our results by considering the role of non-economic control variables in Table 4. Extant research shows that the quality of institutions, such as in anti-corruption policies, may significantly affect CO_2 emissions [53]. Therefore, we include the corruption perceptions index (CPI) from Transparency International (Column 1). Additionally, empirical evidence shows that it is important to account for the human capital when modeling environmental indicators [54,55]. Therefore, we include the education index from the UN in Column 2. Finally, in Column 3, we include the proportion of women in parliament to capture the effect of female political empowerment on environmental degradation [56]. Across all models, renewable energy mediates the effect of industrialization on CO_2 emissions. Therefore, the results confirm that industrialization and renewable energy play an important role in predicting CO_2 emissions in Sub-Saharan African countries.

Table 4. Additional controls.

	I	II	III
$CO_{2_{t-1}}$	0.847266	0.884571	0.858646
	(43.46) ***	(48.64) ***	(49.06) ***
Industry	0.002438	0.002772	0.003614
	(2.04) **	(2.28) **	(4.66) ***
Renewable	0.000283	0.000183	0.000762
	(0.39)	(0.31)	(1.37)
Renewable * Industry	−0.000051	−0.000039	−0.000049
	(2.39) **	(4.02) ***	(3.15) ***
GDP	0.051276	0.033394	0.050422
	(4.84) ***	(3.21) ***	(7.65) ***
GDP squared	−0.116404	−0.074215	−0.110335
	(4.86) ***	(3.04) ***	(7.10) ***
Trade	0.001144	0.001014	0.001154
	(5.60) ***	(6.75) ***	(6.88) ***
Urbanization	0.000859	0.000736	0.000391
	(0.50)	(0.44)	(0.28)
CPI	0.000108		
	(0.08)		
Education		0.247795	
		(0.83)	
Parliament			0.000753
			(0.95)
Constant	−0.526451	−0.511391	−0.538134
	(4.14) ***	(3.07) ***	(6.29) ***
AR(1)	0.000	0.000	0.000
AR(2)	0.969	0.362	0.357
Hansen *p*-value	0.339	0.430	0.218
F-stat	90,807.81	60,291.78	168,723.20
N	539	622	611

* $p < 0.1$; ** $p < 0.05$; *** $p < 0.01$.

5. Conclusions

This study aims to explore the relationship between industrialization, renewable energy, and CO_2 emissions in a sample of 44 Sub-Saharan Africa countries for the period 2000–2015. We relied on the two-step system GMM estimator for this aim, which accounts for endogeneity and omits variable bias. We depart from the EKC framework by incorporating the industry and renewable energy sectors. Our results suggest that industry value adds increased CO_2 emissions while renewable electricity output decreased environmental degradation. If causal, a one percentage point increase in renewable electricity output reduces carbon emissions by 0.22%. Moreover, we find that renewable energy use mediates the relationship between industry value adds and CO_2 emissions.

Our findings have several important policy implications. First, to promote the development of renewables, policymakers can offer low interest loans and tax cuts for purchasing and installing renewable energy generators. In addition, each country can adopt a local renewable energy deployment strategy that outlines the key vision of the government in this sector. Apart from that, the governments can adopt a policy where buildings with an area exceeding a certain threshold are required to replace some of the energy consumption with renewables. It is possible to use subsidies for biogas or hydro power producers in certain countries. Second, it is crucial to institute policies aimed at the promotion of renewable energy technologies across industries. This can be achieved by reducing tax rates for green energy technology adopters, offering low-interest loans and grants to companies and households, and subsidizing green energy.

Moreover, studies show that guaranteed prices act as a potential tool to promote the development of the renewable energy sector [57]. Third, we fail to find the economic presence of the EKC. This highlights that regional economic growth leads to environmental degradation.

Prospective studies can extend our results in many ways. It is essential to assess the role of other factors such as human capital, population, agriculture, or FDI in explaining CO_2 emissions in the Sub-Saharan Africa countries [58–62]. It is important to assess the factors associated with renewable energy adoption [63] and the role of renewable energy in economic growth in the region [64,65].

Author Contributions: Conceptualization, U.M., E.W. and M.E.; methodology, R.S.; software, R.S.; validation, U.M., M.E., E.W. and R.S.; formal analysis, R.S.; and writing—review and editing, R.S., E.W. and U.M. All authors have read and agreed to the published version of the manuscript.

Funding: This research received no external funding.

Institutional Review Board Statement: Not applicable.

Informed Consent Statement: Not applicable.

Data Availability Statement: The data are available from sources cited in the study.

Conflicts of Interest: The authors declare no conflict of interest.

References

1. Iwata, H.; Okada, K.; Samreth, S. Empirical study on the determinants of CO_2 emissions: Evidence from OECD countries. *Appl. Econ.* **2012**, *44*, 3513–3519. [CrossRef]
2. Solarin, S.A. Tourist arrivals and macroeconomic determinants of CO_2 emissions in Malaysia. *Anatolia* **2013**, *25*, 228–241. [CrossRef]
3. Mui-Yin, C.; Chin-Hong, P.; Teo, C.L.; Joseph, J. The determinants of CO_2 emissions in Malaysia: A new aspect. *Int. J. Energy Econ. Policy* **2018**, *8*, 190.
4. Sadikov, A.; Kasimova, N.; Isaeva, A.; Khachaturov, A.; Salahodjaev, R. Pollution, Energy and Growth: Evidence from Post-Communist Countries. *Int. J. Energy Econ. Policy* **2020**, *10*, 656–661. [CrossRef]
5. Chaabouni, S.; Saidi, K. The dynamic links between carbon dioxide (CO_2) emissions, health spending and GDP growth: A case study for 51 countries. *Environ. Res.* **2017**, *158*, 137–144. [CrossRef] [PubMed]
6. Saboori, B.; Sulaiman, J.; Mohd, S. Economic growth and CO_2 emissions in Malaysia: A cointegration analysis of the Environmental Kuznets Curve. *Energy Policy* **2012**, *51*, 184–191. [CrossRef]
7. Jalil, A.; Mahmud, S.F. Environment Kuznets curve for CO_2 emissions: A cointegration analysis for China. *Energy Policy* **2009**, *37*, 5167–5172. [CrossRef]
8. Ahmad, N.; Du, L.; Lu, J.; Wang, J.; Li, H.-Z.; Hashmi, M.Z. Modelling the CO_2 emissions and economic growth in Croatia: Is there any environmental Kuznets curve? *Energy* **2017**, *123*, 164–172. [CrossRef]
9. Pata, U.K. The influence of coal and noncarbohydrate energy consumption on CO_2 emissions: Revisiting the environmental Kuznets curve hypothesis for Turkey. *Energy* **2018**, *160*, 1115–1123. [CrossRef]
10. Lacheheb, M.; Rahim, A.S.A.; Sirag, A. Economic growth and CO_2 emissions: Investigating the environmental Kuznets curve hypothesis in Algeria. *Int. J. Energy Econ. Policy* **2015**, *5*, 1125–1132.
11. Bah, M.M.; Abdulwakil, M.M.; Azam, M. Income heterogeneity and the Environmental Kuznets Curve hypothesis in Sub-Saharan African countries. *GeoJournal* **2019**, *85*, 617–628. [CrossRef]
12. Shahbaz, M.; Hye, Q.M.A.; Tiwari, A.K.; Leitão, N.C. Economic growth, energy consumption, financial development, international trade and CO_2 emissions in Indonesia. *Renew. Sustain. Energy Rev.* **2013**, *25*, 109–121. [CrossRef]
13. Hossain, M.S. Panel estimation for CO_2 emissions, energy consumption, economic growth, trade openness and urbanization of newly industrialized countries. *Energy Policy* **2011**, *39*, 6991–6999. [CrossRef]
14. Mirzaei, M.; Bekri, M. Energy consumption and CO_2 emissions in Iran, 2025. *Environ. Res.* **2017**, *154*, 345–351. [CrossRef]
15. Zuo, Z.; Guo, H.; Cheng, J. An LSTM-STRIPAT model analysis of China's 2030 CO_2 emissions peak. *Carbon Manag.* **2020**, *11*, 577–592. [CrossRef]
16. Nugraha, A.T.; Osman, N.H. CO_2 emissions, economic growth, energy consumption, and household expenditure for Indonesia: Evidence from cointegration and vector error correction model. *Int. J. Energy Econ. Policy* **2019**, *9*, 291–298.
17. Zoundi, Z. CO_2 emissions, renewable energy and the Environmental Kuznets Curve, a panel cointegration approach. *Renew. Sustain. Energy Rev.* **2017**, *72*, 1067–1075. [CrossRef]
18. Shafiei, S.; Salim, R.A. Non-renewable and renewable energy consumption and CO_2 emissions in OECD countries: A comparative analysis. *Energy Policy* **2014**, *66*, 547–556. [CrossRef]
19. Dogan, E.; Seker, F. Determinants of CO_2 emissions in the European Union: The role of renewable and non-renewable energy. *Renew. Energy* **2016**, *94*, 429–439. [CrossRef]

20. Saidi, K.; Omri, A. The impact of renewable energy on carbon emissions and economic growth in 15 major renewable energy-consuming countries. *Environ. Res.* **2020**, *186*, 109567. [CrossRef]
21. Salahuddin, M.; Habib, A.; Al-Mulali, U.; Ozturk, I.; Marshall, M.; Ali, I. Renewable energy and environmental quality: A second-generation panel evidence from the Sub Saharan Africa (SSA) countries. *Environ. Res.* **2020**, *191*, 110094. [CrossRef] [PubMed]
22. Sadorsky, P. Renewable energy consumption, CO_2 emissions and oil prices in the G7 countries. *Energy Econ.* **2009**, *31*, 456–462. [CrossRef]
23. Sebri, M.; Ben-Salha, O. On the causal dynamics between economic growth, renewable energy consumption, CO_2 emissions and trade openness: Fresh evidence from BRICS countries. *Renew. Sustain. Energy Rev.* **2014**, *39*, 14–23. [CrossRef]
24. Danish; Baloch, M.A.; Mahmood, N.; Zhang, J.W. Effect of natural resources, renewable energy and economic development on CO_2 emissions in BRICS countries. *Sci. Total Environ.* **2019**, *678*, 632–638. [CrossRef] [PubMed]
25. Tiwari, A.K. A structural VAR analysis of renewable energy consumption, real GDP and CO_2 emissions: Evidence from India. *Econ. Bull.* **2011**, *31*, 1793–1806.
26. Boontome, P.; Therdyothin, A.; Chontanawat, J. Investigating the causal relationship between non-renewable and renewable energy consumption, CO_2 emissions and economic growth in Thailand. *Energy Procedia* **2017**, *138*, 925–930. [CrossRef]
27. Dong, K.; Hochman, G.; Zhang, Y.; Sun, R.; Li, H.; Liao, H. CO_2 emissions, economic and population growth, and renewable energy: Empirical evidence across regions. *Energy Econ.* **2018**, *75*, 180–192. [CrossRef]
28. Mahmoodi, M. The relationship between economic growth, renewable energy, and CO_2 emissions: Evidence from panel data approach. *Int. J. Energy Econ. Policy* **2017**, *7*, 96–102.
29. Abbasi, K.R.; Adedoyin, F.F.; Abbas, J.; Hussain, K. The impact of energy depletion and renewable energy on CO_2 emissions in Thailand: Fresh evidence from the novel dynamic ARDL simulation. *Renew. Energy* **2021**, *180*, 1439–1450. [CrossRef]
30. Jebli, M.B.; Youssef, S.B. The role of renewable energy and agriculture in reducing CO_2 emissions: Evidence for North Africa countries. *Ecol. Indic.* **2017**, *74*, 295–301. [CrossRef]
31. Waheed, R.; Chang, D.; Sarwar, S.; Chen, W. Forest, agriculture, renewable energy, and CO_2 emission. *J. Clean. Prod.* **2018**, *172*, 4231–4238. [CrossRef]
32. Bhattacharya, M.; Churchill, S.A.; Paramati, S.R. The dynamic impact of renewable energy and institutions on economic output and CO_2 emissions across regions. *Renew. Energy* **2017**, *111*, 157–167. [CrossRef]
33. Nathaniel, S.P.; Iheonu, C.O. Carbon dioxide abatement in Africa: The role of renewable and non-renewable energy consumption. *Sci. Total Environ.* **2019**, *679*, 337–345. [CrossRef] [PubMed]
34. Xu, B.; Lin, B. How industrialization and urbanization process impacts on CO_2 emissions in China: Evidence from nonparametric additive regression models. *Energy Econ.* **2015**, *48*, 188–202. [CrossRef]
35. Liu, X.; Bae, J. Urbanization and industrialization impact of CO_2 emissions in China. *J. Clean. Prod.* **2018**, *172*, 178–186. [CrossRef]
36. Li, K.; Lin, B. Impacts of urbanization and industrialization on energy consumption/CO_2 emissions: Does the level of development matter? *Renew. Sustain. Energy Rev.* **2015**, *52*, 1107–1122. [CrossRef]
37. Dong, F.; Wang, Y.; Su, B.; Hua, Y.; Zhang, Y. The process of peak CO_2 emissions in developed economies: A perspective of industrialization and urbanization. *Resour. Conserv. Recycl.* **2018**, *141*, 61–75. [CrossRef]
38. Shahbaz, M.; Uddin, G.S.; Rehman, I.U.; Imran, K. Industrialization, electricity consumption and CO_2 emissions in Bangladesh. *Renew. Sustain. Energy Rev.* **2014**, *31*, 575–586. [CrossRef]
39. Ullah, S.; Ozturk, I.; Usman, A.; Majeed, M.T.; Akhtar, P. On the asymmetric effects of premature deindustrialization on CO_2 emissions: Evidence from Pakistan. *Environ. Sci. Pollut. Res.* **2020**, *27*, 13692–13702. [CrossRef]
40. Mahmood, H.; Alkhateeb, T.T.Y.; Furqan, M. Industrialization, urbanization and CO_2 emissions in Saudi Arabia: Asymmetry analysis. *Energy Rep.* **2020**, *6*, 1553–1560. [CrossRef]
41. Li, T.; Li, Y.; An, D.; Han, Y.; Xu, S.; Lu, Z.; Crittenden, J. Mining of the association rules between industrialization level and air quality to inform high-quality development in China. *J. Environ. Manag.* **2019**, *246*, 564–574. [CrossRef] [PubMed]
42. Hong, S.; Lee, Y.; Yoon, S.J.; Lee, J.; Kang, S.; Won, E.-J.; Hur, J.; Khim, J.S.; Shin, K.-H. Carbon and nitrogen stable isotope signatures linked to anthropogenic toxic substances pollution in a highly industrialized area of South Korea. *Mar. Pollut. Bull.* **2019**, *144*, 152–159. [CrossRef] [PubMed]
43. Sbia, R.; Shahbaz, M.; Ozturk, I. Economic growth, financial development, urbanisation and electricity consumption nexus in UAE. *Econ. Res.-Ekon. Istraživanja* **2017**, *30*, 527–549. [CrossRef]
44. Ben Jebli, M.; Ben Youssef, S.; Ozturk, I. The Role of Renewable Energy Consumption and Trade: Environmental Kuznets Curve Analysis for Sub-Saharan Africa Countries. *Afr. Dev. Rev.* **2015**, *27*, 288–300. [CrossRef]
45. Lin, B.; Omoju, O.E.; Nwakeze, N.M.; Okonkwo, J.U.; Megbowon, E.T. Is the environmental Kuznets curve hypothesis a sound basis for environmental policy in Africa? *J. Clean. Prod.* **2016**, *133*, 712–724. [CrossRef]
46. Asongu, S.A.; Le Roux, S.; Biekpe, N. Enhancing ICT for environmental sustainability in sub-Saharan Africa. *Technol. Forecast. Soc. Chang.* **2018**, *127*, 209–216. [CrossRef]
47. Khan, H.U.R.; Nassani, A.A.; Aldakhil, A.M.; Abro, M.M.Q.; Islam, T.; Zaman, K. Pro-poor growth and sustainable development framework: Evidence from two step GMM estimator. *J. Clean. Prod.* **2018**, *206*, 767–784. [CrossRef]
48. Muhammad, B. Energy consumption, CO_2 emissions and economic growth in developed, emerging and Middle East and North Africa countries. *Energy* **2019**, *179*, 232–245. [CrossRef]

49. Muhammad, B.; Khan, S. Effect of bilateral FDI, energy consumption, CO_2 emission and capital on economic growth of Asia countries. *Energy Rep.* **2019**, *5*, 1305–1315. [CrossRef]
50. Muhammad, B.; Khan, S. Understanding the relationship between natural resources, renewable energy consumption, economic factors, globalization and CO_2 emissions in developed and developing countries. In *Natural Resources Forum*; Blackwell Publishing Ltd.: Oxford, UK, 2021; Volume 45, pp. 138–156.
51. Essandoh, O.K.; Islam, M.; Kakinaka, M. Linking international trade and foreign direct investment to CO_2 emissions: Any differences between developed and developing countries? *Sci. Total Environ.* **2020**, *712*, 136437. [CrossRef]
52. Ncanywa, T.; Mongale, I.P.; Ralarala, O.; Letsoalo, T.E.; Molele, B.S. Economic complexity to boost the selected sub-Saharan African economies. *J. Econ. Financ. Sci.* **2021**, *14*, 8. [CrossRef]
53. Wang, Z.; Danish; Zhang, B.; Wang, B. The moderating role of corruption between economic growth and CO_2 emissions: Evidence from BRICS economies. *Energy* **2018**, *148*, 506–513. [CrossRef]
54. Omanbayev, B.; Salahodjaev, R.; Lynn, R. Are greenhouse gas emissions and cognitive skills related? Cross-country evidence. *Environ. Res.* **2018**, *160*, 322–330. [CrossRef] [PubMed]
55. Salahodjaev, R. Does Intelligence Improve Environmental Sustainability? An Empirical Test. *Sustain. Dev.* **2015**, *24*, 32–40. [CrossRef]
56. Salahodjaev, R.; Jarilkapova, D. Female parliamentarism and genuine savings: A cross-country test. *Sustain. Dev.* **2019**, *27*, 637–646. [CrossRef]
57. González-Gómez, M. Estimating the long-run impact of guaranteed prices on wind and solar power in Germany. *Energy Sour. Part B Econ. Plan. Policy* **2017**, *12*, 692–698. [CrossRef]
58. Isaeva, A.; Salahodjaev, R.; Khachaturov, A.; Tosheva, S. The Impact of Tourism and Financial Development on Energy Consumption and Carbon Dioxide Emission: Evidence from Post-communist Countries. *J. Knowl. Econ.* **2021**, 1–14. [CrossRef]
59. Salahodjaev, R.; Isaeva, A. Post-Soviet states and CO_2 emissions: The role of foreign direct investment. *Post-Communist Econ.* **2021**, 1–22. [CrossRef]
60. Phiri, J.; Malec, K.; Kapuka, A.; Maitah, M.; Appiah-Kubi, S.N.K.; Gebeltová, Z.; Bowa, M.; Maitah, K. Impact of Agriculture and Energy on CO_2 Emissions in Zambia. *Energies* **2021**, *14*, 8339. [CrossRef]
61. Pachiyappan, D.; Ansari, Y.; Alam, M.S.; Thoudam, P.; Alagirisamy, K.; Manigandan, P. Short and Long-Run Causal Effects of CO_2 Emissions, Energy Use, GDP and Population Growth: Evidence from India Using the ARDL and VECM Approaches. *Energies* **2021**, *14*, 8333. [CrossRef]
62. Huang, Y.; Kuldasheva, Z.; Salahodjaev, R. Renewable Energy and CO_2 Emissions: Empirical Evidence from Major Energy-Consuming Countries. *Energies* **2021**, *14*, 7504. [CrossRef]
63. Eshchanov, B.; Abdurazzakova, D.; Yuldashev, O.; Salahodjaev, R.; Ahrorov, F.; Komilov, A.; Eshchanov, R. Is there a link between cognitive abilities and renewable energy adoption: Evidence from Uzbekistan using micro data. *Renew. Sustain. Energy Rev.* **2021**, *141*, 110819. [CrossRef]
64. Janpolat, K.; Odilova, S.; Nodira, A.; Salahodjaev, R.; Khachaturov, A. Financial Development and Energy Consumption Nexus in 32 Belt and Road Economies. *Int. J. Energy Econ. Policy* **2021**, *11*, 368–373. [CrossRef]
65. Umurzakov, U.; Mirzaev, B.; Salahodjaev, R.; Isaeva, A.; Tosheva, S. Energy Consumption and Economic Growth: Evidence from Post-Communist Countries. *Int. J. Energy Econ. Policy* **2020**, *10*, 59–65. [CrossRef]

energies MDPI

Article

An Analysis of the Use of Energy from Conventional Fossil Fuels and Green Renewable Energy in the Context of the European Union's Planned Energy Transformation

Katarzyna Chudy-Laskowska * and Tomasz Pisula

The Faculty of Management, Department of Quantitative Methods, Rzeszow University of Technology, 35-959 Rzeszow, Poland
* Correspondence: kacha877@prz.edu.pl; Tel.: +48-604077448

Abstract: Over the past few years, considerable emphasis has been put on decarbonization, which, in the context of the recent events in Europe, proves that mixing energy sources is the best strategy. This article discusses ways in which individual EU member states manage their energy source diversification, while comparing their levels of fossil fuels and renewable energy sources (RESs) usage. The research data was acquired from the Eurostat website and comprises of 15 indicators describing the use of energy both from conventional and renewable sources in the European Union, in 2019. The study employs taxonomical methods, such as ranking and cluster analysis. The authors put forward a hypothesis that EU member states approach the use of energy resources in several ways. There are countries which take advantage of both traditional and renewable sources (Netherlands, Germany, Austria, and Italy). However, there is a group of states that relies on a single energy source and exclusively uses either traditional (Poland) or renewable energy resources (Sweden, Finland). The analyses enabled the isolation of country clusters with similar activities and energy strategies.

Keywords: European Union; sustainable energy development; energy transformation; energy strategy; taxonomical analysis; ranking; TOPSIS method

Citation: Chudy-Laskowska, K.; Pisula, T. An Analysis of the Use of Energy from Conventional Fossil Fuels and Green Renewable Energy in the Context of the European Union's Planned Energy Transformation. *Energies* **2022**, *15*, 7369. https://doi.org/10.3390/en15197369

Academic Editors: Surender Reddy Salkuti and Behnam Zakeri

Received: 1 September 2022
Accepted: 4 October 2022
Published: 7 October 2022

1. Introduction

Renewable energy sources, including wind, solar, hydroelectric, ocean thermal, geothermal, biomass, and biofuels, constitute an alternative to fossil fuels and contribute to reductions in greenhouse gas emissions, energy source diversification, and the decreasing dependency on volatile and unstable fossil fuel markets, especially the oil and gas markets. The EU legislation concerning the promotion of renewable energy sources has significantly evolved over the past fifteen years. Moreover, the issue has, considering the recent war in Ukraine, become even more evident owing to certain EU member states' overwhelming reliance on Russian gas and coal supplies.

The European Green Deal sets out guidelines on how to make Europe the first climate-neutral continent by 2050 and provides the most comprehensive package of measures enabling Europe's inhabitants and businesses to benefit from a sustainable ecological transformation. The use of renewable sources based energy offers many potential rewards, including lower greenhouse gas emissions, diversified energy supplies, and the reduced dependency on fossil fuel markets (especially the oil and gas markets). The growth of renewable energy sources in the EU can also stimulate increases in employment by creating jobs in the green technology sector. The share of energy from renewable sources in the EU has nearly doubled from 2004 to 2018, rising from 9.6% in 2004 to 18.9% by 2018. EU member states were obliged to set national energy targets and draw up their 10-year national energy and climate plans (NECPs) as part of the "Horizon 2030" programme. In 2018, Sweden was the European leader in sourcing renewable energy, with more than half (54.6%) of its energy coming from renewable sources. Consequently, Sweden was markedly ahead of Finland (41.2%), Latvia (40.3%), Denmark (36.1%), and Austria (33.4%). At the bottom of the ranking were countries with the lowest share of renewable energy, namely the Netherlands (7.4%), Malta (8.0%), Luxembourg (9.1%), and Belgium (9.4%).

Analyses of consumption patterns of electricity generated from fossil fuels, renewables, and nuclear sources over the 16 years of the enlarged EU (2005–2020), sheds more light on the issue. Figure 1 presents the percentage shares of electricity consumption in EU member states in 2020, by source.

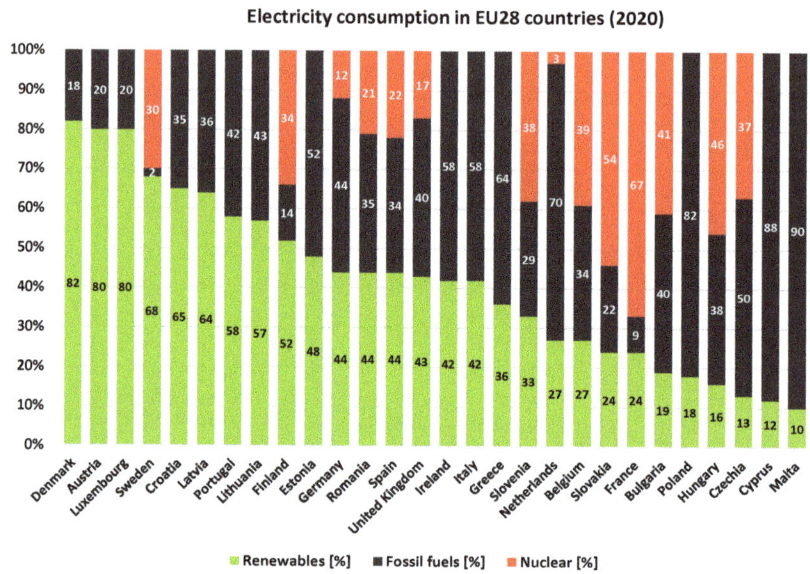

Figure 1. Electricity consumption in the EU countries from major energy sources in 2020. Source: https://ourworldindata.org/grapher/elecmizbar?time=2020&country=AUT~CYP~CZE~DNK~BGR~LUX~HRV~EST~FIN~FRA~DEU~GRC~HUN~ITA~LVA~LTU~GBR~SWE~SVK~SVN~POL~NLD~MLT~PRT~ESP~IRL~BEL~ROU (accessed on 2 July 2022).

While a clear trend for the majority of EU countries to grow their share of electricity consumed from renewable sources is noticeable, a distinctive drop in the share of electricity consumed from conventional fossil fuel sources and nuclear power plants has been witnessed, over the period.

In 2005, the mean annual percentage consumption of electricity from fossil sources (coal, oil, gas) in the EU countries was 61%, dropping to 42% in 2020. As many as 17 member states were, as of that time, still using 50% or more electricity from conventional fossil sources, of which 11 (Malta, Cyprus, Estonia, Poland, United Kingdom, Luxembourg, Ireland, the Netherlands, Greece, Italy, Portugal) consumed 75% or more non-renewable energy. The leaders, Malta and Cyprus, which obtained 100% of their electrical power from conventional sources, as well as Estonia and Poland, in which 99% and 98% of their energy use, respectively, was based on conventional fuels. In 2020, there were only eight countries with a 50% or higher share of electricity consumed from fossil sources, while barely three countries (Malta 90%, Cyprus 88%, Poland 82%) had a share of 75% or higher. The greatest decrease in the share of electricity consumption sourced from fossil sources over the 16-year period was reported in countries such as Luxembourg (by 74%, from 94% to 20%), Denmark (by 55%, from 73% to 18%), and Estonia and Portugal (by 47% and 41%, respectively). Two countries reported a rise in the share of electricity from such sources, namely Lithuania (a 19% increase from 24% in 2005 to 43% in 2020) and Latvia (a 6% increase from 30% to

36%). Other EU member states rolled back their share of fossil fuels in power generation by between 2% and 35%.

A similar downward trend was observed for the electricity generated from nuclear sources in the EU. The tendency reveals a gradual departure from electrical power from nuclear power plants for internal consumption needs. The mean annual percentage share of electricity consumed, in 2005, from nuclear sources in the EU was 21% (highest-ranking countries included France (79%), Lithuania, whose nuclear energy production was abandoned altogether in 2020, (73%), and Slovakia and Belgium (57%)). A distinctive decrease in the aforementioned share consumption is observable over the 16-year period in the vast majority of EU countries. The share percentage was, in 2020, only 16% (a 5% drop) for the EU member states. The share decline was observed in 10 countries that relied more on nuclear energy. The greatest rollbacks were reported in Lithuania (73%), Belgium (17%), Sweden (16%), Germany (15%), and France (12%). However, there were also five countries, who were beneficiaries of electricity from nuclear power plants, where their share had increased, namely Romania (11%), the Czech Republic and Hungary (7%), Spain (2%), and Finland (1%).

The EU has, for over a decade, been gradually moving away from conventional fossil fuel-based energy and nuclear power (in line with the Green Energy strategy), thus resulting in an increased share participation of renewables in the structure of electricity consumption from various sources. In 2005, the mean share of renewable sources-based electricity in the overall electricity consumption in EU countries, was merely 18%, while in 2020, it had grown to as much as 42% (an increase of 24%, i.e., the doubling of its share participation). In early 2005, there were only eight countries with the share of renewable sources-based electricity consumption that was 20% or higher, but with only four countries (Latvia, Austria, Croatia, and Sweden) where the share was 50% or more. The highest percentages were reported for Latvia (70%) and Austria (63%). By 2020, there were 22 countries in which the share of electricity generated from such sources was at least 20%, but as many as nine countries (Denmark, Austria, Luxembourg, Sweden, Croatia, Latvia, Portugal, Lithuania, and Finland) where the share was at least 50%. The share participation was exceptionally high (over 80%) in three countries, namely Denmark (82%), Austria, and Luxembourg (80%). An increase in the share of renewable sources-based electricity consumed can be observed in all countries except for Lithuania, where the share dropped by 6%, from 70% to 64%. The increase was considerably high—more than 50%—in such countries as Luxembourg (by 74%, from 6% to 80%), Denmark (by 55%, from 27% to 82%), and Lithuania (by 54%, from 3% to 57%). In nine other countries, Estonia, Portugal, the United Kingdom, Ireland, Germany, Spain, Italy, Greece, and Belgium, the share rose by about 20% to 50%. However, the expansion was much more limited despite being noticeable (by 9% to 20% in 15 countries. Member states, in which the share of renewable sources-based electricity was still very low (less than 20%) included Malta (only 10%), Cyprus, the Czech Republic, Hungary, Poland, and Bulgaria, with shares of 12%, 13%, 16%, 18%, and 19%, respectively.

The detailed shares of the individual energy sources in total electricity generated in EU countries in 2020 are shown in Figure 2. A detailed analysis of the share participation of the energy sources discussed in this study reveals that countries whose dominant (at least 50% participation) source of electricity is from conventional fossil sources. These countries include Poland and the Czech Republic, where coal contributes 68% and 39%, respectively, of the energy production, Cyprus and Estonia with 88% and 51%, respectively, and which was attributed to oil, Malta, 87%, the Netherlands, 59%, Ireland, 51%, Italy, 48%, and Greece, 40%, who relied on natural gas.

Countries in which a significant (at least 50%) electricity generation is based on renewable sources include Denmark (57% of renewable energy generated from wind power, and 20% from other less common renewable sources), Austria (61% of renewable energy from hydrological sources, and a 10% significant share of energy from wind power), Luxembourg (30% from other renewable sources, 29% from wind energy, and 13% from solar energy, the highest generation in the entire EU). Other countries worthy of note in this group of countries are Sweden with 44% and 17% of renewable energy from hydrological and wind sources, respectively, Croatia with a considerable share of renewable energy from hydrological (43%) and wind sources (13%), Latvia with a significant share of its renewable energy from hydrological (46%) and other renewable sources (15%), Portugal with a sizeable share of renewable energy generated from wind (24%) and hydrological sources (23%), Lithuania with 34% of renewable energy from wind and 13% from other, less common renewable energy sources, as well as Finland with a 23% share of renewable energy generated from hydrological sources, 15% from wind, and 17% from other renewables.

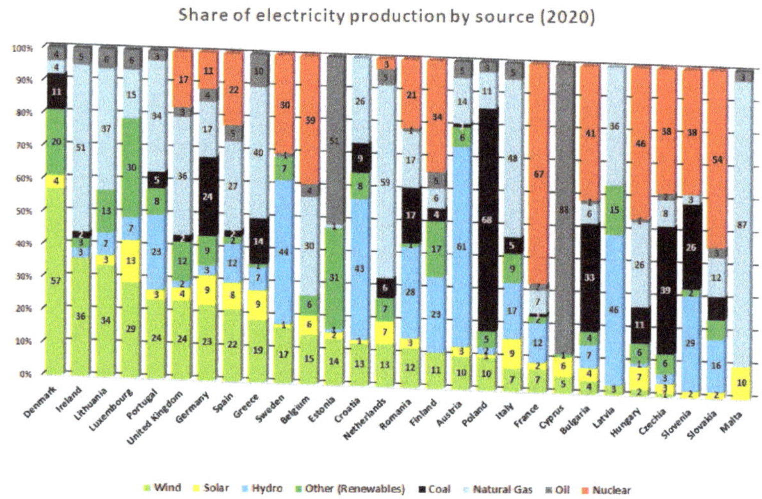

Figure 2. Share of the electricity production in the EU countries by energy source in 2020. Source: https://ourworldindata.org/grapher/share-elec-by-source (accessed on 2 July 2022).

The Table 1 presents a brief history of the EU's activities in promoting and managing the transition from fossil fuels to renewable energy as sources of electricity.

Table 1. Measures intended to boost the share of renewable energy sources in the EU countries.

Date	Activity
April 2009	It was decided that by 2020, 20% of the overall energy consumed in the EU must be sourced from renewable sources. As for the transport sector, member states agreed to attain a 10% share of fuels from renewable sources. Mechanisms to be applied by countries to reach the set targets were also specified, including support systems, common projects, and cooperation between the member states, as well as sustainable growth criteria with regard to biofuels. National renewable energy targets for individual countries to be reached by 2020, were also set, taking into account their starting points and overall renewable energy source potential (from a 10% share of renewables in Malta to 49% in Sweden). Each EU country stated how it intended to attain the individual targets and had to draw up a general action plan [1].
November 2016	The European Commission published the "Clean Energy for All Europeans" package [2].

Table 1. *Cont.*

Date	Activity
December 2018	An amended directive on renewable energy sources [3], as part of the "Clean Energy for All Europeans" package was enacted. The package was aimed at maintaining the EU's position of the global leader in renewable energy sources and assisting the EU in the fulfilment of its emission reduction obligations under the Paris Agreement [4]. The directive has been effective since 2018. A legally binding goal was set, according to which, by 2030 at least 32% of the final energy consumed in the EU should come from renewable sources. In addition, a clause was included allowing the goal to be increased by 2023. EU member states were required to put forward their national energy goals and develop 10-year national energy and climate plans under the "Horizon 2030" programme.
December 2019	The Commission issued the European Green Deal communication [5]. It detailed methods for making Europe a climate-neutral continent by 2050 through the supply of affordable and secure energy.
December 2021	In the package detailing the implementation of the European Green Deal, the European Commission proposed an amendment to the directive on renewable energy sources [6], in order to align its renewable energy target with the new climate goals. The Commission suggested raising the renewable energy source target to 40%. Talks on the energy policy framework for the period after 2030 are under way.
July 2021	The Commission published a new agenda entitled: "Fit for 55: delivering the EU's 2030 Climate Target on the way to climate neutrality" [7]. The review of the renewable energy sources directive proposed an increase in the binding renewable energy share target in the EU's energy basket by up to 40%.

Source: Study based on published EU documents.

Figure 3 shows the national renewable energy targets which the member states agreed to achieve by 2020. However, not all countries were successful in reaching their goals.

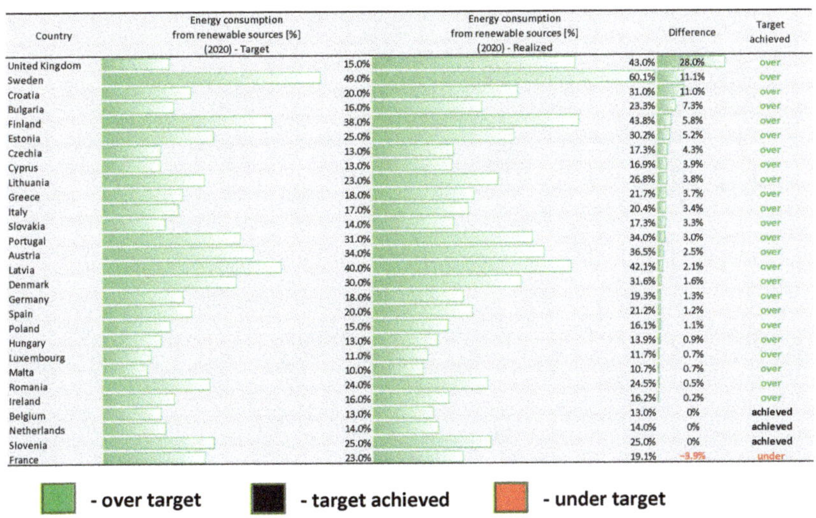

Figure 3. Renewable energy share in the total energy consumption in 2020 vs. the targets set. Source: Eurostat [8,9].

The European Green Deal focuses on three main objectives with regard to the transition to clean energy. To this end, seven main goals were set by the Commission (Table 2).

Table 2. "Green Deal" objectives and goals.

Objectives	Main Goals
1. Providing affordable and secure energy supplies in the EU. 2. Creating a fully integrated, interconnected digitalised EU energy market. 3. Prioritizing energy efficiency, improving energy performance of buildings, and developing an energy sector based largely on renewable sources.	1. Building interrelated energy systems and better integrated networks supporting renewable energy sources. 2. Promoting innovative technology and modern infrastructure. 3. Increasing energy efficiency and promoting eco-projects. 4. Cutting emissions in the gas sector and promoting intelligent integration of all sectors. 5. Empowering consumers and aiding EU countries in counteracting energy poverty. 6. Promoting EU standards and energy technologies globally. 7. Using the entire potential of the European offshore wind power.

Source: Authors' own research.

Having analysed the above objectives and goals, as well as the share of the use of conventional resources and renewables, we formulated the following hypotheses:

Hypothesis 1 (H1). *There is a noticeable tendency towards the sustainable development of national energy systems in the EU countries, based on two pillars: a power industry using conventional fossil fuels and one relying on green renewable energy sources.*

Hypothesis 2 (H2). *There is a group of member states which continues to produce energy predominately from conventional fossil sources and a group of countries which exclusively supports the growth of renewables.*

2. Literature Review

Table 3 presents a detailed review of the literature discussing various aspects of energy sourcing and the potential for its use in various areas of the global economy. The literature we analysed pertained to the period from 2006 to 2021 and discussed the use of energy, both from conventional sources (i.e., fossil fuels) and renewable energy, which was generated from new green renewable energy sources that were neutral for people and the environment.

Relatively few publications are dedicated exclusively to sourcing and using energy from conventional fossil fuels. The study [10] analyses global perspectives and forecasts for hard coal production used in the power industry by means of a logistic forecasting model. The article [11] examines the economic and social aspects of the Polish coal mining industry restructuring in the context of requirements arising from Poland's membership in the EU. Its analysis of the hard coal sales forecast (suitable for use in future energy studies) employs an econometric model of time series developed by the authors. The study [12] forecasts the hard coal consumption for Poland's energy demands. The forecasts are based on an ensemble class model developed by the authors involving a combination of several component models, including adaptive boosting, simulated annealing, and the relevance vector machine (RVM). The main question addressed by the authors is whether Poland is able to fulfil its carbon obligations by 2030. Next, the article [13] analyses the consumption of energy from fossil fuels and the impact of sourcing methods on the natural environment in EU countries.

The second group of publications analysed, focused on discussing mostly or exclusively the production and consumption of energy from renewable sources (the so-called "green energy"). The study [14] addresses the question of whether renewable energy could become the driving force behind a sustainable multifaceted economic development of EU countries. Its analyses employ multivariate comparative methods, in particular the panel vector error correction model (PVEC). The article [15] ranks 11 countries from the so-called EU eastern bloc, according to their renewable energy development levels. To this end, the authors designed a development index for ranking purposes, called the index of renewable energy development (IRED). It also [16] analyses the possible development scenarios and evaluates the function of the EU's common integrated energy system in the context of its sustainable energy development and energy transformation, as well as in terms of its transition to renewable energy. This is of special importance in the event of renewable energy shortages in individual member states (as a consequence of the difference in the green energy development potential and the obvious climatic differences across the EU countries) and the resulting necessity to transfer energy over the integrated power system to other member states across borders. The authors analysed scenarios and applied the correlation analysis using Finland and Italy as examples. Similar issues are discussed in [17], which analyses the possible scenarios for Germany's energy system transformation and decarbonization. The study makes use of a technique for generating simulation scenarios (scenario generation), by applying the GENeSYS-MOD energy modelling system. The publication [18] presents an analysis of the energy efficiency in EU countries in terms of sourcing energy from waste recoveries, i.e., the functioning of the circular economy. Our research method was based on a data envelopment analysis (DEA). Many extant studies focused mainly on broadly defined analyses, for example, of the similarities in the development of potential of energy from renewables across countries (mainly of the EU), examining the similarities between those countries in terms of production, consumption, and the use of renewable energy in various sectors of the national economy. This appears highly significant in the context of energy transformation which is under way in many countries globally and the increasing share of renewables in such countries' energy mix. The research employs simple descriptive analysis methods, straightforward data analysis based on tabular and graphic presentations, dynamics indexes, as well as more advanced statistical methods, such as the comparative cluster analysis, various ranking methods, or simulations designed for generating future development scenarios. We may categorise studies [19–34] as belonging to this group. Several other publications discuss slightly different problems concerning green renewable energy. Study [35] forecasts, among other things, the structure of renewable energy production from various sources and biofuels in Poland, with artificial neural networks being applied in the forecast. As part of study [36], surveys were performed, and opinions were collected from Polish and German citizens, relating to the efficiency and effectiveness of the green energy development management in their countries in terms of the EU's implementation of directives related to the current energy transformation and departure from fossil fuels. A similar subject is discussed in [37], in which surveys were carried out to evaluate the development of solar (photovoltaic) energy in Poland in the ongoing process of energy sources decarbonization. In [38], an evaluation is carried out on the potential of hybrid wind/solar power and a forecast is made about its development in the context of an industry decarbonization strategy worldwide. The forecast employs econometric modelling based on the autoregressive distributed Lag (ARDL) model. In [39], analyses are made for the possible development directions for hydrogen as an eco-friendly fuel (where "eco" stands both for the economy and the environment), which may be successfully used as a source of energy, in transport, for instance.

The third group of publications includes studies discussing the development of energy from fossil fuels and renewable sources. Such publications focus primarily on the comparisons between energy potentials of the countries which rely on conventional fossil fuels and modern renewables. Diverse development perspectives have been analysed for energy systems of the member states, based on two pillars of energy stability. Opportunities, obstacles, and threats to modern energy strategies adopted in various countries worldwide have been discussed. A study [40] is devoted to the forecasts regarding energy demand in Turkey, with the sources analysed in the work using the ARIMA method and it includes various traditional fossil fuels as well as renewables. Another study [41] investigates the main challenges (security of energy supply must be underpinned in the long term, efficient actions must be taken to prevent climate change and investment in replacement, new generation plant and grid installations must be made) and barriers (Emission Trading Scheme (ETS), status of CO_2 capture and storage (CCS)) for EU countries in the implementation of the modern energy mix, based on fossil fuels and renewable sources in the process of supplying energy. Possible scenarios for diverse energy development in EU countries, by 2045, are also analysed and assessed. In [42] the authors offer a critical analysis of the directional changes in the Polish power industry, planned by the government from the perspective of the energy package being implemented in the EU. A study [43] included an analysis of the possible scenarios for the transition to a zero-emissions energy system in seven countries in the Nordic-Baltic region. It creates development scenarios with the use of the Balmorel energy system model for the simulation of the positioning of the energy market in consecutive decades, namely 2020, 2030, 2040, and 2050. Another study [44] contains a compelling, comprehensive historical review of the methods of energy generation, an analysis of global demand, and the use of energy from various (non-renewable and renewable) sources worldwide. It also presents future energy demand forecasts (until 2040) by means of dynamics indexes. Several studies analyse scenarios and possible pathways for the planned energy transformation, as well as problems with the decarbonization of the power industry and the transition to green renewable energy, in search of the optimum energy mix. These subjects are discussed in publications [45–47]. A study [48] contains a comparative analysis of the effectiveness of the energy systems based on varied (fossil and renewable) energy sources. The study compares the efficiency of energy systems in the "new" and "old" EU, using the DEA method. Finally, several other studies [49–53] pertain to highly significant and current problems concerning the detailed comparative analysis of the development levels of energy systems in individual countries and their ranking. Such studies employ various statistical analysis methods, including the taxonomic, cluster analysis (in particular the Hellwig method and the k-means method), as well as various ranking methods (TOPSIS, MULTIMOORA, and the green economy index (GEI)).

Table 3. A review of the literature on energy production and the global energy transformation.

<table>
<tr><th rowspan="2">Region, Country (Temporal Scope of Data)</th><th colspan="2">Energy Sources</th><th rowspan="2">Research Methods Applied</th><th rowspan="2">References</th></tr>
<tr><th>Fossil Fuel Energy</th><th>Renewable (Green) Energy</th></tr>
<tr><td>Turkey
(1950–2004)</td><td colspan="2">[x]</td><td>ARIMA, SARIMA
forecasting methods</td><td>[40]</td></tr>
<tr><td>27 EU countries
(2000–2006)
(Strategic forecasts until 2045)</td><td colspan="2">[x]</td><td>Scenario analysis
Basic statistical descriptive
(Tabular and graph analysis)</td><td>[41]</td></tr>
<tr><td>World
(1950–2008)</td><td>[x]</td><td></td><td>Forecasting methods
(logistic model)</td><td>[10]</td></tr>
<tr><td>28 EU countries
(2003–2014)</td><td></td><td>[x]</td><td>Multivariate econometric analysis
(Panel vector error correction model)</td><td>[14]</td></tr>
<tr><td>Poland
(1995–2014)</td><td>[x]</td><td></td><td>Forecasting methods
(econometric model)</td><td>[11]</td></tr>
<tr><td>11 countries from
the EU's eastern bloc
(2005–2015)</td><td></td><td>[x]</td><td>Ranking methods
Index of Renewable energy
development (IRED)</td><td>[15]</td></tr>
<tr><td>Finland, Italy
(2013)</td><td></td><td>[x]</td><td>Scenario analysis
Correlation analysis</td><td>[16]</td></tr>
<tr><td>Germany
(scenario simulations until 2050)</td><td></td><td>[x]</td><td>Scenario generation
(GENeSYS-MOD
energy modelling system)</td><td>[17]</td></tr>
<tr><td>Poland
(2010–2018)</td><td colspan="2">[x]</td><td>Basic statistical descriptive
(table and graph analysis)</td><td>[42]</td></tr>
<tr><td>EU countries
(2008, 2010, 2012, 2014, 2016)</td><td></td><td>[x]</td><td>Data envelopment analysis (DEA)</td><td>[18]</td></tr>
</table>

Table 3. *Cont.*

Region, Country (Temporal Scope of Data)	Energy Sources		Research Methods Applied	References
	Fossil Fuel Energy	Renewable (Green) Energy		
Nordic-Baltic region (7 countries) (2016, 2017) (scenario simulations for: 2020, 2030, 2040, 2050)	[x]		Scenario generation Balmorel energy system model for energy market simulations	[43]
European Union EU 28 (2016)	[x]		Kruskal–Wallis statistical independence test	[13]
EU countries (2016–2017)		[x]	Scenario simulation Synchronous grids' dynamic Simulink model	[19]
World (1990–2017) (forecasts until 2040)	[x]		Basic statistical descriptive (table and graph analysis) Indexes of dynamics	[44]
Poland (2015–2050)	[x]		Scenario generation MOEM (model of optimal energy mix)	[45]
Poland (1990–2018)		[x]	FCM (fuzzy cognitive maps) Forecasting methods (ANN artificial neural networks)	[35]
EU countries (2017)		[x]	Cluster analysis (k-means method, Ward method)	[20]
Germany EU countries (scenario simulations until 2050)	[x]		Scenario generation (dynELMOD— dynamic electricity model)	[46]
Poland (1965–2018)	[x]		Ensemble time series prediction models (combined adaptive boosting, simulated annealing and relevance vector machine (RVM))	[12]
28 EU countries (2007–2017)		[x]	Panel-data econometric model Cluster analysis (k-means method)	[21]
Czech Republic (with regard to the European Union) (1995–2017)		[x]	Basic analysis of statistical data (visual presentation of data)	[22]
EU countries (2008–2018)		[x]	Ranking methods: TOPSIS method Principal components analysis (PCA)	[23]
Commonwealth of Independent States (12 countries) (2015–2019)		[x]	Index of dynamics (average change index)	[24]
Western EU (6 countries) (2011, 2018, 2020 targets)		[x]	Basic statistical analysis (tabular analysis, charts)	[25]
28 EU countries (2017–2019)	[x]		Ranking methods (Hellwig's measure of the development)	[49]
Visegrad Group (4 countries, compared to 28 EU countries) (1990–2018)	[x]		Cluster analysis (k-means method) Indexes of dynamics	[50]
28 EU countries (2000–2018)	[x]		Cluster analysis (k-means method) Visualization tools (maps, charts)	[51]
28 EU countries (2004–2019)		[x]	Cluster analysis (Hellwig's taxonomic measure of development) Panel econometric model	[26]
Poland, Germany (2018–2020)		[x]	Survey research	[36]
EU countries (especially Poland and Germany) (2011–2021)		[x]	Basis analysis of statistical data (visual presentation of data)	[27]

Table 3. *Cont.*

Region, Country (Temporal Scope of Data)	Energy Sources		Research Methods Applied	References
	Fossil Fuel Energy	Renewable (Green) Energy		
28 EU countries (in comparison to the Visegrad Group) (2009–2019)		[x]	Cluster analysis (Ward's method)	[28]
28 EU countries (2010, 2018)	[x]		Ranking methods Cluster analysis (GEI—Green Economy Index)	[52]
World (1990–2020)		[x]	Econometric modelling Autoregressive distributed lag (ARDL) model	[38]
27 EU countries (2010–2019)		[x]	Indexes of dynamics (simple individual rankings) Correlation coefficients	[29]
Poland (2011–2020) (forecast estimates for 2021–2025)		[x]	Survey research Basic statistical analysis (tabular analysis, charts)	[37]
28 EU countries (2005–2019)		[x]	Cluster analysis (k-means)	[30]
28 EU countries (2010–2019)		[x]	Ranking analysis (TOPSIS method) Cluster analysis (Czekanowski's method)	[31]
(CEE) Central and Eastern EU countries (10 countries) (2008, 2018)		[x]	Cluster analysis (Ward's method)	[32]
Spain (2015–2019)	[x]		Basic statistical analysis (tabular analysis, charts)	[47]
World (until 2020)		[x]	Scenario analysis Basic statistical descriptive (table and graph analysis)	[39]
28 EU countries (2010–2018)	[x]		DEA (data envelopment analysis)	[48]
28 EU countries (2015, 2019)	[x]		Ranking methods Multicriteria decision making (MULTIMOORA method)	[53]
'New' EU member states (10 countries) (2010, 2015, 2019)		[x]	Cluster analysis (Ward method)	[33]
EU countries (2019)		[x]	Cluster analysis (Ward method) Analysis of variance (ANOVA)	[34]

Source: Authors' own research.

3. Materials and Methods

Well-known and commonly applied linear ordering methods for evaluating multi-feature objects were used in order to compile a ranking of the EU countries with regard to the consumption of energy from conventional fossil-based sources and novel methods of generating the so-called green (environment- and climate-friendly) energy. The analyses relied on the TOPSIS method (technique for order preference by similarity to ideal solution) [54], which implemented the generalized distance measure (GDM) [55,56].

The method assumes the known input diagnostic variable matrix $X_{ij}, i = 1, \ldots, m$; $j = 1, \ldots, n$, where n—the number of the diagnostic variables characterising the investigated objects, m—the number of the ranked (ordered) objects (EU countries) and a set weight vector for the diagnostic variables $w_j \in (0, n); \sum_{j=1}^{n} w_j = n$. Our calculations applied identical weights to each diagnostic variable $w_j = 1$.

The algorithm ranking the EU member states by the type of the energy sources used in generating power includes the following steps:

1. It is expected that all diagnostic variables X_j will be treated as stimulants or destimulants. Features characterised as nominants will be converted to the corresponding stimulant values by the following transformation:

$$X_{ij} = \frac{min\left\{nom_j;\ X_{ij}^N\right\}}{max\left\{nom_j; X_{ij}^N\right\}}, \tag{1}$$

where: X_{ij}^N—the value of the j-th nominant observed for the j-th object, nom_j—the nominal value of the j-th variable.

2. A normalised data matrix is created by means of the standardisation procedure according to the formula:

$$Z_{ij} = \frac{X_{ij} - \overline{X_j}}{S_j}, \tag{2}$$

where: $\overline{X_j}$—the mean value of the j-th primary variable, whereas S_j—the standard deviation of the j-th variable.

3. Coordinates for the pattern vector a^+ (ideal solution) for the optimum values of the diagnostic variables and the anti-pattern vector a^- (anti-ideal solution) for the worst values of the diagnostic variables are determined according to the formulas:

$$a^+ = \left(a_1^+, a_2^+, \ldots, a_n^+\right) := \left\{\left(\max_{i=1,\ldots,m} Z_{ij}\middle| j \in J_S\right), \left(\min_{i=1,\ldots,m} Z_{ij}\middle| j \in J_D\right)\right\}, \tag{3}$$

$$a^- = \left(a_1^-, a_2^-, \ldots, a_n^-\right) := \left\{\left(\min_{i=1,\ldots,m} Z_{ij}\middle| j \in J_S\right), \left(\max_{i=1,\ldots,m} Z_{ij}\middle| j \in J_D\right)\right\}, \tag{4}$$

where: J_S—set of stimulants, while J_D—set of destimulants.

4. Calculation of the distance and the i-th object from the pattern GDM_i^+ and the anti-pattern GDM_i^-. The calculations used the GDM (generalized distance measure):

$$GDM_i^+ = \frac{1}{2} - \frac{\sum_{j=1}^n w_j\left(Z_{ij} - a_j^+\right)\left(a_j^+ - Z_{ij}\right) + \sum_{j=1}^n \sum_{l=1,l\neq i,l\neq i_+}^m w_j\left(Z_{ij} - Z_{lj}\right)\left(a_j^+ - Z_{lj}\right)}{2\left[\sum_{j=1}^n \sum_{l=1}^m w_j\left(Z_{ij} - Z_{lj}\right)^2 \cdot \sum_{j=1}^n \sum_{l=1}^m w_j\left(a_j^+ - Z_{lj}\right)^2\right]^{\frac{1}{2}}}, \tag{5}$$

$$GDM_i^- = \frac{1}{2} - \frac{\sum_{j=1}^n w_j\left(Z_{ij} - a_j^-\right)\left(a_j^- - Z_{ij}\right) + \sum_{j=1}^n \sum_{l=1,l\neq i,l\neq i_-}^m w_j\left(Z_{ij} - Z_{lj}\right)\left(a_j^- - Z_{lj}\right)}{2\left[\sum_{j=1}^n \sum_{l=1}^m w_j\left(Z_{ij} - Z_{lj}\right)^2 \cdot \sum_{j=1}^n \sum_{l=1}^m w_j\left(a_j^- - Z_{lj}\right)^2\right]^{\frac{1}{2}}}, \tag{6}$$

where: i_+—pattern object index (number), whereas i_-—anti-pattern object index (number).

5. An aggregate measure (ranking index) corresponding to the degree of similarity of the investigated objects to the ideal solution, is determined according to the formula:

$$TOPSIS\ (GDM)R_i = \frac{GDM_i^-}{GDM_i^- + GDM_i^+}, \tag{7}$$

For $i = 1, \ldots, m$; where: $0 \leq R_i \leq 1$.

6. The objects are placed in a decreasing order depending on the value of measure R_i and the final ranking is generated for the objects (European Union countries). The greater the values of the calculated synthetic index for the country, the higher the country's position in the ranking.

In addition, the analysis made use of an agglomerative clustering—Ward's method—which employs the analysis of variance approach in its procedures [57]. It seeks to minimize the sums of the squared deviations in any pair of clusters which may be formed at any test stag and is one of the most effective clustering methods. The sequence of steps in Ward's method resembles other agglomerative approaches. Significant differences occur in the parameters used in the formula. The sequence is as follows: first, a matrix of taxonomic distances of the dimensions n × n is created, containing the distance between each pair of objects. The matrix is symmetrical in relation to the core diagonal, which is composed of zeros only. Next, the procedure involves searching for pairs of objects (and then clusters) for which the mutual distance is the shortest. The objects are labelled "p" and "q", with p < q. Then, "p" and "q" are combined into a single cluster, which occupies the position labelled with number "p". At the same time, object "q" (cluster) is removed, and the numbers of clusters higher than the "q" object are incremented by one. In this way, the dimension of the matrix is decreased by 1. Next, the distance of the new cluster to each remaining one is calculated according to the formula:

$$D_{pr} = a_1 \cdot d_{pr} + a_2 \cdot d_{qr} + b \cdot d_{pq}, \tag{8}$$

where: D_{pr}—distance from the new cluster to cluster "r", d_{pr}—distance of original cluster "p" to cluster "r", d_{qr}—distance to original cluster "q" from cluster "r", d_{pq}—relative distance between the original clusters "p" and "q", a_1, a_2, b—parameters calculated in Ward's method by formulas:

$$a_1 = \frac{n_p + n_r}{n_p + n_q + n_r}, \quad a_2 = \frac{n_q + n_r}{n_p + n_q + n_r}, \quad b = \frac{-n_r}{n_p + n_q + n_r}. \tag{9}$$

n_P, n_q, n_r—means the quantity of single objects in each group.

The group means method was used to describe the newly formed clusters. An analysis of the group means was performed for the resulting clusters, with the aim of obtaining indicators (diagnostic features) dominant in a given group. For the numerical data matrix, the overall arithmetic means of indicators (without grouping) were calculated and labelled by $\overline{W}_i$. Next, the group arithmetic means of the indicators in resulting clusters were calculated and labelled as $\overline{w}_i$. The structural index of each cluster is the quotient $\frac{\overline{w}_i}{\overline{W}_i}$. High values of the structural index of the means offer information about the dominance of a specific feature in the resulting group. If the mean level of a phenomenon in the group is identical to its mean level across the entire population of objects, then the quotient of means equals 1 (or 100%). Values in excess of 1 (>100%) demonstrate that the mean level of the factor in the group is significantly above its overall mean, while values lower than 1 (<100%) indicate that the mean level of the factor is lower in the investigated group, comparatively to the entire analysed population.

4. Results

An analysis was performed based on the selected indicators, allowing us to create two rankings pertaining to the use of fossil fuels and renewables.

4.1. Data Characteristics

The taxonomic analyses relied on information from the databases kept by Eurostat—the European Statistical Office. The statistical data processed in the study were related to the generation, use, and consumption of energy from various sources (traditional fossil fuels and green renewables), resource levels, as well as the consumption of various types of fuel for power generation purposes in the process of electricity production. The analyses were performed for 2019 (latest available Eurostat data at the time of drafting this article).

Original statistical data retrieved mainly from the Eurostat database (https://ec.europa.eu/eurostat/web/energy/data/database (accessed on 9 July 2022)) were used in the statistical analyses.

Certain data in the statistical-taxonomic analyses were pre-processed on the basis of the original data values and expressed as intensity indicators for the primary diagnostic variables, representing their values converted per 1 million inhabitants of a given country.

A total of 15 indicators characterising the use of fossil fuels and renewables in 28 EU countries, in 2019, were selected for the purpose of this study. The indicators selected for further analysis were labelled (X_1 to X_{15}), with the following meaning and interpretation:

X_1—Stock levels for oil products (fuel oil) [thousand tonnes]/1 [million inhabitants] [58,59];

X_2—Share of fossil fuels in gross available energy [%] [60];

X_3—Final consumption—energy use (hard coal) [thousand tonnes]/1 [million inhabitants] [59,61];

X_4—Gross electricity production (fossil fuels) [thousand tonnes of oil equivalent]/1 [million inhabitants] [59,62];

X_5—Stock levels for natural gas [million cubic meters]/1 [million inhabitants] [59,63];

X_6—Share of fuels in final energy consumption—energy use (natural gas) [%] [64];

X_7—Gross electricity production (natural gas) [thousand tonnes of oil equivalent]/1 [million inhabitants] [59,62];

X_8—Energy available for final consumption (oil and petroleum products excluding biofuels) [thousand tonnes of oil equivalent]/1 [million inhabitants] [59,65];

X_9—Solar thermal collector surface [square meters]/1 [million inhabitants] [59,66];

X_{10}—Share of fuels in final energy consumption—energy use (renewables and biofuels) [%] [64];

X_{11}—Liquid biofuels production capacities (pure biodiesels) [thousand tonnes per year]/1 [million inhabitants] [59,67];

X_{12}—Electricity production capacities for renewables and wastes (hydro) [megawatt]/1 [million inhabitants] [59,68];

X_{13}—Total energy supply (geothermal) [thousand tonnes of oil equivalent]/1 [million inhabitants] [59,69];

X_{14}—Total energy supply (wind) [thousand tonnes of oil equivalent]/1 [million inhabitants] [59,69];

X_{15}—Total energy supply (solar—photovoltaic) [thousand tonnes of oil equivalent]/1 [million inhabitants] [59,69].

Table 4 lists the indicators analysed in the study and their basic descriptive statistics.

Table 4. Descriptive statistics of selected variables.

Variable	Mean	Min	Max	σ	V_z	S
Variables used in the country ranking for the development of a fossil fuel-based energy industry.						
X_1	35.5	0	152.8	39.4	110.9	1.3
X_2	72.3	31.8	96.7	14.7	20.3	−0.8
X_3	45.6	0	400.5	75.1	164.7	4.2
X_4	62.5	0	301.4	85.9	137.5	1.7
X_5	184.1	0	959.3	260.4	141.5	1.7
X_6	16.4	0	37.3	10.2	62.3	0.2
X_7	100.5	0	350.7	96.5	96.0	1.2
X_8	1025.1	475.8	3825.2	611.4	59.6	3.8
Variables used in the country ranking for the development of a green renewable energy industry.						
X_9	146.6	0	1237.7	253.5	172.9	3.4
X_{10}	12.5	4.3	27.4	6.4	50.8	0.9
X_{11}	39.2	0	122.9	35.9	91.4	0.7
X_{12}	449.1	0	2167.3	542.3	120.8	1.9
X_{13}	6.4	0	90.2	17.2	267.7	4.6
X_{14}	61.6	0	239.2	59.5	96.7	1.4
X_{15}	15.8	0.1	48.0	12.5	79.2	0.8

σ—standard deviation, V_z—coefficient of variation, S—skewness. Source: Authors' own research.

The table splits the data into two categories: the features which were used to rank the countries by the use of fossil fuel energy sources (X_1–X_8), and the features which were applied to rank the countries by the use of renewables (X_9–X_{15}). Meanwhile, the full set of indicators (X_1–X_{15}) was used for the classification by means of Ward's method.

All indicators selected for the purpose of the present study are highly variable, as demonstrated by the standard deviation and the coefficient of variation. The highest variability across countries was reported for three indicators: X_{13}—Total energy supply (geothermal), X_9—Solar thermal collector surface, and X_3—Final consumption—energy use (hard coal).

The coefficient of variation for X_{13} was 267.7%. As many as eight countries (Estonia, Ireland, Latvia, Lithuania, Luxembourg, Malta, Finland, and Sweden) lack any supply of geothermal energy. In contrast, in Italy, the index reaches the highest value of 90.2 [thousand tonnes of oil equivalent] per 1 [million people].

The use of solar collectors 7 (variable X_9) also reveals major differences across the EU, with a coefficient of variation of 172.9%. In Estonia, Lithuania, and Slovakia, the index has a value of 0. In contrast, it is the highest for Cyprus—specifically 1237.7 [square meters] per 1 [million inhabitants].

In the fossil fuel group, index X_3 is characterised by the highest variability, as its coefficient of variation equals 164.7%. The lowest value (0) of the index is reported for Malta, and the highest is in Poland, namely 400.5 [thousand tonnes] per 1 [million inhabitants].

All indicators except for one reveal a right-hand skewness, i.e., for most countries the values of the indicators are below the mean level for the specific feature. The only index which stands out in this respect is: X_2—Share of fossil fuels in gross available energy. It displays a left-hand skewness, which means that in most of the countries analysed in this study, the share of fossil fuels in gross available energy is higher than the calculated mean value.

4.2. EU Countries' Rankings Depending on the Energy Sources Used

Two rankings were compiled (Figure 4) on the basis of the variables investigated in this study: one with regard to the use of fossil fuels and another with regard to the use of renewables. The results reveal that the consumption of fossil fuels and renewables is strongly varied across the analysed member states. Poland proves to be the most controversial case, with the second position in the fossil fuel ranking and the last position in the renewable sources ranking.

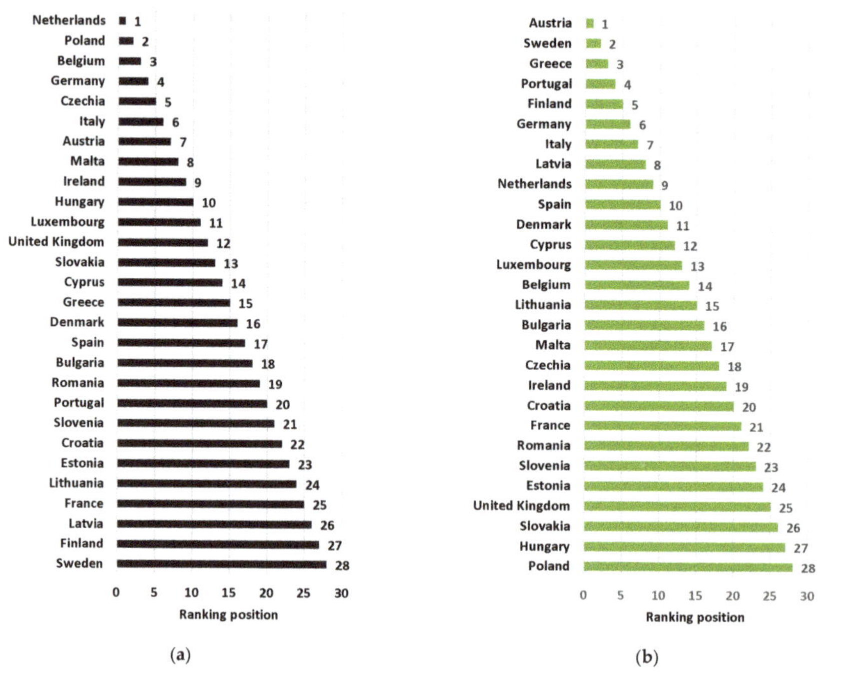

Figure 4. EU countries' ranking in terms of: (**a**) fossil fuels; (**b**) renewables. Source: authors' own research.

The best situation is observed for countries with an optimal mixed approach to the use of conventional fossil fuels and green renewables.

The largest share in the consumption of fossil fuels is observed for the Netherlands (1st position), Poland (2nd position), and Belgium (3rd position). In contrast, the smallest share of fossil fuels is reported for Lithuania (26th position), Finland (27th position), and Sweden (28th position). As for the renewable energy sources, Austria leads the ranking (1st position), followed by Sweden (2nd position), and Greece (3rd position). At the bottom of the ranking are Slovakia (26th position), Hungary (27th position), and Poland (28th position).

4.3. Analysis of the Similarities in Obtaining Energy from Various Sources in EU Countries with the Use of Cluster Analysis Methods

Based on features selected for the purpose of this study, the countries were grouped according to their use of both fossil and renewable fuels in order to facilitate the ranking analysis. The study made use of the taxonomic grouping method (Ward's method), which enabled a more detailed analysis of the country groups identified, thus characterising them in terms of the utilisation of various energy sources. Based on the scree plot (Figure 5a), a decision was made to split the Ward's diagram at the linkage distance of 8.4, as at that point on the scree plot, a marked surge in the linkage distance is clearly visible. There is also a significant value on level 4 of the linkage distance, although a split at that length would generate too many clusters (multiple single-element clusters), and as such would prove far from helpful in drawing conclusions. For this reason, a split into five clusters was preferred.

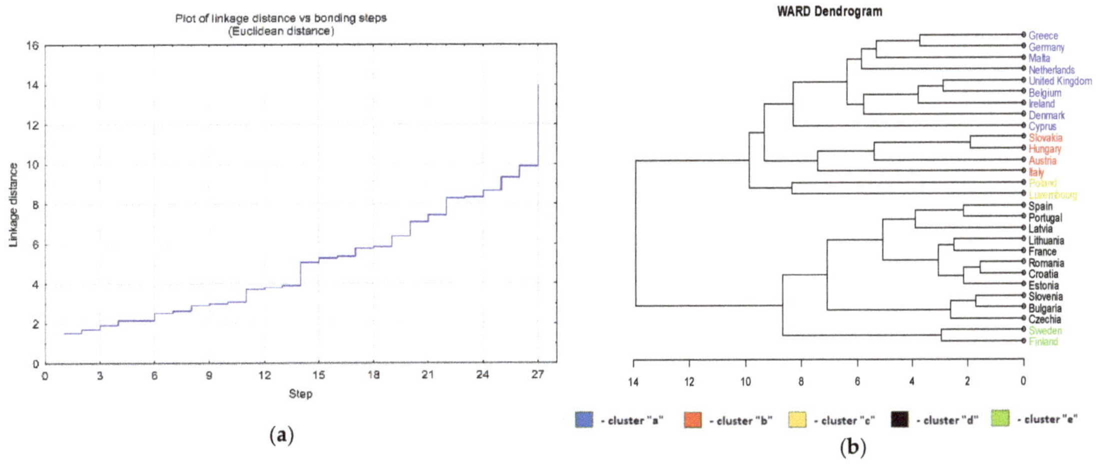

Figure 5. Results of the cluster analysis using Ward's method: (**a**) scree plot; (**b**) dendrogram—Ward's method (Euclidean distance). Source: Authors' own research.

The first cluster, labelled "a", included nine countries: Belgium, United Kingdom, Ireland, Denmark, Germany, Greece, Malta, the Netherlands, and Cyprus. These countries had relatively solid positions in the rankings due to the use of renewables but ranked highly also in terms of their use of fossil fuels. A conclusion which can be drawn from the analysis of the quotient of the group mean indicators (Figure 6) is that the cluster is characterized by high values for X_1—Stock levels for oil products (fuel oil) and X_7—Gross electricity production (natural gas), as well as elevated use of thermal energy, with the more-than-average values of index X_9—Solar thermal collector surface. The solar energy indicator X_{15} also reaches above-average values (the highest of all clusters). The same holds true for the wind energy index, although not to such a considerable extent as in the countries from cluster "e".

Labelled "b", the second cluster groups together four countries, namely Austria, Hungary, Slovakia, and Italy. Austria and Italy rank high in terms of the exploitation of both fossil and renewable energy sources, whereas Hungary and Slovakia also have high positions in fossil fuels, yet also extremely low positions with regard to the use of renewables. An analysis of the quotient of the group mean indicators indicates that the use of geothermal energy reaches radically elevated levels (above-average values of indicator X_{13}). Remarkably high values are also reached by indicators X_5 and X_6, which correspond to energy production from natural gas. What is more, the countries in the cluster also show a considerable consumption of solar (photovoltaic) and water power (hydropower), as demonstrated by the above-average mean quotients for the diagnostic variables X_9, X_{15}, and X_{12}.

The third cluster, named "c", included two countries: Luxembourg and Poland. This pair appears quite peculiar, although this study is concerned with the use of energy sources rather than economic development. Consequently, the fact that the two countries were placed in the same cluster should be seen as perfectly natural. Indicator X_3, Final consumption—energy use (hard coal), has definitely the highest quotient of means in the group. The feature corresponds to the consumption of energy from fossil fuels, particularly hard coal. High values of the quotient of means were also found for indicators such as: X_4—Gross electricity production (fossil fuels) and X_8—Energy available for final consumption (oil and petroleum products excluding biofuels).

The fourth cluster, coded "d", contains the largest number of objects (11 countries), namely Bulgaria, Slovenia, the Czech Republic, Croatia, Romania, Estonia, France, Lithuania, Latvia, Portugal, and Spain. These are mostly countries which ranked last in terms of the use of renewables or non-renewables. The Czech Republic is the only member state with a high (5th) position in the fossil fuel ranking in this group, while Portugal and Latvia perform quite well in terms of the exploitation of renewable energy sources. Although they are countries with an average consumption of both types of sources, they generate a considerable portion of energy from fossil fuels.

The last cluster, labelled "e", consists of two countries: Sweden and Finland. They have high (2nd and 5th) positions in the renewables ranking. In addition, they are at the bottom of the list of countries which use fossil fuels (27th and 28th position). However, the group mean quotient analysis reveals that these countries are characterised by an above-average level of indicators corresponding to the use of renewables. Specifically, variables with a remarkably high share include X_{12}—Electricity production capacities for renewables and wastes (hydro), X_{14}—Total energy supply (wind) and X_{10}—Share of fuels in final energy consumption—energy use (renewables and biofuels), as well as X_{11}—Liquid biofuels production capacities (pure biodiesels).

Figure 7 presents countries plotted at coordinates corresponding to the computed ranking positions and divided into four groups. The EU countries may be categorised as: (1) countries with a very high position in both energy sources rankings, (2) those with the lowest position in both rankings, (3) and (4) those with only one well-developed area of power industry, i.e., relying either on fossil fuels or renewables.

From Figure 7, we may conclude that the countries which occupy the top positions in both rankings include the Netherlands, Germany, Austria, and Italy. They are leaders in both areas, with high fossil fuels' and renewables' consumption levels. Countries at the bottom of both rankings include France, Estonia, Slovenia, and Croatia.

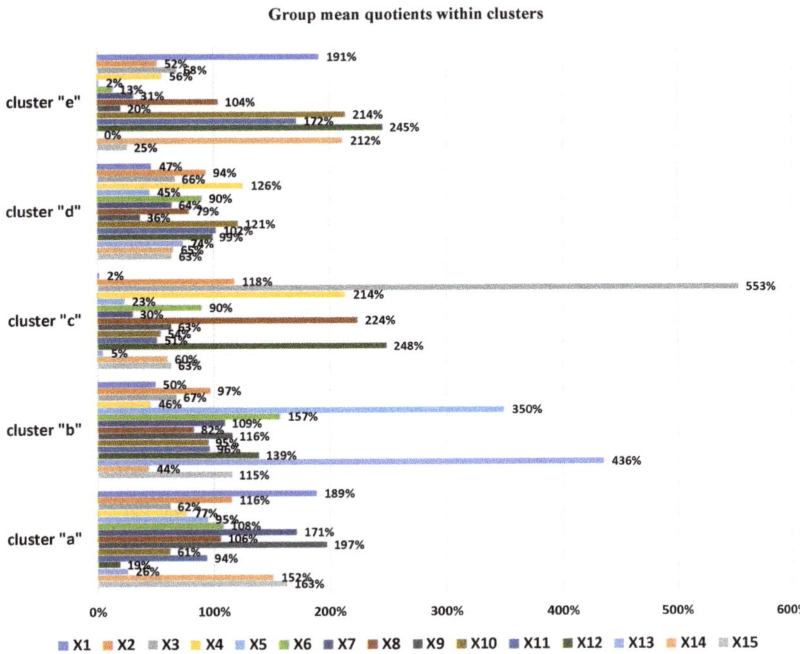

Figure 6. Percentage share of the group means comparatively to the overall mean for the selected diagnostic variables in the clusters obtained by Ward's method. Source: Authors' own research.

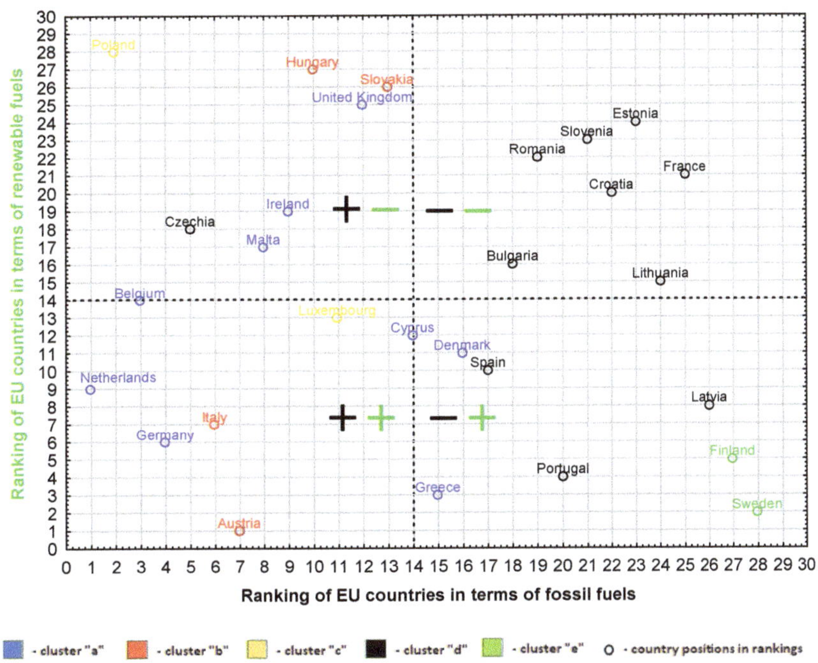

Figure 7. A visualisation of the two rankings and four groups of the EU countries. Source: Authors' own research.

5. Discussion

With global energy demand on a constant increase, ensuring current production levels and creating favourable conditions for future economic growth means that energy must be easily available and inexpensive, and the supply system must be resistant to various disruptions. Power outages may lead to substantial financial losses and wreak havoc in its functioning. Equally important are the energy availability and the ability to purchase it at reasonable and acceptable prices. Energy security, understood as guaranteed energy supply, is usually defined in terms of the energy system's resistance to exceptional and unpredictable events which may threaten the physical integrity of energy flow or cause an uncontrollable price surge, regardless of its economic rationale [70]. Energy security ensures that consumers' current and future demands for fuels and energy are met in a technically and economically justified way, with a minimum negative effect of the energy sector, on the environment and living conditions.

The predicted increase in the demand for energy necessitates its supply from multiple sources. A diversified energy mix guarantees security for the energy system thanks to its elasticity in meeting the requirements of a given country. Most renewable energy sources are intermittent [71], which opens spatial and temporal gaps between their availability and consumption by end users. To tackle this issue, the employment of large-scale energy storage systems may greatly improve the utilisation rate and stability of the use of renewable energy. The pumped hydro energy storage technology, for example, can quickly balance the load and adjust the frequency to meet the requirements of the power system [72], as well as the redox flow batteries being a representative of electrochemical energy storage technique [73].

Energy supply security depends on many factors, amongst which, several key determinants can be highlighted:

- diversity (diversification) of capabilities—sustainable and well-balanced energy production systems including various power generation technologies, together with suitable production capabilities help to make maximum use of advantages offered by a specific technology;
- interchangeability of fuels—diversity in the consumption of fuels may be a crucial factor to enhancing energy security. Conversion of fuels such as coal into gas, gas into liquid fuel, and coal gasification facilitate meeting the demand even if conventional fuel supplies are disrupted;
- political threats—the energy supply system may be exposed to threats and disorganisation due to various, often conflicting political interests of different countries, or due to terrorist attacks

The invasion of Ukraine forced the European countries to look for immediate responses involving the replacement of raw materials which had been imported from the East barely some months before. At present, Europe is amidst the decarbonization process with gas as a transition fuel—ironically, a fuel imported from Russia.

There are many signs that a domestic mining industry and access to coal deposits will remain a diversifying factor relevant for the energy security. From the power industry's standpoint, an ideal solution would be to rely on many diverse sources to enhance security and guarantee a continuous energy supply—a sine qua non of the existence and constant growth of a modern society.

The analysis performed in this article demonstrated that EU countries use conventional and renewable sources to a varying degrees. The choice of the strategy depends on the state's resources, geographical and climate factors, sea access, sunlight, wind intensity, and the presence of fossil fuel deposits.

The research results presented in the article, despite being obtained only for the 15 indicators (determinants of energy use based fossil and renewable sources) selected for the study are comparable with the findings of other publications dealing with renewable energy sources' development in EU countries. However, a full comparison with such findings is not possible due to the use of various sets of diagnostic indicators and different research methods by the authors. It is worthy of note that much of the aforementioned studies dealt with the use of renewable energy only, while a few others dealt with energy use from both energy sources (fossil and renewable). Our study covers both approaches and also indicates which approaches are dominant in a given EU country.

Similar synthetic measures confirming the more favourable situation in terms of renewable sources based (solar, wind, hydro, and bio) electricity production in the rich old EU members, while highlighting problems with the greening of electricity production in the large group of new EU member states were presented in article [26]. The observations contained in the article corroborate our findings, rankings in development of renewable energy development in highly developed old EU countries, including Austria, Sweden, Finland, Portugal, and Germany and also the relatively poorer use of renewable sources based energy in poorer Eastern European countries despite observable changes.

In [20], the authors showed an elaborate variety of sources from which renewable energy is obtained in some EU countries. This is attributable to their geographical locations, financial capabilities, traditions, as well as economic potentials and social awareness. Consequently, there are great opportunities for cooperation and exchange of experience in this area between individual countries, which should result in a more comprehensive and effective use of countries' individual renewable energy sources.

Similarly, in [23] the results of the author's study showed that EU countries are characterized by significant differences in the development of renewable energy sources, in 2018. The unquestionable leaders in this respect are Sweden, Austria, Finland, and Latvia, which is also confirmed by study findings.

A group of member states, including Germany, Italy, Austria, and the Netherlands, have a highly diversified strategy for their power industry, as they take advantage of both conventional and renewable energy sources. Greece, Portugal, Finland, and Sweden currently rely mostly on renewables, while the power industries of Poland, Belgium, the Czech Republic, and Malta, are based largely around fossil fuels. Countries with the lowest level of energy security make limited use of both types of energy sources, and these include Estonia, France, Slovenia, and Croatia.

The European Union has, for long, been applying strategies and actions, aimed at motivating member states to increase their share in obtaining renewable sources based energy. Such strategies and actions by the EU have already brought tangible effects, currently visible even in countries such as Poland that traditionally base their energy industry on fossil sources. The data analysis, presented in Figures 1 and 2 in the introduction section, provides such a conclusion.

However, these measures seem insufficient in the context of the observed climate changes. Therefore, the EU ought to further stimulate more intense country involvement in renewable energy sources development through actions such as:

- financing development investments in wind, solar, hydro, geothermal, and bio-energy,
- applying tax breaks for companies in the renewable energy sector,
- introducing higher restrictions on the required share of renewable energy in the total value of energy obtained,
- increasing country and energy consumers' awareness that renewable energy is an unavoidable and important source of energy.

Taking cognizance of the fact that obtaining renewable energy at similar levels due to factors earlier mentioned and the previously signalled discontinuities in the renewable energy sources supply is impossible in all countries, the European Union should also reinforce support for initiatives in co-financing investments in ecological energy production from fossil sources. Basing a country's energy sector solely on one pillar, such as renewable energy, or on any other source, such as gas, is very risky and adversely affects the country's energy stability as has recently been confirmed by the current war in Ukraine.

6. Conclusions

The analyses were based on quantitative indicators, describing energy derived from both fossil and renewable sources, which is rarely undertaken in previous similar studies, where predominant analytical approaches were solely based on the developments in countries' green energy sector and the use of energy from renewable sources.

Our study used methods of multivariate comparative analysis, such as Ward's cluster analysis method and the ranking method based on the synthetic measure of the TOPSIS development.

The findings of this study corroborate our research hypotheses formulated at the beginning, namely that EU countries display a tendency to foster sustainable development of their energy systems based on two pillars (both energy from conventional fossil fuels, and green renewable energy sources). Such countries include Austria, Germany, Italy, and the Netherlands. On the other hand, there is a group of member states which continue to produce energy predominately from conventional fossil sources (Poland and the Czech Republic), as well as a group of those which exclusively support the growth of their power industry based on renewables, such as Finland and Sweden.

The research was carried out relying on statistical data from 2019. Hence, it is planned that future long-term (2022–2030) studies should be undertaken in order to compare future results with current findings.

Given the present uncertainties in terms of energy supply, emphasis must be put on the diversification of energy sources to ensure the highest possible level of energy security, a guarantor of further socio-economic development.

The conclusions drawn from this study can serve as recommendations for decision makers in developing energy strategies and can help identify countries which require assistance in choosing appropriate development paths in terms of their energy strategy.

The research on the use of hydrogen as a clean energy source seems hopeful in the pursuit for improving EU countries' energy balance. In terms of accelerating the development of renewable energy, the EU hopes to achieve its goal by producing 10 million tons of green hydrogen, while importing another 10 million tons by 2030. The current energy prices in Europe are relatively high globally, and the use of renewable energy will further increase the cost of green hydrogen production in Europe, although large-scale long-distance transportation technology of hydrogen is yet to mature.

In addition, European Commission decision makers have proposed to double the solar PV capacity by 2025 and install 600 GW by 2030. In order to support the solar energy industry in Europe and retain and regain its technological and industrial leading position in the field of solar energy, the European Union has decided to set up the "EU Solar Energy Industry Alliance", hoping that in the process of energy transformation, European enterprises can assail their desired market position.

Author Contributions: Conceptualization, K.C.-L. and T.P.; methodology, K.C.-L. and T.P.; validation, K.C.-L. and T.P.; investigation, K.C.-L. and T.P.; resources, K.C.-L. and T.P.; data curation, K.C.-L. and T.P.; writing—original draft preparation, K.C.-L. and T.P.; writing—review and editing, K.C.-L. and T.P.; visualization, K.C.-L. and T.P.; supervision, K.C.-L. and T.P. All authors have read and agreed to the published version of the manuscript.

Funding: This research received no external funding.

Institutional Review Board Statement: Not applicable.

Informed Consent Statement: Not applicable.

Data Availability Statement: Publicly available data.

Conflicts of Interest: The authors declare no conflict of interest.

References

1. European Commission. Available online: https://ec.europa.eu/info/energy-climate-change-environment/implementation-eu-countries/energy-and-climate-governance-and-reporting/national-energy-and-climate-plans_en#documents (accessed on 10 April 2022).
2. EU Monitor. Available online: https://www.eumonitor.eu/9353000/1/j4nvke1fm2yd1u0_j9vvik7m1c3gyxp/vkcwedk9layr/v=s7z/f=/com(2016)860_en.pdf (accessed on 10 April 2022).
3. EUR-LEX. Available online: https://eur-lex.europa.eu/legalcontent/EN/TXT/PDF/?uri=CELEX:32018L2001&from=EN (accessed on 10 April 2022).
4. United Nations Climate Change. Available online: https://unfccc.int/sites/default/files/english_paris_agreement.pdf (accessed on 10 April 2022).
5. EUR-LEX. Available online: https://eur-lex.europa.eu/resource.html?uri=cellar:b828d165-1c22-11ea-8c1f-01aa75ed71a1.0002.02/DOC_1&format=PDF (accessed on 10 April 2022).
6. EUR-LEX. Available online: https://eur-lex.europa.eu/resource.html?uri=cellar:dbb7eb9c-e575-11eb-a1a5-01aa75ed71a1.0001.02/DOC_1&format=PDF (accessed on 10 April 2022).
7. EUR-LEX. Available online: https://eur-lex.europa.eu/legal-content/EN/TXT/PDF/?uri=CELEX:52021DC0550&from=PL (accessed on 10 April 2022).
8. EUROSTAT. Available online: https://ec.europa.eu/eurostat/documents/2995521/10335438/8-23012020-AP-EN.pdf/292cf2e5-8870-4525-7ad7-188864ba0c29 (accessed on 10 April 2022).
9. EUROSTAT. Available online: https://ec.europa.eu/eurostat/statistics-explained/index.php?title=Renewable_energy_statistics (accessed on 10 April 2022).
10. Hook, M.; Zittel, W.; Schindler, J.; Aleklett, K. Global coal production outlooks based on a logistic model. *Fuel* **2010**, *89*, 3546–3558. [CrossRef]
11. Manowska, A.; Tobór Osadnik, K.; Wyganowska, M. Economic and social aspects of restructuring Polish coal mining: Focusing on Poland and the EU. *Resour. Policy* **2017**, *52*, 192–200. [CrossRef]
12. Li, Y.; Zhang, H.; Kang, Y. Will Poland fulfil its coal commitment by 2030? An answer based on a novel time series prediction method. *Energy Rep.* **2020**, *6*, 1760–1767. [CrossRef]
13. Martins, F.; Felgueiras, C.; Smitkova, M.; Caetano, N. Analysis of fossil fuel energy consumption and environmental impacts in European Countries. *Energies* **2019**, *12*, 964. [CrossRef]
14. Armeanu, D.S.; Vintila, G.; Gherghina, S.C. Does renewable energy drive sustainable economic growth? Multivariate panel data evidence for EU-28 countries. *Energies* **2017**, *10*, 381. [CrossRef]
15. Momete, D.C. Measuring Renewable Energy Development in the Eastern Bloc of the European Union. *Energies* **2017**, *10*, 2120. [CrossRef]
16. Wang, H.; Di Pietro, G.; Wu, X.; Lahdelma, R.; Verda, V.; Haavisto, I. Renewable and sustainable energy transitions for countries with different climates and renewable energy sources potentials. *Energies* **2018**, *11*, 3528. [CrossRef]
17. Bartholdsen, H.-K.; Eidens, A.; Loffler, K.; Seehaus, F.; Wjda, F.; Burandt, T.; Oei, P.-Y.; Kemfert, C.; von Hirschhausen, C. Pathways for Germany's low-carbon energy transformation towards 2050. *Energies* **2019**, *12*, 2988. [CrossRef]
18. Halkos, G.; Petrou, K.N. Analysing the energy efficiency of EU member states: The potential of energy recovery from waste in the circular economy. *Energies* **2019**, *12*, 3718. [CrossRef]
19. Agathokleous, C.; Ehnberg, J. A quantitative study on the requirement for additional inertia in the European Power System until 2050 and the potential role of wind power. *Energies* **2020**, *13*, 2309. [CrossRef]
20. Brodny, J.; Tutak, M. Analyzing similarities between the European Union Countries in terms of the structure and volume of energy production from renewable energy sources. *Energies* **2020**, *13*, 913. [CrossRef]
21. Simionescu, M.; Strielkowski, W.; Tvaronaviciene, M. Renewable energy in final energy consumption and income in the EU-28 countries. *Energies* **2020**, *13*, 2280. [CrossRef]
22. Sivek, M.; Jirasek, J.; Kavina, P.; Vojnarova, M.; Kurkova, T. Divorce after hundreds of years of marriage: Prospects for coal mining in the Czech Republic with regard to the European Union. *Energy Policy* **2020**, *142*, 111524. [CrossRef]
23. Brodny, J.; Tutak, M.; Bindzar, P. Assessing the level of renewable energy development in the European Union member states. A 10-year perspective. *Energies* **2021**, *14*, 3765. [CrossRef]
24. Chowaniak, M.; Gródek-Szostak, Z.; Kotulewicz-Wisińska, K.; Luc, M.; Suder, M.; Szeląg-Sikora, A. The RES in the Countries of the Commonwealth of Independent States: Potential and Production from 2015 to 2019. *Energies* **2021**, *14*, 1856. [CrossRef]
25. Cucchiella, F.; Condemi, A.; Rotilio, M.; Annibaldi, V. Energy transitions in Western European Countries: Regulation comparative analysis. *Energies* **2021**, *14*, 3940. [CrossRef]
26. Huterski, R.; Huterska, A.; Zdunek-Rosa, E.; Voss, G. Evaluation of the level of electricity generation from renewable energy sources in European Union countries. *Energies* **2021**, *14*, 8150. [CrossRef]
27. Jałowiec, T.; Wojtaszek, H. Analysis of the RES potential in accordance with the energy policy of the European Union. *Energies* **2021**, *14*, 6030. [CrossRef]
28. Kacperska, E.; Łukasiewicz, K.; Pietrzak, P. Use of renewable energy sources in the European Union and the Visegrad Group countries—Results of cluster analysis. *Energies* **2021**, *14*, 5680. [CrossRef]

29. Komarnicka, A.; Murawska, A. Comparison of consumption and renewable sources of energy in European Union Countries—Sectoral indicators, economic conditions and environmental impacts. *Energies* **2021**, *14*, 3714. [CrossRef]
30. Matuszewska-Janica, A.; Żebrowska-Suchodolska, D.; Ala-Karvia, U.; Hozer-Koćmiel, M. Changes in electricity production from renewable energy sources in the European Union countries in 2005–2019. *Energies* **2021**, *14*, 6276. [CrossRef]
31. Musiał, W.; Zioło, M.; Luty, L.; Musiał, K. Energy policy of European Union Member states in the context of renewable energy sources development. *Energies* **2021**, *14*, 2864. [CrossRef]
32. Pakulska, T. Green energy in Central and Eastern European (CEE) countries: New challenges on the path to sustainable development. *Energies* **2021**, *14*, 884. [CrossRef]
33. Wałachowska, A.; Ignasiak-Szulc, A. Comparison of renewable energy sources in 'New' EU Member States in the context of national Energy transformations. *Energies* **2021**, *14*, 7963. [CrossRef]
34. Włodarczyk, B.; Firoiu, D.; Ionescu, G.H.; Ghiocel, F.; Szturo, M.; Markowski, L. Assessing the sustainable development and renewable energy sources relationship in EU countries. *Energies* **2021**, *14*, 2323. [CrossRef]
35. Brodny, J.; Tutak, M.; Saki, S.A. Forecasting the structure of energy production from renewable energy sources and biofuels in Poland. *Energies* **2020**, *13*, 2539. [CrossRef]
36. Jałowiec, T.; Wojtaszek, H.; Miciuła, I. Green energy management through the implementation of RES in the EU. Analysis of the opinions of Poland and Germany. *Energies* **2021**, *14*, 8097. [CrossRef]
37. Lew, G.; Sadowska, B.; Chudy-Laskowska, K.; Zimon, G.; Wójcik-Jurkiewicz, M. Influence of photovoltaic development on decarbonization of power generation—Example of Poland. *Energies* **2021**, *14*, 7819. [CrossRef]
38. Khan, H.R.; Awan, U.; Zaman, K.; Nassani, A.A.; Haffar, M.; Abro, M.M.Q. Assessing hybrid solar-wind potential for industrial decarbonization strategies: Global shift to green development. *Energies* **2021**, *14*, 7620. [CrossRef]
39. Sharma, S.; Agarwal, S.; Jain, A. Significance of Hydrogen as Economic and Environmentally Friendly Fuel. *Energies* **2021**, *14*, 7389. [CrossRef]
40. Ediger, V.S.; Akar, S. ARIMA forecasting of primary energy demand by fuel in Turkey. *Energy Policy* **2007**, *35*, 1701–1708. [CrossRef]
41. Kavouridis, K.; Koukouzas, N. Coal and sustainable energy supply challenges and barriers. *Energy Policy* **2008**, *36*, 693–703. [CrossRef]
42. Gawlik, L.; Mokrzycki, E. Changes in the structure of electricity generation in Poland in view of the EU climate package. *Energies* **2019**, *12*, 3323. [CrossRef]
43. Lund, P.D.; Skytte, K.; Bolwig, S.; Bolkesjo, T.F.; Bergaentzle, C.; Gunkel, P.A.; Kirkerud, J.G.; Klitkou, A.; Koduvere, H.; Gravelsins, A.; et al. Pathway analysis of a zero-emission transition in the Nordic-Baltic region. *Energies* **2019**, *12*, 3337. [CrossRef]
44. Ahmad, T.; Zhang, D. A critical review of comparative global historical energy consumption and future demand: The story told so far. *Energy Rep.* **2020**, *6*, 1973–1991. [CrossRef]
45. Antosiewicz, M.; Nikas, A.; Szpor, A.; Witajewski-Baltvilks, J.; Doukas, H. Pathways for the transition of the Polish power sector and associated risks. *Environ. Innov. Soc. Transit.* **2020**, *35*, 271–291. [CrossRef]
46. Kittel, M.; Goeke, L.; Kemfert, C.; Oei, P.-Y.; von Hirschhausen, C. Scenarios for coal-exit in Germany—A model-based analysis and implications in the European context. *Energies* **2020**, *13*, 2041. [CrossRef]
47. Pena-Ramos, J.A.; del Pino-Garcia, M.; Sanchez-Bayon, A. The Spanish energy transition into the EU Green Deal: Alignments and paradoxes. *Energies* **2021**, *14*, 2535. [CrossRef]
48. Simeonovski, K.; Kaftandzieva, T.; Brock, G. Energy efficiency management across EU countries: A DEA approach. *Energies* **2021**, *14*, 2619. [CrossRef]
49. Daroń, M.; Wilk, M. Management of energy sources and the development potential in the energy production sector—A comparison of EU countries. *Energies* **2021**, *14*, 685. [CrossRef]
50. Gostkowski, M.; Rokicki, T.; Ochnio, L.; Koszela, G.; Wojtczuk, K.; Ratajczak, M.; Szczepaniuk, H.; Bórawski, P.; Bełdycka-Bórawska, A. Clustering Analysis of Energy Consumption in the Countries of the Visegrad Group. *Energies* **2021**, *14*, 5612. [CrossRef]
51. Halkos, G.; Tsilika, K. Visual exploration of energy use in EU 28: Dynamics, patterns, policies. *Energies* **2021**, *14*, 7532. [CrossRef]
52. Kasztelan, A. On the road to a Green Economy: How do European Union countries 'Do their homework'? *Energies* **2021**, *14*, 5941. [CrossRef]
53. Sipa, M.; Gorzeń-Mitka, I. Assessment of the progress towards the management of renewable energy consumption in the innovativeness context—A country approach. *Energies* **2021**, *14*, 5064. [CrossRef]
54. Hwang, C.-L.; Yoon, K. *Multiple Attribute Decision Making*, 1st ed.; Springer-Verlag: Berlin/Heidelberg, Germany, 1981; pp. 128–132.
55. Jajuga, K.; Walesiak, M.; Bak, A. On The General Distance Measure. In *Exploratory Data Analysis in Empirical Research;* Schwaiger, M., Opitz, O., Eds.; Springer: Berlin/Heidelberg, Germany, 2003; pp. 104–109.
56. Walesiak, M. *Uogólniona Miara Odległości GDM w Statystycznej Analizie Wielowymiarowej z Wykorzystaniem Programu R*, 2nd ed.; Wydawnictwo Uniwersytetu Ekonomicznego we Wrocławiu: Wrocław, Poland, 2016; p. 43.
57. Grabiński, T. *Metody Taksonometrii;* Wydawnictwo Akademii Ekonomicznej w Krakowie: Kraków, Poland, 1992; pp. 101–111.
58. Eurostat Database. Stock Levels for Oil Products. Available online: https://ec.europa.eu/eurostat/databrowser/view/nrg_stk_oil/default/table?lang=en (accessed on 1 October 2021).

59. Eurostat Database. Population on 1 January by Age and Sex. Available online: https://ec.europa.eu/eurostat/databrowser/view/demo_pjan/default/table?lang=en (accessed on 1 October 2021).
60. Eurostat Database. Share of Fossil Fuels in Gross Available Energy. Available online: https://ec.europa.eu/eurostat/databrowser/view/nrg_ind_ffgae/default/table?lang=en (accessed on 1 October 2021).
61. Eurostat Database. Supply, Transformation and Consumption of Solid Fossil Fuels. Available online: https://ec.europa.eu/eurostat/databrowser/view/NRG_CB_SFF__custom_2742493/default/table?lang=en (accessed on 1 October 2021).
62. Eurostat Database. Production of Electricity and Derived Heat by Type of Fuel. Available online: https://ec.europa.eu/eurostat/databrowser/view/NRG_BAL_PEH__custom_2742647/default/table?lang=en (accessed on 1 October 2021).
63. Eurostat Database. Stock Levels for Gaseous and Liquefied Natural Gas. Available online: https://ec.europa.eu/eurostat/databrowser/view/nrg_stk_gas/default/table?lang=en (accessed on 1 October 2021).
64. Eurostat Database. Share of Fuels in Final Energy Consumption. Available online: https://ec.europa.eu/eurostat/databrowser/view/NRG_IND_FECF__custom_2743065/default/table?lang=en (accessed on 1 October 2021).
65. Eurostat Database. Simplified Energy Balances. Available online: https://ec.europa.eu/eurostat/databrowser/view/NRG_BAL_S__custom_3041423/default/table?lang=en (accessed on 1 October 2021).
66. Eurostat Database. Solar Thermal Collectors' Surface. Available online: https://ec.europa.eu/eurostat/databrowser/view/nrg_inf_stcs/default/table?lang=en (accessed on 1 October 2021).
67. Eurostat Database. Liquid Biofuels Production Capacities. Available online: https://ec.europa.eu/eurostat/databrowser/view/nrg_inf_lbpc/default/table?lang=en (accessed on 1 October 2021).
68. Eurostat Database. Electricity Production Capacities for Renewables and Wastes. Available online: https://ec.europa.eu/eurostat/databrowser/view/nrg_inf_epcrw/default/table?lang=en (accessed on 1 October 2021).
69. Eurostat Database. Complete Energy Balances. Available online: https://ec.europa.eu/eurostat/databrowser/view/NRG_BAL_C__custom_2743696/default/table?lang=en (accessed on 1 October 2021).
70. Government of the Republic of Poland. Available online: https://www.gov.pl/web/polski-atom/bezpieczenstwo-energetyczne-podstawa-rozwoju-spoleczenstwa (accessed on 1 July 2022).
71. Sun, C.; Negro, E.; Vezzu, K.; Pagot, G.; Cavinato, G.; Nale, A.; Bang, Y.H.; di Noto, V. Hybrid inorganic-organic proton-conducting membranes based on SPEEK doped with WO3 nanoparticles for application in vanadium redox flow batteries. *Electrochim. Acta* **2019**, *309*, 311–325. [CrossRef]
72. Wang, W.; Tai, G.; Pei, J.; Pavesi, G.; Yuan, S. Numerical investigation of the effect of the closure law of wicket gates on the transient characteristics of pump-turbine in pump mode. *Renew. Energy* **2022**, *194*, 719–733. [CrossRef]
73. Sun, C.; Zhang, H. Review of the Development of First-Generation Redox Flow Batteries: Iron-Chromium System. *ChemSusChem* **2022**, *15*, e202101798. [CrossRef]

energies MDPI

Review

Green and Renewable Energy Innovations: A Comprehensive Bibliometric Analysis

Grzegorz Mentel [1,*], Anna Lewandowska [2], Justyna Berniak-Woźny [2] and Waldemar Tarczyński [3]

1 Department of Quantitative Methods, Rzeszow University of Technology, 35-959 Rzeszow, Poland
2 Department of Management, University of Information Technology and Management, 35-225 Rzeszów, Poland
3 Department of Sustainable Finance and Capital Markets, Institute of Economics and Finance, University of Szczecin, 71-101 Szczecin, Poland
* Correspondence: gmentel@prz.edu.pl

Abstract: Taking into account factors such as unfavorable climate changes, shrinking fossil fuel resources, low energy efficiency, and the pace of population growth, the transformation towards green and renewable energy is one of the most important goals and challenges facing the world. The energy sector is the source of about 75% of global greenhouse gas emissions and energy-related emissions are reaching new record levels. For the energy transition to succeed, innovation at the level of technology, business processes and policies (local, national, and international) are necessary. Therefore, the aim of this article is to analyze the size, structure, and dynamics of research on innovations in the field of green and renewable energy in the last decade in order to identify the main topics and research trends in this field. The authors conducted a bibliometric review based on the PRISMA guidance together with visualization analysis based on the VOSviewer software. For this purpose, the Web of Science Core Collection (WoS CC) database was used, and based on defined inclusion criteria, the authors selected 1144 records for bibliographic analysis. The database was subjected to a performance analysis from the perspective of the number of publications per year, dominant countries, and journals. Further, science mapping was employed to analyze such features of the publications as co-citations, co-occurrences, and bibliometric coupling. Based on the results, gaps in green and renewable energy innovations were identified and issues for future research were defined and recommended.

Keywords: green energy; renewable energy; innovation; trends; bibliometric analysis

Citation: Mentel, G.; Lewandowska, A.; Berniak-Woźny, J.; Tarczyński, W. Green and Renewable Energy Innovations: A Comprehensive Bibliometric Analysis. *Energies* **2023**, *16*, 1428. https://doi.org/10.3390/en16031428

Academic Editor: Benjamin McLellan

Received: 7 January 2023
Revised: 19 January 2023
Accepted: 28 January 2023
Published: 1 February 2023

1. Introduction

Population growth and the accompanying socio-economic development increase the demand for energy and energy-related services, as all societies need such services to meet basic human needs and improve human well-being, and to develop and support production processes [1]. In 2014, the total global consumption of primary energy amounted to approximately 160,310 million MWh, and it was predicted that this value would increase to 240,318 million MWh in 2040 [2]. Global reserves of fossil fuels are depleting very quickly: based on average global usage forecasts, it is assumed that oil and gas may last for 50 more years, and coal and uranium for about 100 more years [3]. Due to the long-term and very intensive use of non-renewable energy sources, there are also harmful effects on the environment, such as air pollution, climate change, and irreversible loss of natural resources.

The UN states that to keep global warming below 1.5 °C, emissions must be reduced by 45% until 2030 and to net zero by 2050 [4]. It is estimated that the energy sector is the source of about three-quarters of global greenhouse gas emissions [5]. Thus, the energy sector plays a fundamental and growing role in achieving the decarbonization of the economy [6]. Therefore, in May 2021, the International Energy Agency (IEA) published its monumental

Net Zero by 2050 report, which assumes that the electricity sector must move from being the highest emitting sector in 2020 to the first sector to achieve net zero emissions by 2040 in the world [7]. At the same time, the IEA emphasizes that achieving this goal requires unprecedented efforts by all actors.

Decarbonization of the electricity sector is at the heart of the response to the threat of climate change. Therefore, the Paris Agreement underlines the urgent need for action in this area—in particular, the shift from fossil fuels to renewable energy and increasing energy efficiency [8]. This problem has also been included in the UN Sustainable Development Goals [9]. Goal 7 is to achieve universal access to affordable, reliable, and modern energy and to significantly increase the share of renewable energy in the energy mix. People's dependence on energy has increased significantly, but energy poverty remains an international development challenge. As energy is a prerequisite for many aspects of modern life, lack of access reflects social and economic inequalities. However, it must be based on low-emission technological solutions that reduce greenhouse gas emissions, and thus improve living and health conditions and ensure sustainable development of life on the planet [9].

However, as the International Renewable Energy Agency (IRENA) highlights in its 2022 Tracking Report on Sustainable Development Goal 7 [10], the latest available data and selected energy scenarios show that at the current pace of progress, the world does not stand much chance of achieving any of the goals and metrics under SDG 7. This is especially true for the most vulnerable groups of countries, which from the beginning have been significantly lagging behind in this area. Worldwide, 91% of the population had access to electricity in 2020, still leaving 733 million people without access to electricity. According to forecasts, the problem of lack of access to electricity in 2030 will still affect 630 million people [11,12].

Renewable energy is defined by the IPCC [13] as:

'(. . .) any form of energy from solar, geophysical, or biological sources that is replenished by natural processes at a rate that equals or exceeds its rate of use. Renewable energy is obtained from the continuing or repetitive flows of energy occurring in the natural environment and includes low carbon technologies such as solar energy, hydropower, wind, tide and waves and ocean thermal energy, as well as renewable fuels such as biomass'.

In many cases, the terms *renewable energy* and *green energy* are used interchangeably, but there are important differences between the terms. Green energy is a subset of renewable energy and includes resources that provide more environmental benefits as they come from natural sources such as the sun, water, and wind [14].

The International Energy Agency (IEA) proposes the following classification of renewable energy technologies based on their stage of development:

- First generation technologies, i.e., hydropower, biomass combustion, geothermal energy and heat;
- Second generation technologies, i.e., solar heating and cooling, wind energy, bioenergy, solar photovoltaics;
- Third generation technologies, i.e., thermal solar energy, ocean energy, geothermal energy, integrated bioenergy systems [15].

Globally, it is estimated that in 2008 renewable energy accounted for 12.9% of the total 492 exajoules (EJ) of primary energy supply [14]. Moreover, as the WHO report [16] shows, in 2010–2019 the share of renewable energy sources in total final energy consumption increased by only 2.7%. It was a good sign that although the COVID-19 pandemic had a bad effect on the energy transformation and stopped many green energy projects, in 2020 the use of renewable energy sources continued to grow and accounted for over 80% of all new electricity capacity added that year [17]. However, in 2021, the share of renewable energy sources in global electricity production reached only 28.7% (after a slight increase of 0.4%) resulting from the global electricity demand reaching its all-time high, the slowdown in economic activity caused by COVID-19, and drought in several regions that reduced hydropower production [18]. According to Our World in Data [7], global primary energy consumption is still largely based on fossil fuels such as oil (33.1%), coal (27%), and gas (24.3%). The dominant renewable energy sources include hydropower (6.45%), wind (92.2%), and solar (1.1%). Nuclear power, delivering 4.3% of energy, is not classified as a renewable source as it uses radioactive fuel, but it is defined as a low-carbon source [19].

If we are to reach the milestones of a net zero emissions scenario by 2050, thus increasing the share of renewables in power generation from almost 29% in 2021 to over 60% by 2030, then global efforts must be doubled and annual production of green and renewable energy must grow by more than 12% on average between 2022 and 2030, which is twice the 2019–2021 average [20]. This requires a complete change in the way energy is produced, transported, and consumed, and a significant acceleration in the implementation of available technologies, as well as the development and dissemination of technologies that are not yet on the market. We need innovative techniques to respond to climate change. On the one hand, we need innovation to reduce costs, improve efficiency, and enable the integration of renewable technologies in energy systems. On the other hand, innovation is also needed to discover and develop fourth-generation renewable energy technologies that will revolutionize the entire energy system. In other words, incremental improvements are necessary and will continue to make significant progress but may not be enough to drive a complete transformation of the energy sector that requires disruptive technologies and processes. However, this is a large undertaking that requires huge investments and a significant acceleration of innovation and implementation work. Innovations are also necessary in the entire energy system, such as in market design and accompanying policies and regulations, infrastructure integrating renewable energy sources with energy systems, and new business models [21].

Growing interest in energy transformation accelerated by the climate change and energy crises has led to the emergence of a research stream on green and renewable energy innovations. Scientists are working on many innovations in the field of renewable energy aiming to achieve high efficiency with less pollution and to integrate existing and new technologies. They are also analyzing political and financial frameworks as well as new business models necessary to achieve the energy transformation required and define critical areas. Therefore, the aim of this article is to analyze the size, structure, and dynamics of research on innovations in the field of green and renewable energy in the last decade in order to identify the main topics and research trends in this field. More detailed knowledge on this issue will give a helicopter view of the current state of the art. Scientific relationships on the international and authorship level will also prompt research agendas for the years to come that would support the green transformation. There are three key research questions (RQs) regarding this aim:

RQ1. What is the structure of the research on green and renewable energy innovations?

RQ2. What are the research dynamics and the most important activities being carried out within green and renewable energy innovation?

RQ3. What are the future research directions related to green and renewable energy innovation?

The article is structured as follows. In Section 2 the methodology and data sets are defined. In Section 3, the main findings of reviews and the results of additional analyses are presented. Section 4 discusses the implications of the empirical results and concludes the paper.

2. Materials and Methods

In order to achieve the aim and answer the research questions posed in the previous section, the authors used a bibliometric review with visualization analysis, well-established methods presenting an overview of research trends in a selected area [22].

The bibliometric review is based on the use of various approaches to the characterization of quantitative, qualitative, and structural changes in scientific research and the profile of publications in a selected subject area [23]. The main publication classification factors used by bibliometrics are journals, authors, institutions, countries, keywords, and references [24]. According to Bjork et al. [25] the key advantage of bibliometric analysis is gaining a helicopter view of a specific research field. Bibliometric research, very popular in the information and natural sciences, has become equally popular in other fields over time, including the social sciences, and its popularity can be attributed to the development of scientific databases such as Scopus, Web of Science, and Google Scholar, as well as the advancement, accessibility, and affordability of bibliometric visualization software such as Gephi, Leximancer, and VOSviewer. According to Corsini et al. [26], there are several arguments in favor of using bibliometric methods, such as the possibility of obtaining a review of extensive scientific literature, objective results (as opposed to narrative reviews based on a critical and subjective summary of selected scientific papers), and the possibility of using modern solutions in the field of large database analysis.

Bibliometric research techniques can be divided into three groups: review, evaluation, and relational techniques [27].

1. Review techniques include systematic literature review, meta-analyses, and qualitative research.
2. Evaluative techniques assess the level of academic impact and are divided into three groups of metrics [28]. (1) Productivity metrics—number of articles in a given period or per author; (2) Impact measures—the number of citations in a given period, per document, author, or journal; (3) Hybrid metrics—average number of citations per article, and metrics that include productivity and impact. Evaluative techniques are limited by the lack of network identification between authors, publications, journals or countries and institutions [29].
3. Relational techniques identify the relationship between units of analysis within a research field and provide information about the structure of that field by identifying authors or affiliations and discovering leading themes and research methods. Relational techniques are based on four types of analysis: co-citation analysis, co-authorship analysis, bibliographic coupling, and co-word analysis [30,31].

Zupic and Cater [32] recommend the following workflow guidelines for mapping selected research filed with bibliometric methods: (1) research design, (2) compilation of bibliometric data, (3) analysis, (4) visualization, and (5) interpretation. The workflow of this study can be mapped as follows: The research design is outlined as Section 2.1; the compilation of bibliometric data comprises Section 2.2; and the analysis, visualization, and interpretation are detailed in Section 3.

2.1. Research Design

At the research design stage, we define research questions and select optimal bibliometric methods. In this paper, three research questions are put forward.

RQ1. *What is the structure of the research on green and renewable energy innovations?*

RQ1.1. *What is the volume of published articles on the topic of green and renewable energy innovations?*

RQ1.2. *What are the most impactful journals, authors, and countries in the field?*

RQ2. *What are the research dynamics and the most important activities carried out within the green and renewable energy innovations?*

RQ2.1. *What is the intellectual structure in the field?*

RQ2.2. *Which are the research topics most addressed in the domain?*

RQ3. *What are the future research directions related to green and renewable energy innovations?*

To answer RQ1, we used performance analysis, which aims to examine the impact of selected metrics on a given field [33], explore and define leading research actors in a field [34], and recognize the importance of different factors in that research field. The most common analysis metrics are the number of publications and citations per year, or the source of publications, with the number of publications being an indicator of productivity, while citation being an indicator of impact.

To answer RQs 2 and 3 we have employed the following three mapping science techniques:

1. Co-citation analysis to explore the relationship between cited publications and deepen understanding of the development of leading topics in the research field.
2. Co-word analysis to explore existing or future relationships between themes in a given research field.
3. Coupling analysis (of countries and sources) for establishing a similarity relationship between sample documents.

2.2. Compilation of Bibliographic Data

The compilation of bibliometric data covers the selection of the database(s), filtering the bibliographic data, and refining them. The next step was the selection of analytical software and the decision on how to visualize the results of the analysis. We followed the PRISMA (Preferred Reporting Items for Systematic Reviews and Meta-Analyses) guidelines and procedures [35]. We decided to use the Web of Science Core Collection (WoS CC), which is considered as the leading database for classifying scientific research. The Web of Science Core Collection (WoS CC) contains over 21,100 peer-reviewed journals published worldwide in over 250 scientific disciplines.

The search was based on the keyword combination "green energ*" or "renewable energ*" and "innovat*". We searched for articles that have this phrase in the title. A total of 5897 records were found by searching the WoS CC database. In the screening phase we excluded papers:

- Published before 2012—N = 1231
- Document types other than articles—N = 1472
- Papers in languages other than English—N = 40
- Limited to four databases: SCI-EXPANDED, SSCI, ESCI, A&HCI—N = 102

The WoS CC database was limited to open access papers only, as a result of which another 1908 papers were excluded as not relevant for this research. After the data search and selection process was completed, the qualified records were subjected to bibliometric analysis as shown in Figure 1, and the final database consisted of 1144 documents.

The database was downloaded in TXT format because the authors planned to use it for visualization in VOSviewer software, which requires CSV or TXT files. Since WoS CC has built-in analyzer functions, an initial descriptive analysis was performed using these functions and then the analysis functions of Excel were used. Tables were created to provide quantitative data. Additionally, VOSviewer software, version 1.6.18, was employed to quantitatively and visually analyze selected publications on green and renewable energy innovations, using such techniques as co-citations, co-occurrences, and bibliometric coupling maps.

The analysis, visualization, and interpretation are presented in the next section, Results.

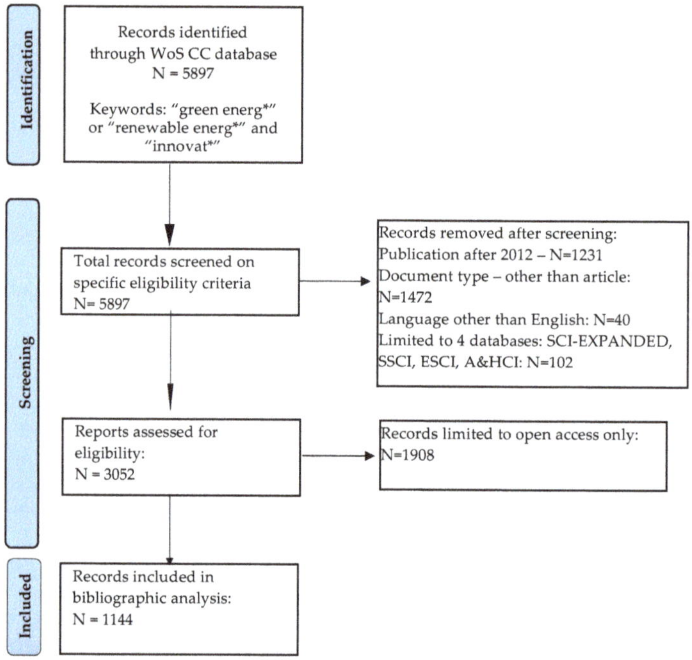

Figure 1. PRISMA flowchart [36].

3. Results

As already signaled in the Introduction, innovation is one of the key elements of achieving SDG 7 and accelerating the development of green and renewable energy. Both from the perspective of science and practice, an in-depth understanding of scientific achievements in the field of green and renewable energy is of key importance [37]. In order to make an in-depth analysis of this research area and identify current topics and directions of scientific research, bibliometric research was used [38], under which 1144 documents selected from the WoS CC database were evaluated.

The basic criterion for assessing the interest of researchers in a given scientific area and the dynamics of this interest is the number of publications and its distribution over the years [39–43]. Figure 2 shows the number of published documents on green and renewable energy and innovation in 2012–2022. It is worth noting that there is an increasing trend in the number of scientists researching the topic of innovation in the field of green and renewable energy. This increase is particularly noticeable in the last three years, i.e., during the period of intensified discussions on the climate and energy crisis and the green transformation. As the problem of climate change and dwindling fossil resources becomes more urgent, the authors predict that this trend will continue into the future.

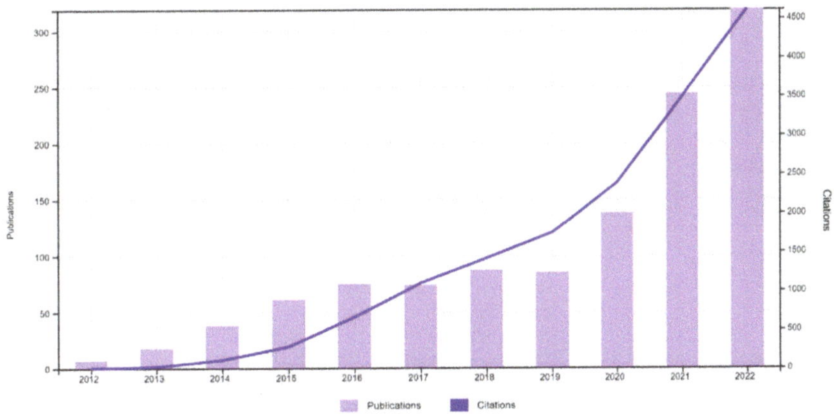

Figure 2. The number of documents on green and renewable energy and innovation and the number of citations of documents (2012–2022).

Table 1 presents the researchers who published at least four papers on green and renewable energy and innovation. The most productive authors in this rating with six articles are Jahid (University of Ottawa, Canada), Anser (Super Univ, Pakistan), and Alsharif (Sejong University, North Korea), followed by Cabeza (University of Lleida, Spain), Ahmad (Shandong University of Technology, China), Dincer (Instanbul Medipol University, Turkey), and Yüksel (Instanbul Medipol University, Turkey) with five papers on green and renewable energy innovation. Meanwhile, twelve authors each published four studies.

Table 1. The leading authors of documents on green and renewable energy and innovation.

Authors	No. of Papers	% of 1144	Citations
Jahid, Abu	6	0.5245	56
Anser, Muhammad Khalid	6	0.5245	32
Alsharif, Mohammed H.	6	0.5245	27
Cabeza, Luisa F.	5	0.4371	390
Ahmad, Mahmood	5	0.4371	20
Dincer, Hasan	5	0.4371	11
Yuksel, Serhat	5	0.4371	11
Coma, Julia	4	0.3497	361
Perez, Gabriel	4	0.3497	361
Stucki, Tobias	4	0.3497	156
Sarkodie, Samuel Asumadu	4	0.3497	121
Asif, Muhammad	4	0.3497	119
Kharel, Rupak	4	0.3497	109
Zgid, Dominika	4	0.3497	106
Kaiwartya, Omprakash	4	0.3497	94
Sun, Huaping	4	0.3497	42
Olah, Judit	4	0.3497	26
Nassani, Abdelmohsen A.	4	0.3497	7
Zaman, Khalid	4	0.3497	7

As Table 2 shows, the most popular journals in the field of green and renewable energy and innovation are *Energies* (MDPI, IF: 3.252) with 19 publications and 961 citations, *Sustainability* (MDPI, IF: 3.889) with 91 publications and 650 citations, and *IEEE Access* (IEEE, IF: 3.476) with 40 published articles and 551 citations. They are followed by the *Environmental Science and Pollution Research* (Springer Heidelberg, IF: 5.19) with 36 articles and 281 citations and the *Frontiers in Environmental Science* (Frontiers Media SA, IF: 5.411) with 31 papers and 83 citations.

Table 2. Top 20 journals with articles on green and renewable energy and innovation.

Journal	No. of Papers	% of 1144	Citations
Energies	119	10.402	961
Sustainability	91	7.955	650
IEEE Access	40	3.497	551
Environmental Science and Pollution Research	36	3.147	281
Frontiers in Environmental Science	31	2.710	83
Energy Reports	26	2.273	79
Energy Policy	19	1.661	739
Economic Research-Ekonomska Istrazivanja	16	1.399	80
Frontiers in Energy Research	16	1.399	48
Applied Energy	15	1.311	845
Applied Sciences-Basel	11	0.962	73
IEEE Journal on Selected Areas in Communications	9	0.787	444
Renewable Energy	9	0.787	305
Journal of Cleaner Production	9	0.787	262
Energy Economics	7	0.612	290
International Journal of Environmental Research and Public Health	7	0.612	47
Renewable and Sustainable Energy Reviews	6	0.524	147
Energy Strategy Reviews	6	0.524	58
Mathematical Problems in Engineering	6	0.524	21
Heliyon	5	0.437	27

The essential subjects of the research are publications by country, including their social networks [44–46]. Table 3 presents the most important countries in accordance with the number of published papers on green and renewable energy innovations. The most productive authors in this area are from the People's Republic of China with 363 papers (which is 31% of all papers on green and renewable energy innovations). Moreover, important study in this area has been carried out in the USA and England (approximately 10% and 8% of review articles, respectively). Figure 3 shows a geographical map of 93 countries in accordance with the number of published articles.

Table 3. Documents on green and renewable energy and innovation by country.

Country	No. of Papers	% of 1144	Citations
Peoples Republic of China	363	31.731	4182
USA	118	10.315	3009
England	102	8.916	2619
Italy	71	6.206	867
Pakistan	63	5.507	814
Germany	60	5.245	1607
Poland	55	4.808	343
Saudi Arabia	54	4.720	970
India	51	4.458	525
Canada	49	4.283	1197
Spain	45	3.934	937
South Korea	41	3.584	667
Turkey	40	3.497	643
Malaysia	38	3.322	425
France	36	3.147	603
Taiwan	33	2.885	190
Netherlands	29	2.535	658
Australia	28	2.448	365
Sweden	27	2.360	357
Switzerland	23	2.010	921

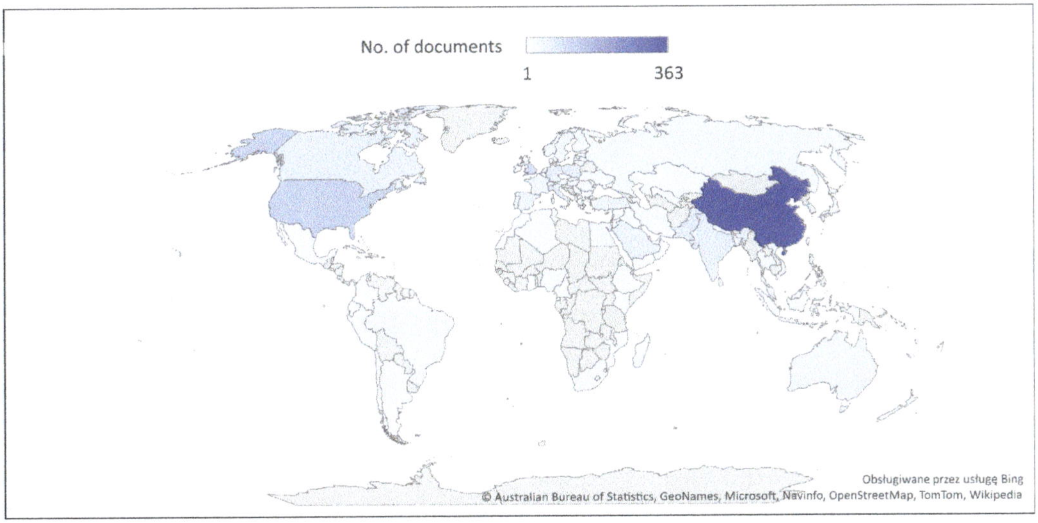

Figure 3. A map of countries in accordance with the number of published articles on green and renewable energy and innovation.

The analysis of the organization's bibliographic links is very interesting due to its complementarity with the results of the citation analysis of authors, institutions, and countries. Figure 4 shows data with a minimum of three documents and 100 institutions meeting the thresholds.

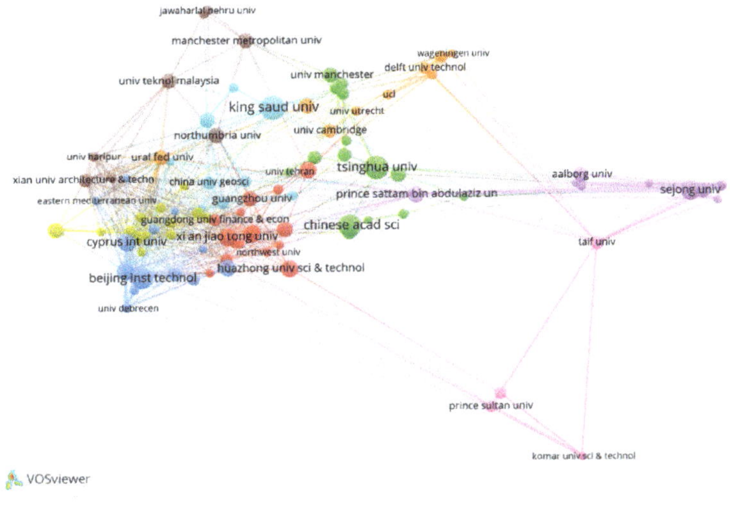

Figure 4. Bibliographic coupling of organizations.

Table 4 presents the most frequently cited articles from the research area under study. It is worth noting that there is a very good representativeness of studies published in the last decade (2012–2022) [47–56]. This is confirmed by the fact that the topic of innovation in the field of green and renewable energy is a development topic. Moreover, the importance of some studies, such as Richter [54] and Lehr et al. [55], indicate that this topic began to appear in the 2000s, but a particularly dynamic increase in the number of studies in this field could be observed from 2015.

Table 4. Most cited articles.

Article	Year	Author(s)	Citations	Citations per Year
Green preparation of reduced graphene oxide for sensing and energy storage applications	2014	Bo et al.	390	48.75
Energy, land-use and greenhouse gas emissions trajectories under a green growth paradigm	2017	van Vuuren et al.	324	64.80
Tailoring the energy landscape in quasi-2D halide perovskites enables efficient green-light emission	2017	Quan et al.	316	63.2
Energy-efficient base-stations sleep-mode techniques in green cellular networks: A survey	2015	Wu et al.	260	37.14
Tubular graphitic-C3N4: A prospective material for energy storage and green photocatalysis	2013	Tahir et al.	207	23.00
Renewable energy-driven innovative energy-efficient desalination technologies	2014	Ghaffour et al.	191	23.87
Green scheduling of a two-machine flowshop: Trade-off between makespan and energy consumption	2016	Mansouri et al.	175	29.16
Business model innovation for sustainable energy: German utilities and renewable energy	2013	Richter	168	18.67
Green jobs? Economic impacts of renewable energy in Germany	2012	Lehr et al.	166	16.60
Vertical greenery systems for energy savings in buildings: A comparative study between green walls and green facades	2017	Coma et al.	159	31.80

Based on the co-citations of references analysis, it was possible to define the network of citations. This technique, as reported by Albort-Morant et al. [39] describes the key authors, often cited with other authors. Figure 5 shows a citation map where each circle represents a document, and the size of the circle corresponds to the number of citations [57].

Figure 5. Co-citation of references map.

The analysis defined four clusters of papers that cited each other. The red cluster shows research that mainly concerns the environment–competitiveness relationship [58], renewable energy and technological innovations [59], and energy-saving innovations [60]. The green cluster shows studies that relate to the panel unit root test, as described in [61], and diagnostic tests for cross-sectional relationships in panels, as in [62]. The blue cluster covers the topics of green public finance and sustainable green finance [63], CO_2 emissions in the context of economic growth and renewable energy production [64] and energy poverty and energy efficiency [65]. The yellow cluster focuses on research methods [66,67].

In the next step, the technique of bibliographic coupling of authors was employed to identify active researchers in the field and provides a view of the research structure of the front of the field. Of the 4194 authors, we selected those with a minimum of three publications and 10 citations. Only 44 authors met the threshold. Figure 6a presents the authors bibliographic coupling map. To make the map more readable we created two more showing the four defined clusters more precisely. Figure 6b presents the first three clusters, and Figure 6c presents the fourth cluster, namely:

- The red cluster with 12 authors led by Anser M.K. (Shandong University of Technology, China), Ahmad M. (Super Univ, Pakistan), and Olah J. (University of Johannesburg, South Africa) focuses on financial, technological, and social aspects of green transformation.
- The blue cluster represented by nine authors, such as Dincer H. (Istanbul Mediapol University, Turkey), Stucki T. (Bern University of Applied Sciences, Switzerland), and Sun H. (Tianjin University of Commerce, China), focuses research on green and renewable energy policy and economics.
- The yellow cluster covers eight authors including Asif M. (Glasgow Caledonian University, UK), Cabez L.F. (University of Lleida, Spain), and Coma J. (University of Lleida, Spain), whose research focuses on energy aspects of sustainable and green buildings as well as technological aspects of the green transformation.
- The green cluster includes 11 authors led by Ansharif MA (Sejong University, South Korea), Ansari N. (New Jersey Institute of Technology, US), and Kaiwartya O (Nottingham Trent University, UK), who focus their research on such themes as mobile and cellular networks or sensors for green computing.

Following Phillips et al. [67], in order to illustrate the geographic allocation of knowledge, bibliographic coupling of countries was used. Figure 7 shows three clusters in relation to the network of countries defined as a result of the analysis. The green cluster includes Asian countries under the strong leadership of the People's Republic of China supported by Latin America (Mexico and Brazil), South Africa, and Canada. Another cluster, red, combines research from the United States and European Union (mostly Italy, Germany, Poland, and France) with contribution from Israel and Egypt. In the blue cluster, the most productive countries are England, Saudi Arabia, Spain, Malaysia, and Australia.

The analysis of the author's keywords is also very important due to the fact that the keywords correspond to the context of the publication and thus define the main topics and research trends [68].

Figure 8 shows the most common author keywords and the author keyword network in articles. The minimum number of author keywords for a co-occurrence map is five and the minimum cluster size is specified as 12. The author's most popular keywords are *renewable energy* (131), *energy efficiency* (90), *green energy* (57), *energy consumption* (40), and *sustainable development* (34). The identified 92 author's keywords were classified into four clusters.

(a)

(b)

(c)

Figure 6. (**a**) Authors bibliographic coupling map. (**b**) Authors bibliographic coupling map—clusters 1–3. (**c**) Authors bibliographic coupling map—cluster 4.

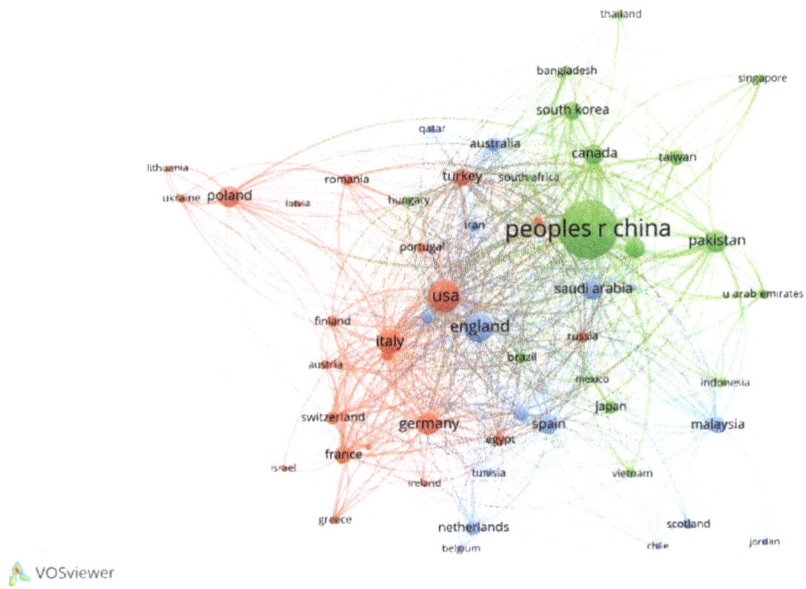

Figure 7. Bibliographic coupling of countries.

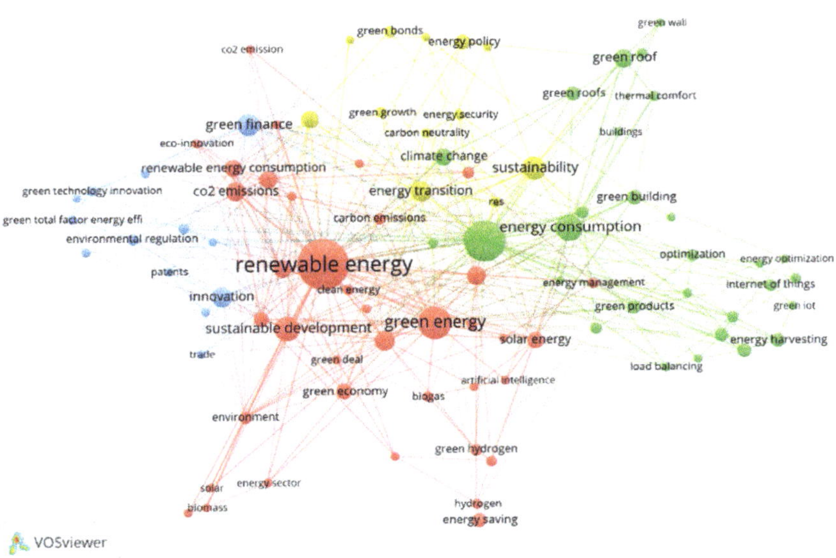

Figure 8. Co-occurrence map (keywords network).

- The red cluster (38 keywords) encapsulates such terms as *renewable energy* and *green energy* followed by *sustainable development*, *CO_2 emissions*, *solar energy*, *green economy*, *environmental sustainability*, *green innovation*, *renewable energy sources*, and *technological innovation*, which led to the naming of this cluster as "renewable technology innovations".

- The green cluster (29 keywords) revolves around the terms *energy efficiency* and *energy consumption*, followed by *the green roof, green products, climate change, green building, energy harvesting, smart grid, green communication,* and *optimization*, which led to the naming of this cluster as "energy consumption efficiency and sustainability".
- The blue cluster (13 keywords) revolves around the terms *green finance, innovation,* and *environmental regulation*, followed by *patents, green technology innovation, financial development, energy intensity,* and *environmental policy*, which led to the naming of this cluster as "renewable energy innovations ecosystem".
- The yellow cluster (12 keywords) revolves around the terms *sustainability, energy transition,* and *economic growth*, followed by *energy policy, green bonds, green growth, energy security,* and *carbon neutrality*, which led to the naming of this cluster as "renewable energy innovation investments".

Using the same map, Figure 9 shows the chronological evolution of research topics. Thanks to this, it is possible to identify the most current topics in the researched area, which turn out to be green finance, green innovations, renewable energy consumption, CO_2 emission, and carbon neutrality, while energy efficiency related to green construction or energy harvesting have been the main topics of interest in the previous decade.

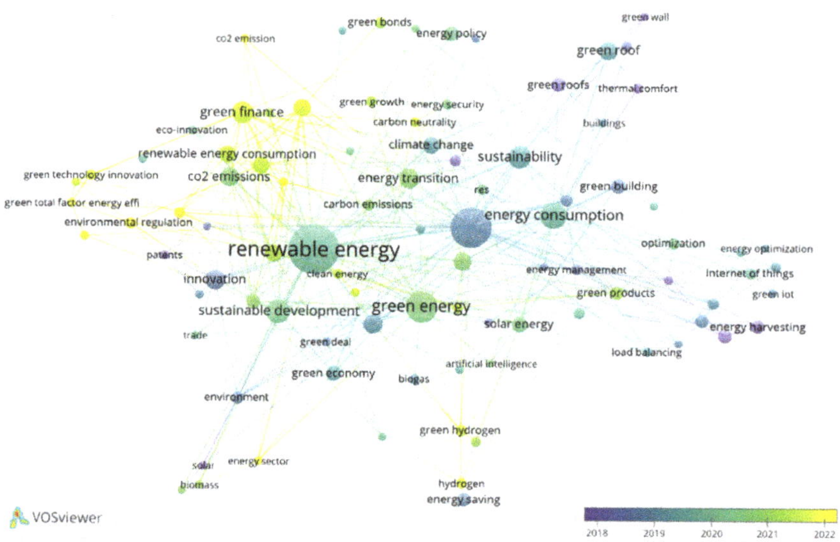

Figure 9. Co-occurrence map (overlay visualization).

Figure 10 presents the most cited journals. The best three journals in the researched area are *Energies, Sustainability,* and *IEEE Access*. These journals appear as the most important sources in green and renewable energy research (in the innovation context) and can be helpful for further research and also to the identification of consistent journals on green energy and renewable energy.

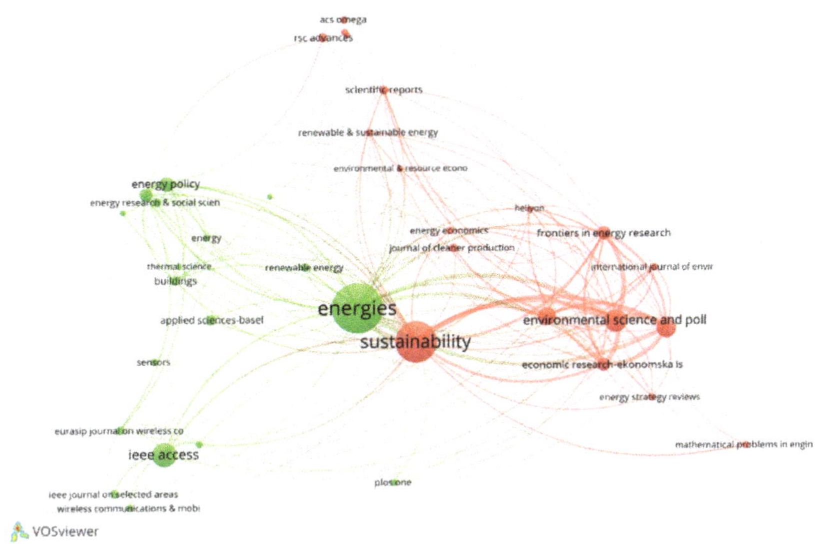

Figure 10. Bibliometric coupling of sources.

4. Discussion

The amount of research into green and renewable energy innovation has been increasing over the last decade, especially in the last three years. Until 2019, the number of publications in this field was about 70 per year, after which there was a significant increase to about 130 published works in 2020 and even 320 in 2022. The increase in 2020–2022 may, among others, be related to changes in EU legislation [69]. This concerns the amended renewable energy directive (Directive (EU) 2018/2001) as part of the "Clean energy for all Europeans" package. EU Member States were obliged to transpose it into national law by June 2021. EU Member States were also obliged to propose a national energy target and establish ten-year national energy and climate plans under the "Horizon 2030" program. The Directive establishes a binding target according to which at least 32% of the final energy must come from renewable sources by 2030 and includes a clause to increase this target by 2023, as well as an increase in the 14% share of renewable energy in transport by 2030. In July 2021, in accordance with the EU's new climate ambitions, the co-legislators were tasked with changing the target to 40% by 2030. An important element that will affect the scale of the use of renewable energy sources will be technological progress, both in terms of currently known methods of generating energy and in completely new technologies, including energy storage technologies [70–72]. The strong increase in the number of publications in 2022 was also significantly influenced by the war in Ukraine and the related energy and fuel crises. This factor caused huge turbulence and caused opposite trends—strengthening the development of green and renewable energy technologies and the return to fossil fuels and nuclear energy. However, the assessment of the impact of the war requires further and in-depth research, which will undoubtedly translate into a further increase in scientific publications. However, regardless of the development of the situation on the energy market, this huge increase in scientific activity in the last decade indicates the importance of the problem and the need for further research in this field.

The study of the most cited references shows that the green and renewable energy innovation bibliography has been defined by various related research areas (such as energy fuels, environmental sciences, ecology, science technology, business economics, telecommunications, and chemistry). The most frequently cited articles concern research methods, the unit root panel test and diagnostic tests of cross-sectional relationships in panels [61,66,67], the environment-competitiveness relationship [62], and green energy solutions in construction [73] as well as technological innovations in the field of green energy [74]. The most productive authors are Jahid (University of Ottawa, Canada), Anser (Super Univ, Pakistan), and Alsharif (Sejong University, North Korea), followed by Cabeza (University of Lleida, Spain), Ahmad (Shandong University of Technology, China), Dincer (Instanbul Medipol University, Turkey), and Yüksel (Instanbul Medipol University, Turkey). Their origin is differentiated both from a geographical perspective and from an institutional perspective. However, globally the most productive scientists in this area are from China, USA, and UK. The research geographical context is very important because green energy innovation is greatly reliant on the macro environment and business ecosystem (including financial support).

According to the analysis of bibliographic coupling of countries, the leading countries are China, USA, UK, Italy, Pakistan, Germany, Poland, Saudi Arabia, and India. In some of countries, especially in Africa, green and renewable energy innovations are barely analyzed. Moreover, the relatively few studies in Southern Europe (except for Italy) or Eastern Europe and Balkans suggests the need for such research and wider cooperation between regional scholars. Additional studies at a regional level are missing to understand green and renewable energy innovation nature.

Some researchers have suggested wider and more intense collaboration in their study on green and renewable energy innovations to share the resulting benefits and costs [75]. Accordingly, the study not only indicated a great deal of cooperation among the different clusters, but also collaboration between Canada and China in the first cluster, collaboration between the USA, Italy, Germany, and Poland in the second cluster, and collaboration between England, Spain, Malaysia, and Australia in the third cluster. Overall, cooperation is happening mainly between developed countries whereas cooperation in developing countries is very poor.

The analysis of the co-occurrence of the author's keywords identified four clusters dominated by four research topics. The first concerns the innovative capacity of green technologies (cluster one), the second relates to energy efficiency (cluster two), the third concerns the macro-environment (cluster three), and the last concerns policies, sources and tools for financing investments in green and renewable energy sources (cluster four).

4.1. Limitations of the Study

The results of the conducted research are burdened with a number of limitations. First, science mapping is a comprehensive, quantitative method of assessing the structure of the knowledge base, but the emphasis on quantitative information related to publications in a given area is no substitute for review methods that qualitatively analyze the substantive content of scientific publications. Therefore, the review presented in the article is only a helicopter view on research on innovations in green and renewable energy, and thus a starting point for further, in-depth research in this area. In addition, the interpretation of maps prepared with the VOSviewer software is subjective. Second, one specific WoS CC database was selected for the study, while others such as Scopus and Google Scholar were excluded. Therefore, we cannot guarantee that our search strategy was able to cover all relevant articles. Finally, due to our inclusion criteria, we included publications in English only in this study. Undoubtedly, there are valuable studies and publications in the native language of researchers who do not publish in English.

4.2. Future Research Directions

The results of the analysis of the co-occurrence of the author's keywords show some similarities and links between innovation and green and renewable energy, which requires further research. The trends in the use of the author's keywords show that many current keywords are *green finance, renewable energy consumption,* CO_2 *emissions, innovations in the field of green technologies,* and *green and renewable energy*, i.e., in the area of climate change and financing the development of green and renewable energy sources.

Further research is needed to assess whether this trend in the use of the author's keywords will continue in the future. In particular, the importance of green buildings for the development of energy efficiency has long been recognized in the literature [76,77]. Indeed, energy consumption is a significant criterion of analysis and measure of innovation activities [78–80]. Finally, as an important topic for various organizations, climate change and environmental regulation have become of particular interest to innovation scientists [81–83].

It is worth mentioning that the essential features of green and renewable energy and innovation are not sufficiently emphasized in the clusters. The author's co-cited keywords focused mainly on technology, knowledge, research and development, and management. When it comes to innovations in the field of green technologies or eco-innovations, these are topics that are much less popular among researchers. Thus, in further research scientists could be more focused on terms such as "green technologies innovation" or "eco-innovations".

Moreover, further analysis of the investments of various funds is needed to better understand innovation as a pilar of climate change policy and to explain the logic of government funding priorities. In future research, it is worth assessing innovation in green and renewable energy from other perspectives, such as public/private sector, financial support, competitiveness. In addition, this study could be supplemented with qualitative methods, e.g., expert panels or focus groups with representatives of various entities.

Finally, this research approach could be continued by, for example, inclusion in the analysis of other databases such as Scopus, Google Scholar, and inclusion of other types of publications, including reviews and conference proceedings.

4.3. Contributions

The obtained research results filled the knowledge gap in several aspects. In the first place, to the authors' knowledge, this is the first study in which a comprehensive bibliometric analysis was carried out, which: (1) analyzed the number of published articles on innovation in the field of green and renewable energy; and (2) identified the most influential authors, journals, and countries in the field of green and renewable energy innovation.

We answered the research questions posed:

- RQ1: The structure of research on innovation in the field of green and renewable energy was characterized and the intellectual structure was defined and the most frequently discussed research topics in this field were identified.
- RQ2: The dynamics of research and the most important activities carried out as part of innovations in the field of green and renewable energy were characterized, and connections between authors, journals, countries, and institutions active in the field of research on innovations in green and renewable energy sources were defined.
- RQ3: Future research directions related to green and renewable energy innovations were recommended.

5. Conclusions

The aim of the article was to analyze the size, structure, and dynamics of research on innovations in the field of green and renewable energy in the last decade in order to identify the main topics and research trends in this area. To achieve this goal, a comprehensive bibliometric analysis of leading authors, journals, countries, and the most frequently cited documents in the field of innovation in green and renewable energy in the years 2012–2022 was carried out. The analysis was based on 1144 articles selected from WoS CC. Performance analysis of the records was supplemented with mapping using the VOSviewer software. The results of the review presented in the article provide scientists with knowledge about the structure and dynamics of research on innovations in green and renewable energy and identify knowledge gaps and future research directions. The most important conclusions from the research are as follows:

1. A significant increase in research in the analyzed area has been observed in the last decade, from 7 articles in 2012 to 320 articles (with 4642 citations) in 2022.
2. The most productive authors are Jahid (University of Ottawa, Canada), Anser (Super Univ, Pakistan), Alsharif (Sejong University, North Korea), Cabeza (University of Lleida, Spain), Ahmad (Shandong University of Technology, China), Dincer (Instanbul Medipol University, Turkey), and Yüksel (Instanbul Medipol University, Turkey).
3. The results showed that innovations in the field of green and renewable energy, and in particular their external environment (ecosystem, financing) are an important problem and research topic in developed countries but should be studied even more intensively in developing countries.
4. The most popular journals in the field of green and renewable energy and innovation are *Energies* (MDPI, IF: 3.252), *Sustainability* (MDPI, IF: 3.889), *IEEE Access* (IEEE, IF: 3.476), *Environmental Science and Pollution Research* (Springer Heidelberg, IF: 5.19) and *Frontiers in Environmental Science* (Frontiers Media SA, IF: 5.411).
5. The analysis of the co-occurrence of the author's keywords identified four clusters dominated by four research topics. The first concerns the innovative capacity of green technologies (cluster one), the second relates to energy efficiency (cluster two), the third concerns the macro-environment (cluster three), and the last concerns policies, sources and tools for financing investments in green and renewable energy sources (cluster four).

Author Contributions: Conceptualization, J.B.-W., A.L., G.M. and W.T.; Data Curation, J.B.-W. and A.L.; Formal Analysis, J.B.-W. and A.L.; Investigation, J.B.-W., A.L. and G.M., Methodology, J.B.-W. and A.L., Writing—Original Draft Preparation, J.B.-W. and A.L.; Writing—Review and Editing, W.T., J.B.-W., A.L. and G.M. All authors have read and agreed to the published version of the manuscript.

Funding: This research received no external funding.

Institutional Review Board Statement: Not applicable.

Informed Consent Statement: Not applicable.

Data Availability Statement: Not applicable.

Conflicts of Interest: The authors declare no conflict of interest.

References

1. Bernai, R.R. Renewable Energy Sources and Climate Change Mitigation, Special Report of the Intergovernmental Panel on Climate Change (IPCC). *Econ. Energy Environ. Policy* **2013**, *2*, 101–112.
2. Bilgili, M.; Ozbek, A.; Sahin, B.; Kahraman, A. An overview of renewable electric power capacity and progress in new technologies in the world. *Renew. Sustain. Energy Rev.* **2015**, *49*, 323–334. [CrossRef]
3. Kiciński, J. Green energy transformation in Poland. Bulletin of the Polish Academy of Sciences. *Tech. Sci.* **2021**, *69*, e136213.
4. Ogunbode, C.A.; Doran, R.; Böhm, G. Exposure to the IPCC special report on 1.5 C global warming is linked to perceived threat and increased concern about climate change. *Clim. Chang.* **2020**, *158*, 361–375. [CrossRef]
5. Ritchie, H.; Roser, M.; Rosado, P. CO_2 and Greenhouse Gas Emissions, Published Online at OurWorldInData.org. 2020. Available online: https://ourworldindata.org/co2-and-other-greenhouse-gas-emissions (accessed on 1 December 2022).
6. Barresi, G.; Marisca, C. Energy policy and environmental sustainability in regional development strategy: The regional environmental energy plan in Sicily (PEARS). In *Moving from the Crisis to Sustainability: Emerging Issues in the International Context*; Calabro, G., D'Amico, A., Lafranchi, M., Moschella, G., Pulejo, L., Salomone, R., Eds.; Franco Angeli Editore: Milano, Italy, 2011; pp. 209–218. ISBN 9788856847055.
7. Bouckaert, S.; Pales, A.F.; McGlade, C.; Remme, U.; Wanner, B.; Varro, L.; D'Ambrosio, D.; Spencer, T. Net Zero by 2050: A Roadmap for the Global Energy Sector. 2021. Available online: https://trid.trb.org/view/1856381 (accessed on 1 December 2022).
8. Delbeke, J.; Runge-Metzger, A.; Slingenberg, Y.; Werksman, J. The Paris agreement. In *Towards a Climate-Neutral Europe*; Routledge: London, UK, 2019; pp. 24–45.
9. United Nations General Assembly: Transforming Our World: The 2030 Agenda for Sustainable Development: Draft Resolution Referred to the United Nations Summit for the Adoption of the Post-2015 Development Agenda by the General Assembly at Its 69th Session. UN Doc. A/70/L.1 of 18 September 2015. Available online: https://digitallibrary.un.org/record/803352 (accessed on 1 December 2022).

10. IRENA, Tracking SDG 7: The Energy Progress Report, Publ.-SDG-7-2021. 2021. Available online: https://www.irena.org/publications/2021/Jun/Tracking-SDG-7-2021 (accessed on 1 December 2022).
11. Salman, M.; Zha, D.; Wang, G. Assessment of energy poverty convergence: A global analysis. *Energy* **2022**, *255*, 124579. [CrossRef]
12. Piwowar, A. Energy Poverty as a Current Problem in the Light of Economic and Social Challenges. *Energies* **2022**, *15*, 8554. [CrossRef]
13. Change, IPCC Climate. Mitigation of climate change. *Contrib. Work. Group III Fifth Assess. Rep. Intergov. Panel Clim. Chang.* **2014**, *1454*, 147.
14. Perathoner, S.; Centi, G. New Energy Sources and CO_2 Treatment. *Glob. Chang. Energy Issues Regul. Policies* **2013**, *2*, 143–160. [CrossRef]
15. International Energy Agency. *World Energy Outlook*; OECD/IEA: Paris, France, 2009; p. 17.
16. World Health Organization, Tracking SDG 7: The Energy Progress Report 2022. Available online: https://trackingsdg7.esmap.org/ (accessed on 1 December 2022).
17. Zhongming, Z.; Linong, L.; Xiaona, Y.; Wangqiang, Z.; Wei, L. World Adds Record New Renewable Energy Capacity in 2020. 2021. Available online: http://resp.llas.ac.cn/C666/handle/2XK7JSWQ/321399 (accessed on 1 December 2022).
18. Zhongming, Z.; Linong, L.; Xiaona, Y.; Wangqiang, Z.; Wei, L. Global Energy Review. 2021. Available online: https://www.iea.org/reports/global-energy-review-2021/renewables (accessed on 1 December 2022).
19. Jawerth, N. What is the clean energy transition and how does nuclear power fit in? *IAEA Bull. (Online)* **2020**, *61*, 4–5.
20. Nijs, W.; Tarvydas, D.; Toleikyte, A. EU Challenges of Reducing Fossil Fuel Use in Buildings. 2021. Available online: https://www.google.pl/url?sa=t&rct=j&q=&esrc=s&source=web&cd=&ved=2ahUKEwjN0YfNp4r8AhWitYsKHekVAy8QFnoECA8QAQ&url=https%3A%2F%2Fpublications.jrc.ec.europa.eu%2Frepository%2Fbitstream%2FJRC127122%2FJRC127122_01.pdf&usg=AOvVaw2brUp-avWpRxkbxosL3vBJ (accessed on 1 December 2022).
21. Adkin, L.E. Technology Innovation as a Response to Climate Change: The Case of the Climate Change Emissions Management Corporation of Alberta. *Rev. Policy Res.* **2019**, *36*, 603–634. [CrossRef]
22. Laengle, S.; Lobos, V.; Merigó, J.M.; Herrera-Viedma, E.; Cobo, M.J.; De Baets, B. Forty years of Fuzzy Sets and Systems: A bibliometric analysis. *Fuzzy Sets Syst.* **2021**, *402*, 155–183. [CrossRef]
23. Rey-Martí, A.; Ribeiro-Soriano, D.; Palacios-Marqués, D. A bibliometric analysis of social entrepreneurship. *J. Bus. Res.* **2016**, *69*, 1651–1655. [CrossRef]
24. Abramo, G.; D'Angelo, C.A.; Di Costa, F. National research assessment exercises: A comparison of peer review and bibliometrics rankings. *Scientometrics* **2011**, *89*, 929–941. [CrossRef]
25. Bjork, S.; Offer, A.; Söderberg, G. Time series citation data: The Nobel Prize in economics. *Scientometrics* **2014**, *98*, 185–196. [CrossRef]
26. Corsi, S.; Ruggeri, G.; Zamboni, A.; Bhakti, P.; Espen, L.; Ferrante, A.; Noseda, M.; Varanini, Z.; Scarafoni, A. A Bibliometric Analysis of the Scientific Literature on Biostimulants. *Agronomy* **2022**, *12*, 1257. [CrossRef]
27. Benckendorff, P.; Zehrer, A. A network analysis of tourism research. *Ann. Tour. Res.* **2013**, *43*, 121–149. [CrossRef]
28. Koseoglu, M.A.; Rahimi, R.; Okumus, F.; Liu, J. Bibliometric studies in tourism. *Ann. Tour. Res.* **2016**, *61*, 180–198. [CrossRef]
29. Hall, C.M. Publish and perish? Bibliometric analysis, journal ranking and the assessment of research quality in tourism. *Tour. Manag.* **2011**, *32*, 16–27. [CrossRef]
30. Fabregat-Aibar, L.; Barberà-Mariné, M.G.; Terceño, A.; Pié, L. A Bibliometric and Visualization Analysis of Socially Responsible Funds. *Sustainability* **2019**, *11*, 2526. [CrossRef]
31. Feng, Y.; Zhu, Q.; Lai, K.-H. Corporate social responsibility for supply chain management: A literature review and bibliometric analysis. *J. Clean. Prod.* **2017**, *158*, 296–307. [CrossRef]
32. Zupic, I.; Čater, T. Bibliometric methods in management and organization. *Organ. Res. Methods* **2015**, *18*, 429–472. [CrossRef]
33. Cobo, M.J.; López-Herrera, A.G.; Herrera-Viedma, E.; Herrera, F. Science mapping software tools: Review, analysis, and cooperative study among tools. *J. Am. Soc. Inf. Sci. Technol.* **2011**, *62*, 1382–1402. [CrossRef]
34. Donthu, N.; Reinartz, W.; Kumar, S.; Pattnaik, D. A retrospective review of the first 35 years of the International Journal of Research in Marketing. *Int. J. Res. Mark.* **2021**, *38*, 232–269. [CrossRef]
35. Moher, D.; Liberati, A.; Tetzlaff, J.; Altman, D.G.; Altman, D.; Antes, G.; Atkins, D.; Barbour, V.; Barrowman, N.; Berlin, J.A.; et al. Preferred reporting items for systematic reviews and meta-analyses: The PRISMA statement (Chinese edition). *J. Chin. Integr. Med.* **2009**, *7*, 889–896. [CrossRef]
36. Cortese, T.T.P.; de Almeida, J.F.S.; Batista, G.Q.; Storopoli, J.E.; Liu, A.; Yigitcanlar, T. Understanding Sustainable Energy in the Context of Smart Cities: A PRISMA Review. *Energies* **2022**, *15*, 2382. [CrossRef]
37. Page, M.J.; McKenzie, J.E.; Bossuyt, P.M.; Boutron, I.; Hoffmann, T.C.; Mulrow, C.D.; Shamseer, L.; Tetzlaff, J.M.; Akl, E.A.; Brennan, S.E.; et al. The PRISMA 2020 statement: An updated guideline for reporting systematic reviews. *Int. J. Surg.* **2021**, *88*, 105906. [CrossRef]
38. Van Oorschot, J.A.; Hofman, E.; Halman, J.I. A bibliometric review of the innovation adoption literature. *Technol. Forecast. Soc. Chang.* **2018**, *134*, 1–21. [CrossRef]
39. Albort-Morant, G.; Henseler, J.; Leal-Millán, A.; Cepeda-Carrión, G. Mapping the field: A bibliometric analysis of green innovation. *Sustainability* **2017**, *9*, 1011. [CrossRef]

40. Ahmed, S.; Huang, B. Control engineering practice in 25 years: A bibliometric overview. *Control Eng. Pract.* **2019**, *88*, 16–20. [CrossRef]
41. Udomsap, D.; Hallinger, P. A bibliometric review of research on sustainable construction, 1994–2018. *J. Clean. Prod.* **2020**, *254*, 120073. [CrossRef]
42. Hernández-Torrano, D.; Ibrayeva, L. Creativity and education: A bibliometric mapping of the research literature (1975–2019). *Think. Ski. Creat.* **2020**, *35*, 100625. [CrossRef]
43. Xie, H.; Zhang, Y.; Duan, K. Evolutionary overview of urban expansion based on bibliometric analysis in Web of Science from 1990 to 2019. *Habitat Int.* **2020**, *95*, 102100. [CrossRef]
44. Zou, Y.; Luo, Y.; Zhang, J.; Xia, N.; Tan, G.; Huang, C. Bibliometric analysis of oncolytic virus research, 2000 to 2018. *Medicine* **2019**, *98*, e16817. [CrossRef]
45. Veloutsou, C.; Mafe, C.R. Brands as relationship builders in the virtual world: A bibliometric analysis. *Electron. Commer. Res. Appl.* **2020**, *39*, 100901. [CrossRef]
46. Peng, R.Z.; Zhu, C.; Wu, W.P. Visualizing the knowledge domain of intercultural competence research: A bibliometric analysis. *Int. J. Intercult. Relat.* **2020**, *74*, 58–68. [CrossRef]
47. Bo, Z.; Shuai, X.; Mao, S.; Yang, H.; Qian, J.; Chen, J.; Yan, J.; Cen, K. Green preparation of reduced graphene oxide for sensing and energy storage applications. *Sci. Rep.* **2014**, *4*, 4684. [CrossRef] [PubMed]
48. Van Vuuren, D.P.; Stehfest, E.; Gernaat, D.E.; Doelman, J.C.; van den Berg, M.; Harmsen, M.; De Boer, H.S.; Bouwman, L.F.; Daioglou, V.; Edelenbosch, O.Y.; et al. Energy, land-use and greenhouse gas emissions trajectories under a green growth paradigm. *Glob. Environ. Chang.* **2017**, *42*, 237–250. [CrossRef]
49. Na Quan, L.; Zhao, Y.; de Arquer, F.P.G.; Sabatini, R.; Walters, G.; Voznyy, O.; Comin, R.; Li, Y.; Fan, J.Z.; Tan, H.; et al. Tailoring the Energy Landscape in Quasi-2D Halide Perovskites Enables Efficient Green-Light Emission. *Nano Lett.* **2017**, *17*, 3701–3709. [CrossRef]
50. Wu, J.; Zhang, Y.; Zukerman, M.; Yung, E.K.-N. Energy-Efficient Base-Stations Sleep-Mode Techniques in Green Cellular Networks: A Survey. *IEEE Commun. Surv. Tutor.* **2015**, *17*, 803–826. [CrossRef]
51. Tahir, M.; Cao, C.; Butt, F.K.; Idrees, F.; Mahmood, N.; Ali, Z.; Aslam, I.; Tanveer, M.; Rizwan, M.; Mahmood, T. Tubular graphitic-C3N4: A prospective material for energy storage and green photocatalysis. *J. Mater. Chem. A* **2013**, *1*, 13949–13955. [CrossRef]
52. Ghaffour, N.; Lattemann, S.; Missimer, T.; Ng, K.C.; Sinha, S.; Amy, G. Renewable energy-driven innovative energy-efficient desalination technologies. *Appl. Energy* **2014**, *136*, 1155–1165. [CrossRef]
53. Mansouri, S.A.; Aktas, E.; Besikci, U. Green scheduling of a two-machine flowshop: Trade-off between makespan and energy consumption. *Eur. J. Oper. Res.* **2016**, *248*, 772–788. [CrossRef]
54. Richter, M. Business model innovation for sustainable energy: German utilities and renewable energy. *Energy Policy* **2013**, *62*, 1226–1237. [CrossRef]
55. Lehr, U.; Lutz, C.; Edler, D. Green jobs? Economic impacts of renewable energy in Germany. *Energy Policy* **2012**, *47*, 358–364. [CrossRef]
56. Coma, J.; Pérez, G.; de Gracia, A.; Burés, S.; Urrestarazu, M.; Cabeza, L.F. Vertical greenery systems for energy savings in buildings: A comparative study between green walls and green facades. *Build. Environ.* **2017**, *111*, 228–237. [CrossRef]
57. Vošner, H.B.; Kokol, P.; Bobek, S.; Železnik, D.; Završnik, J. A bibliometric retrospective of the Journal Computers in Human Behavior (1991–2015). *Comput. Human Behav.* **2016**, *65*, 46–58. [CrossRef]
58. Porter, M.E.; van der Linde, C. Toward a New Conception of the Environment-Competitiveness Relationship. *J. Econ. Perspect.* **1995**, *9*, 97–118. [CrossRef]
59. Johnstone, N.; Haščič, I.; Popp, D. Renewable Energy Policies and Technological Innovation: Evidence Based on Patent Counts. *Environ. Resour. Econ.* **2010**, *45*, 133–155. [CrossRef]
60. Popp, D. Induced Innovation and Energy Prices. *Am. Econ. Rev.* **2002**, *92*, 160–180. [CrossRef]
61. Pesaran, M.H. A simple panel unit root test in the presence of cross-section dependence. *J. Appl. Econ.* **2007**, *22*, 265–312. [CrossRef]
62. Pesaran, M.H. General diagnostic tests for cross-sectional dependence in panels. *Empir. Econ.* **2021**, *60*, 13–50. [CrossRef]
63. Zhang, D.; Mohsin, M.; Rasheed, A.K.; Chang, Y.; Taghizadeh-Hesary, F. Public spending and green economic growth in BRI region: Mediating role of green finance. *Energy Policy* **2021**, *153*, 112256. [CrossRef]
64. Charfeddine, L.; Kahia, M. Impact of renewable energy consumption and financial development on CO_2 emissions and economic growth in the MENA region: A panel vector autoregressive (PVAR) analysis. *Renew. Energy* **2019**, *139*, 198–213. [CrossRef]
65. Li, W.; Chien, F.; Hsu, C.-C.; Zhang, Y.; Nawaz, M.A.; Iqbal, S.; Mohsin, M. Nexus between energy poverty and energy efficiency: Estimating the long-run dynamics. *Resour. Policy* **2021**, *72*, 102063. [CrossRef]
66. Pesaran, M.H.; Shin, Y.; Smith, R.J. Bounds testing approaches to the analysis of level relationships. *J. Appl. Econom.* **2001**, *16*, 289–326. [CrossRef]
67. Phillips, P.C.B.; Perron, P. Testing for a unit root in time series regression. *Biometrika* **1988**, *75*, 335–346. [CrossRef]
68. Sakata, I.; Sasaki, H.; Akiyama, M.; Sawatani, Y.; Shibata, N.; Kajikawa, Y. Bibliometric analysis of service innovation research: Identifying knowledge domain and global network of knowledge. *Technol. Forecast. Soc. Chang.* **2013**, *80*, 1085–1093. [CrossRef]

69. Directive (EU) 2018/2001 of the European Parliament and of the Council of 11 December 2018 on the Promotion of the Use of Energy from Renewable Sources (Recast). Dz.U. L 328 z 11.12.2018. Available online: https://eur-lex.europa.eu/legal-content/EN/TXT/PDF/?uri=CELEX:32018L2001&from=EN (accessed on 1 December 2022).
70. Xiao, X.; Roh, B.-M.; Hamilton, C. Porosity management and control in powder bed fusion process through process-quality interactions. *CIRP J. Manuf. Sci. Technol.* **2022**, *38*, 120–128. [CrossRef]
71. Damak, C.; Leducq, D.; Hoang, H.M.; Negro, D.; Delahaye, A. Liquid Air Energy Storage (LAES) as a large-scale storage technology for renewable energy integration—A review of investigation studies and near perspectives of LAES. *Int. J. Refrig.* **2020**, *110*, 208–218. [CrossRef]
72. Mentel, G.; Tarczyński, W.; Azadi, H.; Abdurakmanov, K.; Zakirova, E.; Salahodjaev, R. R&D Human Capital, Renewable Energy and CO_2 Emissions: Evidence from 26 Countries. *Energies* **2022**, *15*, 9205. [CrossRef]
73. Sailor, D. A green roof model for building energy simulation programs. *Energy Build.* **2008**, *40*, 1466–1478. [CrossRef]
74. Huang, X.; Han, T. and Ansari N. On Green-Energy-Powered Cognitive Radio Networks. *IEEE Commun. Surv. Tutor.* **2015**, *17*, 827–842. [CrossRef]
75. Wurlod, J.-D.; Noailly, J. The impact of green innovation on energy intensity: An empirical analysis for 14 industrial sectors in OECD countries. *Energy Econ.* **2018**, *71*, 47–61. [CrossRef]
76. Khan, K.I.; Nasir, A.; Rashid, T. Green Practices: A Solution for Environmental Deregulation and the Future of Energy Efficiency in the Post-COVID-19 Era. *Front. Energy Res.* **2022**, *10*, 878670. [CrossRef]
77. Coma, J.; Pérez, G.; Solé, C.; Castell, A.; Cabeza, L.F. Thermal assessment of extensive green roofs as passive tool for energy savings in buildings. *Renew. Energy* **2016**, *85*, 1106–1115. [CrossRef]
78. Zhao, Y.; Ramzan, M.; Adebayo, T.S.; Oladipupo, S.D.; Adeshola, I.; Agyekum, E.B. Role of Renewable Energy Consumption and Technological Innovation to Achieve Carbon Neutrality in Spain: Fresh Insights From Wavelet Coherence and Spectral Causality Approaches. *Front. Environ. Sci.* **2021**, *9*, 769067. [CrossRef]
79. Borràs, J.G.; Lerma, C.; Mas, Á.; Vercher, J.; Gil, E. Energy Efficiency Evaluation of Green Roofs as an Intervention Strategy in Residential Buildings in the Field of Spanish Climate. *Buildings* **2022**, *12*, 959. [CrossRef]
80. Saharkhiz, Y.; Valizadeh, M. Energy consumption reduction in buildings through applying green roofs. *Turk. Online J. Des. ART Commun.* **2016**, *6*, 1692–1696. [CrossRef]
81. Wang, Y.; Hu, J.; Hu, Y.; Wang, Y. Which is more effective: The carrot or the stick? Environmental policy, green innovation and enterprise energy efficiency—A quasi-natural experiment from China. *Front. Environ. Sci.* **2022**, *10*, 870713. [CrossRef]
82. Pegels, A.; Lütkenhorst, W. Is Germany's energy transition a case of successful green industrial policy? Contrasting wind and solar PV. *Energy Policy* **2014**, *74*, 522–534. [CrossRef]
83. Pettifor, H.; Wilson, C.; Chryssochoidis, G. The appeal of the green deal: Empirical evidence for the influence of energy efficiency policy on renovating homeowners. *Energy Policy* **2015**, *79*, 161–176. [CrossRef]

MDPI
St. Alban-Anlage 66
4052 Basel
Switzerland
Tel. +41 61 683 77 34
Fax +41 61 302 89 18
www.mdpi.com

Energies Editorial Office
E-mail: energies@mdpi.com
www.mdpi.com/journal/energies

www.ingramcontent.com/pod-product-compliance
Lightning Source LLC
Chambersburg PA
CBHW061819210926
51850CB00006B/752
* 9 7 8 3 0 3 6 5 6 8 0 0 3 *